Ectomycorrhizal Symbioses in Tropical and Neotropical Forests

Ectomycorrhizal Symbioses in Tropical and Neotropical Forests

Editors

Amadou M. Bâ
Université Antilles-Guyane
Guadeloupe (French West Indies)
France

Krista L. McGuire
Barnard College
Columbia University
New York
USA

Abdala G. Diédhiou
Université Cheikh Anta Diop
Dakar
Sénégal

CRC Press
Taylor & Francis Group
Boca Raton London New York

CRC Press is an imprint of the
Taylor & Francis Group, an **informa** business
A SCIENCE PUBLISHERS BOOK

CRC Press
Taylor & Francis Group
6000 Broken Sound Parkway NW, Suite 300
Boca Raton, FL 33487-2742

Cover Illustrations

Photograph of Pink sporocarp: Reproduced by kind courtesy of Ducousso Marc, CIRAD
Photograph of Yellow ectomycorrhia: Reproduced by kind courtesy of the first editor, Amadou M. Bâ

Printed in the United States of America on acid-free paper

International Standard Book Number: 978-1-4665-9468-5 (Hardback)

Library of Congress Cataloging-in-Publication Data

Ectomycorrhizal symbioses in tropical and neotropical forests / editors: Amadou M. Bâ, Krista L. McGuire, Abdala G. Diédhiou.
pages cm
Includes bibliographical references and index.
ISBN 978-1-4665-9468-5 (hardcover : alk. paper) 1. Ectomycorrhizal fungi. 2. Ectomycorrhizas. 3. Rain forest ecology. I. Bâ, Amadou M. (Amadou Mustapha) II. McGuire, Krista L. III. Diédhiou, Abdala G. (Abdala Gamby)

QK604.2.M92E27 2014
577.34--dc23

2013047501

Visit the Taylor & Francis Web site at
http://www.taylorandfrancis.com

CRC Press Web site at
http://www.crcpress.com

Science Publishers Web site at
http://www.scipub.net

Foreword

***Ectomycorrhizal symbioses in tropical and neotropical forests*: A major step toward to a neglected research imperative.**

In a 2009 paper in the *New Phytologist*, Ian Alexander and myself called the mycorrhizal research in the tropics '**a neglected research imperative**'. Although Janse's '*Les endophytes radicaux de quelques plantes javanaises*' in 1896 was published shortly after the description of ectomycorrhizas by Frank in 1885, the research on tropical mycorrhizas attracted too limited attention in the next 100 years. Although papers dealing with mycorrhiza and tropic continuously accumulated in the last decade (*ca.* +10% per year, according to ISI Web of Knowledge), they represented only, each year between 2009 and 2012, a remarkably constant 0.6% fraction of the papers dealing with mycorrhiza.

Yet, one should wish more than such a constant progression. Tropical ecosystems represent more than 0.6% of land ecosystems and have pivotal role in the Earth's biogeochemical cycle and climate; moreover, threats on biodiversity in tropical forests should encourage faster study in tropical latitudes. In this framework, a book devoted to '***Ectomycorrhizal symbioses in tropical and neotropical forests***' is very timely.

Indeed, we are far from the simple view that the tropics are dominated by arbuscular mycorrhizas, as popularized by Malloch et al. in *PNAS* in 1975. However, what determines the success of ectomycorrhizas in tropical ecosystems remains unclear. One of the most striking features of tropical ectomycorrhizal symbioses is their frequent occurrence in 'monodominant' forests, where a single species dominates 60 to 100% of the canopy. The present book investigates these exciting monodominant forests by comparing their mycorrhizal diversity with that of more host-diverse ectomycorrhizal forests (Diédhiou et al., Chapter 1), and by questioning the role of ectomycorrhizas among the positive feedbacks that maintain the dominant species over generations (McGuire, Chapter 10).

Beyond such questions, basic descriptions of these insufficiently studied ectomycorrhizal symbioses are still required. This book nicely describes ectomycorrhizas of Nyctaginaceae (Haug et al., Chapter 2) and

Gnetum (a liana probably related to Pinaceae; Bechem, Chapter 8) on the plant side, and of Sebacinales (Moyersoen, Chapter 5) on the fungal side, in a timely revival of often overlooked morphological and microscopic investigations. The *Gnetum* chapter illustrates a more functional approach for studying ECM fungi based on physiology and ecophysiology. These functional aspects are also central in the chapter evaluating the role of the ectomycorrhizal symbiosis in protection of *Coccoloba uvifera* against sea salt (Bâ et al., Chapter 9), and in plant adaptation to soils highly concentrated in heavy metals (Jourand et al., Chapter 6).

More ecological questions, such as the response of the ectomycorrhizal community to disturbance (Onguene et al., Chapter 3), open the door to the use of ectomycorrhizal symbioses for reforestation, especially in physiologically constraining environments. The possibility of facilitation by 'nurse plants', which pre-cultivates a fungal community suitable for installing a target tree species, receives supports in Madagascarian sclerophyllous forests (Baohanta et al., Chapter 4) and the tree performances after inoculation of *Afzelia* by *Scleroderma* and *Thelephora* raises good hopes in Western Africa (Sanon et al., Chapter 7). In Asiatic dipterocarps forests, inoculation, using mycoflora from soil sampled under conspecific parents is promising for forest rehabilitation (Tata, Chapter 11). In an overview of the inoculation methods and results, Duponnois et al. (Chapter 12) rightly point out that ectomycorrhizal symbioses are very promising tools—yet, more long-term assessments of inoculant survival and of host growth promotion are required, especially after out planting to field sites. Encouraging results obtained from some models should not dissimulate that the method needs independent optimization for each plant-fungus model. Anarchic use of any inoculant or any method could introduce undesirable exotic fungi, and discredit ectomycorrhizal inoculations by yielding low results. In this framework, the careful case studies reported in this book are very relevant.

Finally, beyond trees and forests, edible fungal fruit bodies represent often neglected and potentially threatened resources, with high cultural and economical importance (Yorou et al., Chapter 13). '***Ectomycorrhizal symbioses in tropical and neotropical forests***' is a rich milestone in our knowledge of fungal and ectomycorrhizal functioning diversity over all the tropics, from Africa to America, including Pacific islands, with rich insights into functional and applied questions. The reader clearly feels emergent models for further studies and practices: *Scleroderma* and *Pisolithus* spp. on the fungal side, *Coccoloba, Afzelia, Gilbertiodendron, Uapaca, Dicymbe* and *Shorea* on the plant side. Researches must now go on, in order to develop robust methods ensuring environmental protection and sustainability of ecosystems services.

As a 'primarily temperate' mycorrhizologist, I have always been fascinated by the lessons and exquisite diversity offered by tropical mycorrhizal symbioses, and I feel greatly honoured to write these introductory lines. Closing the pages of ***'Ectomycorrhizal symbioses in tropical and neotropical forests'***, I feel more fascinated than ever, and hope all readers will share this great excitement!

Marc-André SELOSSE
Professor at Muséum national d'Histoire naturelle and
President of the Société botanique de France
Observatoire de la Structure et de l'Evolution de la Biodiversité
(UMR7205 CNRS – Muséum), Département Systématique et Evolution,
45 rue Buffon, 75005 Paris, France

Preface

Ectomycorrhizal (EcM) symbioses are mutualistic plant-fungus associations that play a major role in function, maintenance and evolution of biodiversity and ecosystems stability and productivity. EcM associations are integral to the biology and ecology of forest trees, affecting growth, water and nutrient absorption and protection against pathogens. They are also a research imperative in tropical and neotropical forest ecosystems because they concern ecologically and economically important tree species (e.g., Ceasalpinioid subfamily in Africa, family of Dipterocarpaceae in Asia and South America). The book is an overview of the knowledge of EcM symbioses in tropical and neotropical ecosystem forests. The contents address diversity and function of ectomycorrhiza (ECM) associated with forest plants, impacts of ectomycorrhiza on plant diversity and composition, regeneration and dynamics of ecosystems, and biomass production in forestry, adaptation of EcM plants to nutrient deficient, salted and ultramafic soils.

Diédhiou et al. (Chapter 1) has provided valuable information on the diversity and community structure of EcM fungi from mixed and monodominant forests in tropical Africa. They found that monodominant forest tends to harbor more EcM-forming lineages than the mixed forest, while there was no significant difference in terms of richness of EcM fungal species between these two forest types. In the monodominant *Dicymbe corymbosa* forest in Guyana, it appears that positive plant-soil feedbacks function to maintain EcM monodominance through seedling facilitation and alterations of decomposer communities and nutrient cycling pathways. It is also possible that the EcM monodominant trees are able to support a greater diversity of EcM fungi compared to non-dominant EcM trees, which may partially explain why these trees attain such extreme dominance. With a greater diversity of symbionts, trees have access to more diverse pools of soil resources (McGuire, Chapter 10). EcM forest communities could contribute not only to the preservation of monodominant clump refuges in South Cameroon with endemic species, but also to the protection of the biodiversity hot spots of Africa's rain forests. Maintenance of such EcM forest communities with their associated edible mushrooms also helps to preserve a valuable source of alternative protein-rich food for local communities who depend on non-timber forest products (Onguene et al., Chapter 3).

Different approaches exist to characterize EcM. Morpho-anatomical features are important information, together with ultrastructure and DNA sequences of EcM are used to confirm the EcM status of Sebacinales species. These results are not only important for the taxonomy of Sebacinales EcM but also for the understanding of Sebacinales ecology (Moyersoen, Chapter 5). Haug et al. (Chapter 2) showed that although the common ancestor of Nyctaginaceae may have been involved with one thelephoroid taxon, coevolution of Nyctaginaceae and ectomycorrhizal Thelephoraceae in different habitats may have led to the evolution of several different species.

On the other hand, EcM fungi are an integral part of plant physiology. For example, there is a possibility that the fungus *Scleroderma sinnamariense* can access some organic P sources in nature when inorganic P is limiting, an ability which would be beneficial to the *Gnetum africanum*, a notable liana for its edible leaves (Bechem, Chapter 8). Bâ et al. (Chapter 9) that *Scleroderma bermudense* possess considerable resistance to salinity. Tolerance to salt stress was considerably enhanced by *S. bermudense*. Otherwise, the focus on EcM *Pisolithus albus* isolated from soils in New Caledonia highlighted the identification of an ultramafic nickel-tolerant ecotype, showing specific and adaptive molecular response to this metal. In thus, this fungus plays a key role in plant host adaptation to toxic nickel concentrations as found in these soils (Jourand et al., Chapter 6).

From ecological point of view, the EcM fungi could be from an adult tree of *Uapaca bojeri*, an endemic tree of the Malagasy sclerophyllous forest, where various putative EcM fungi were observed, or from endemic and pioneer shrub species (*Leptolenabojeriana* and *Sarcolaenaoblongifolia*) that persist on disturbed sites and facilitate the survival of EcM fungi propagules that could potentially infect roots of *U. bojeri* seedlings (Baohanta et al., Chapter 4). Sanon et al. (Chapter 7) already propose to use two effective EcM fungi, *S. dictyosporum* and *Thelephora* sp., to inoculate *Afzelia africana*, a timber tree for the reforestation programme in West Africa. Similar results were obtained in Asian forests, where diptercarp seedlings are inoculated by mycoflora from soil sampled under conspecific parents in order to rehabilitate forests (Tata, Chapter 11). Duponnois et al. (Chapter 12) present an overview of the inoculation methods and results rightly point to ectomycorrhizal symbioses as very promising tools, but more long-term assessments of inoculant survival and of host growth promotion are required, especially after outplanting to field sites. Finally, beyond trees and forests, edible fungal fruitbodies represent often neglected and potentially threatened resources, of high cultural and economic importance (Yorou et al., Chapter 13).

From this finding, it should be evident that further research is necessary in tropical and neotropical forests to (i) assess and compare the morphological and phylogenetic community composition between monodominant and mixed forests, (ii) to determine the role of EcM fungi in the recruitment and establishment of seedlings of EcM tree species, (iii) to use EcM symbioses for reforestation programmes.

Amadou M. Bâ
Krista L. McGuire
Abdala G. Diédhiou

Contents

CHAPTER 1

Diversity and Community Structure of Ectomycorrhizal Fungi in Mixed and Monodominant African Tropical Rainforests

Abdala Gamby Diedhiou,[1,2,*] *Helvyne Christelle Michaella Ebenye,*[2,3,4] *Marc-André Selosse,*[3,4] *Nérée Onguene Awana*[5] and *Amadou Mustapha Bâ*[2,6]

1. Introduction

Mycorrhizal symbioses play a prominent role in the biology and ecology of forest trees. They involve soil fungi and roots of trees, which together as a symbiosis provide the fungi with carbohydrates and enhance the uptake of water and nutrients for the trees, and also have a major protective role for the roots (Smith and Read 2008). Forest trees are primarily associated with two types of mycorrhizas: arbuscular mycorrhizas which include fungi from the phylum of Glomeromycota and ectomycorrhizas (EcMs) mainly formed by members of Ascomycota and Basidiomycota. It is currently estimated that 6,000–10,000 plant species (Smith and Read 2008, Brundrett 2009) and 20,000–25,000 fungal species (Rinaldi et al. 2008) are involved in

Authors' affiliations given at the end of the chapter.

ectomycorrhizal symbioses. EcMs are the most widespread mycorrhizal type in the forests of cool-temperate and boreal latitudes (Molina et al. 1992). Conversely, tropical forests are dominated by arbuscular mycorrhizas (Malloch et al. 1980), a situation that once constrained the introduction of EcM *Pinus* species in the tropics (see Pringle et al. 2009 for review). EcMs being found only in a minority of ecologically and economically important tree species that can form stands where they dominate (Hart et al. 1989, Alexander 1989, 2006, Alexander and Lee 2005, McGuire 2007). Moreover, the species richness of EcM fungi appears to have a unimodal relationship with latitudinal gradient (Tedersoo and Nara 2010, Tedersoo et al. 2012).

In tropical Africa, EcMs are mainly distributed in open forests, gallery forests, and rainforests of the Guineo-Congolian basin, Zambezian Miombo woodlands of East and South-Central Africa, and Sudanian savannah woodlands of the sub-Sahara (Fig. 1, Bâ et al. 2012).

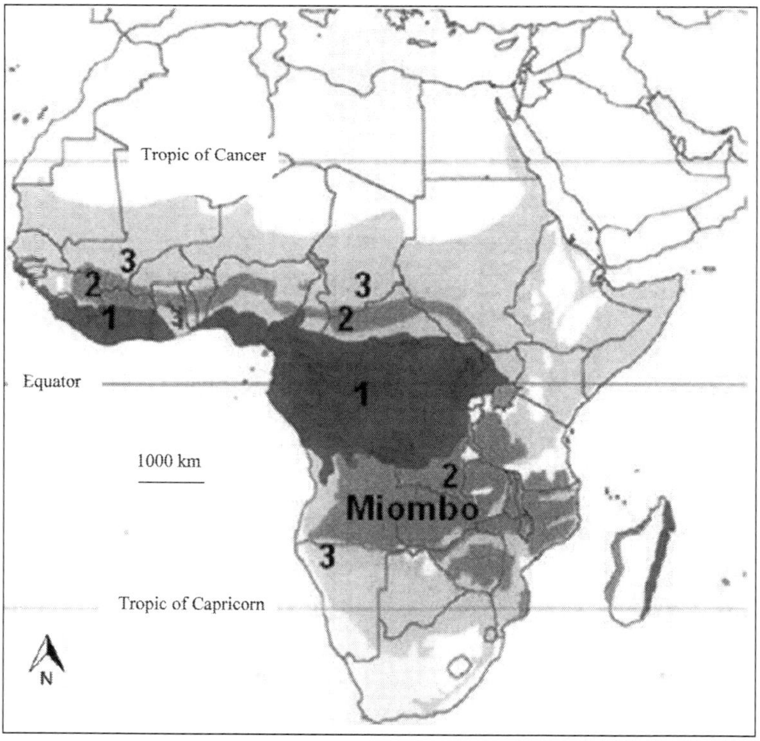

Fig. 1. Distribution of EcM trees in tropical Africa: (1) rainforests in the Guinea-Congo region; (2) open forests in the Sudanian and Zambezian regions; (3) savanna woodlands in the Sudanian and Zambezian regions (Bâ et al. 2012).

Color image of this figure appears in the color plate section at the end of the book.

EcM associations are found mainly on Caesalpinioideae, Dipterocarpaceae, Phyllantaceae, Sarcolaenaceae, Asteropeiaceae, Sapotaceae, Papilionoideae, Gnetaceae (now considered to be close to Pinaceae; Burleigh and Mathews 2004), and Proteaceae (Bâ et al. 2012). In their natural habitats, some African EcM tree species tend to aggregate in patches forming monodominant or mixed stands where they significantly contribute to the forest basal area (Newbery et al. 2004). For instance, *Microberlinia bisulcata*, *Tetraberlinia bifoliolata*, and *T. moreliana* form up to 70% of local patches in the Korup National Park in Cameroon (Newbery et al. 1997). Similarly, in the Southern Guinea rainforests (West Africa), Caesalpinioideae and Phyllanthaceae trees are the most abundant native EcM species, growing in mixed patches with high regeneration of seedlings (Diédhiou et al. 2010). In the Congo basin, *Gilbertiodendron dewevrei* forms monodominant forests where it represents more than 90% of trees in some stands, with abundant seedlings and saplings (Hart et al. 1989, Hart 1995).

It has been documented that EcM fungi contribute to the establishment and structure of these African EcM plant communities (Högberg and Piearce 1986, Newbery et al. 2000, Onguene and Kuyper 2001). However, the diversity and specificity of African EcM fungi have remained inadequately understood. Until recently, available African EcM fungi data was mainly obtained from sporocarp collections, *ex situ* fungal baiting, inoculation of seedlings, and pure culture synthesis trials (Thoen and Bâ 1989, Thoen and Ducousso 1989, Bâ and Thoen 1990, Bâ et al. 1991, Buyck et al. 1996, Sanon et al. 1997, Diédhiou et al. 2004, 2005). The use of molecular tools, such as barcoding with ITS (Nilsson et al. 2008) for *in situ* identification and phylogenetic analysis substantially improved our knowledge of the diversity and specificity of African EcM fungi (Rivière et al. 2007, Tedersoo et al. 2007, 2010a, 2011, Diédhiou et al. 2010, Jairus et al. 2011).

Indeed, a high diversity of EcM fungal species belonging to more than 25 phylogenetic lineages has been revealed from Continental Africa, Madagascar, and the Seychelles. As in temperate forests, the /russula–lactarius and /tomentella–thelephora lineages dominated the EcM fungal flora in these African forests. Furthermore, a low level of host preference and dominance of multi-host fungal species have been revealed from mixed EcM forests (Diédhiou et al. 2010, Tedersoo et al. 2011), suggesting that some EcM tree species may facilitate the recruitment and establishment of conspecific and non-conspecific seedlings. However, little is known about EcM fungal diversity in African tropical monodominant forests. Based on these different observations we addressed the following question: Do mixed forests and monodominant forests in tropical Africa have similar diversity and community structure patterns of EcM fungi? In a monodominant forest of South-Eastern Cameroon we described the EcM and sporocarp diversity, and compared these results to those found from a mixed forest of Southern Guinea (Rivière et al. 2007, Diédhiou et al. 2010).

2. Site Description

The mixed forest was located in Southern Guinea, one of the last regions of West Africa retaining a primary tropical rainforest. Above- and below-ground EcM fungal diversity surveys were conducted in typical evergreen rainforests covering hills and mountains ranging in altitude from 500 m in the Ziama forest (8°51′N, 9°31′W) to 1,752 m on the Mount Nimba forest (7°60′N, 8°49′W). The evergreen rainforests are characterized by a mean annual rainfall of 2,500–3,000 mm and a dry season (mean rainfall <15 mm) from January to March. Temperatures are generally above 24°C with relative humidity up to 80%. The soils are generally poor, lateritic and prone to heavy leaching (McGinley 2008). Canopy trees are at least 30 m tall, with some emergent individuals reaching 50–60 m in height. The canopy is dominated by *Cryptosepalum tetraphyllum* (EcM Caesalpinioideae, Fabales), as well as the AM trees *Erythrophleum ivorense* (Fabaceae) and *Heritiera utilis* (Malvaceae, Malvales). The other EcM trees are represented by Caesalpinioideae (*Afzelia bella, Anthonotha fragans, A. macrophylla, Gilbertiodendron limba, Paramacrolobium coeruleum* and *Pelligriniodendron diphyllum*), and Phyllanthaceae (a family formerly classified within Euphorbiaceae: *Uapaca esculenta, U. chevalieri, U. guineensis* and *Uapaca heudelotii*). Some of these Caesalpinioideae and Phyllanthaceae tree species grow in mixed patches where they dominate (Rivière et al. 2007, Diédhiou et al. 2010). Sporocarps (fruit-bodies) and ECMs of mature trees and seedlings were collected within 3 plots of 0.5–1 ha each in the mixed forest. For the mature trees, root tips were sampled around trees by tracing roots from the base of trunk whenever possible.

The monodominant forest of *G. dewevrei* is located in the Dja Faunal Reserve (2°49′ to 3°23′N; 12°25′ to 13°35′E) in the Southeastern region of Cameroon, which extends from Mouloundou to south of Bertoua. The climate is characterized by two wet seasons from March to June and from August to November with a mean annual rainfall varying from 1600 to 1700 mm. The two dry seasons are December–February (18–59 mm) and July (54 mm). Mean annual temperature varies from 23.7° to 24.3°C. Soils are clayey ferralitic, acidic, and poor in nutrients (Peh et al. 2011). The vegetation has a main canopy of 30–40 m with highest trees rising to 60 m (Letouzey 1985). The monodominant forests are surrounded by mixed forests including many tree and shrub species, which form mainly arbuscular mycorrhiza (e.g., *Afrostyrax lepidophyllus, Afzelia bipendensis, Anthonotha ferruginea, Baphia pubescens, Beilschmiedia louisii, Cryptosepalum congolum, Drypetes paxii,*

Entangrophragma congoensi, Erismadelphus exsul var. *platyphyllus, Fernandoa ferdinandii* and *Hesteria trillesiana*). Sporocarps and EcMs of mature trees, saplings and seedlings of *G. dewevrei* were collected within 3 plots of 0.5 ha each in a monodominant stand. EcMs from mature trees and saplings were sampled by extracting cylindrical soil cores from around trees.

3. General Traits of the EcM Fungal Communities from the Two Forest Types

A total of 856 samples, 463 from the monodominant forest and 393 from the mixed forest, were collected and analyzed by morphological and molecular approaches. Of the 856 samples 343 (145 and 198 for the monodominant forest and mixed forest, respectively) were represented by sporocarps and 513 (354 and 159 for the monodominant forest and mixed forest, respectively) were represented by EcM tips. Most of the morphologically identified sporocarps sampled from the mixed forest were vouchered and stored in the Museum National d'Histoire Naturelle, France (Rivière et al. 2007). The voucher specimens of sporocarps from the monodominant forest are kept at the herbarium of the Institute of Agricultural Research for Development (IRAD), Yaounde, Cameroon. EcMs were classified into morphotypes (MTs) for molecular analyses based on distinctive macroscopic and microscopic features: branching, color and texture of the mantle, presence or absence of emanating hyphae, rhizomorphs, and sclerotia linked to EcMs (Diédhiou et al. 2004, 2009). The molecular identification of EcM fungi from sporocarps and MTs was conducted by DNA sequencing of the internal transcribed spacer (ITS) region (Diédhiou et al. 2010, H.C.M. Michaella Ebenye unpublished data) and a portion of mitochondrial rDNA large subunit (mtLSU) (Rivière et al. 2007). Hence, from the 856 samples collected, 613 were successfully sequenced and the sequences were aligned with the Clustal X program (Thompson et al. 1997). Alignments were manually optimized with the Genedoc program (Nicholas et al. 1997). BlastN searches of obtained sequences against the INSD were conducted to determine their taxonomical affiliation (Altschul et al. 1990). ITS sequences (OTUs, operational taxonomic units) displaying >90% and >97% full-length similarity to identified fungal taxa were considered congeneric and conspecific, respectively. These sequence similarities are arbitrary, but were deemed as adequate screening for DNA barcoding DNA barcoding thresholds for fungal taxonomy at genus and species levels, respectively (Smith et al. 2007, Nilsson et al. 2008, Hughes et al. 2009, Tedersoo et al. 2010b).

Hence, 625 samples were successfully identified at the family, genus or species level by combining DNA barcoding and morphological analyses of sporocarps and EcMs collected from the two forest types. The identified fungi fell into three phyla: Ascomycota, Basidiomycota and Zygomycota (Table 1). The Basidiomycota was the most abundant phylum and the Zygomycota the least abundant. The identified fungi were then assigned to the phylogenetic lineages of EcM fungi predefined by Tedersoo et al. (2010b); however, we excluded from the analysis the genera *Chalciporus*, *Geastrum* and *Leptodontidium* as their EcM status is still debatable (see Rinaldi et al. 2008, Tedersoo et al. 2010b, Comandini et al. 2012). The names of the identified EcM-forming lineages are written in lower case, non-italicized font with a slash (e.g., /russula-lactarius; Moncalvo et al. 2002). Thus, 18 EcM-forming lineages (Table 1) accounting for 88% of the total abundance of identified samples were recorded from both African tropical forests. This result is consistent with those obtained from other tropical forests where a considerable number of fungi of unknown trophic status have been revealed from EcMs (Tedersoo et al. 2010a, Peay et al. 2010, Smith et al. 2011, 2013). The fungi of unknown trophic status included members of Agaricomycetes, Sordariomycetes, Zygomycetes, Leotiomycetes, Dothideomycetes and Tremellomycetes (Table 1). Although these latter fungi were molecularly identified from EcMs, they are traditionally considered to be parasitic, saprophytic, endophytic, or of unknown trophic status. However, some of these fungi, particularly Mortierellaceae (Zygomycetes), Hypocreaceae (Sordariomycetes) and Polyporaceae (Agaricomycetes) found on many healthy EcMs (data not shown) may represent new EcM-forming lineages in tropical ecosystems (Peay et al. 2010, Smith et al. 2011, 2013) or even on larger scales. Further work is thus necessary to clearly determine the EcM lifestyle of not only these tropical fungal taxa but also of many unexplored fungi which were often revealed in EcMs by DNA barcoding. Indeed, it was only ten years ago that important EcM taxa such as Sebacinales were confirmed to associate with tree roots (Selosse et al. 2002). The EcM-forming lineage inventory is a continuous challenge and requires extensive careful work from field to laboratory, particularly in the fungal taxa where EcM associations are uncommon (de Roman et al. 2005, Rinaldi et al. 2008, Tedersoo et al. 2010b, Comandini et al. 2012).

Among the 18 phylogenetic lineages of EcM fungi revealed from the mixed and monodominant forests, only one (/tuber-helvella) has not yet been reported from other African tropical forests (Tedersoo et al. 2007, 2010a, 2011, Jairus et al. 2011, Bâ et al. 2012). This corroborates the hypothesis of a lower diversity of phylogenetic lineages of EcM fungi in tropical forests relative to temperate forests where individual sites usually support more than 20 lineages (Tedersoo and Nara 2010). Besides the genus *Tuber*, known only from the Holarctic realm (Bonito et al. 2010), the /tuber-helvella lineage includes many separate taxa which are widely distributed in the Austral regions (Tedersoo et al. 2010b). However, the members of this EcM-forming lineage would be poorly represented in the African tropical forests

Table 1. Fungal species identified from EcMs and sporocarps collected from the two forest types (EcM-forming lineages are written in lower case, non-italicized font preceded with a slash (e.g.,/russula-lactarius)).

Type of forest:			Mixed		Monodominant	
Fungi:						
Phylum	**Class/EcM lineage**	**Family/Genus**	**sporocarps**	**EcMs**	**sporocarps**	**EcMs**
Basidiomycota	/russula-lactarius	*Russula*	+	+	+	+
		Lactarius	+	+	+	+
	/tomentella-thelephora	*Tomentella*	+	+	+	+
		Thelephora		+		+
	/boletus	*Boletus*	+	+	+	+
		Leccinum	+	+		
		Strobilomyces	+	+		
		Xerocomus	+		+	
		Tubosaeta	+			
	/sebacina	*Sebacina*				+
	/pisolithus-scleroderma	*Scleroderma*	+	+	+	
	/amanita	*Amanita*	+	+	+	+
	/tricholoma	*Tricholoma*	+	+		
		Pseudobaeospora			+	
	/clavulina	*Clavulina*		+	+	+
	/coltricia	*Coltricia*	+			
	/cantharellus	*Cantharellus*	+		+	
		Sistotrema				+
	/cortinarius	*Cortinarius*			+	
	/inocybe	*Inocybe*	+		+	

Table 1. contd....

Table 1. contd.

Type of forest:			Mixed		Monodominant	
Fungi:						
Phylum	**Class/EcM lineage**	**Family/Genus**	**sporocarps**	**EcMs**	**sporocarps**	**EcMs**
	/ramaria-gautieria	*Ramaria*			+	
	/suillus-rhizopogon	*Truncocolumella*			+	
	Agaricomycetes	*Chalciporus*	+		+	
		Geastrum			+	
		Polyporaceae				+
	Tremellomycetes	*Trichosporon*				+
		Tremellaceae				+
Ascomycota	/marcelleina-peziza gerardii	*Marcelleina*				+
	/tuber-helvella	*Helvella*			+	
	/helotiales	Helotiales				+
	/elaphomyces	*Elaphomyces*				+
	Sordariomycetes	Nectriaceae				+
		Chaetosphaeria				+
		Cercophora				+
		Sordariomycete				+
		Hypocreaceae				+
	Leotiomycetes	*Leptodontidium*				+
		Leohumicola				+
		Hyaloscyphaceae				+
	Dothideomycetes	Botryosphaericeae				+
		Leptosphaeria				+
Zygomycota	Zygomycetes	Mortierellaceae				+

as observed in the suillus-rhizopogon lineage assumed to be specific to Pinaceae (Molina et al. 1992, Bruns et al. 2002). On the other hand, there is a noticeable absence of some Holarctic and Austral EcM-forming lineages such as /cenococcum and /laccaria, and some panglobal EcM-forming lineages (e.g., /entoloma, /hebeloma-alnicola and /hysterangium) as well.

The/russula-lactarius, and /tomentella-thelephora, were the most abundant EcM-forming lineages, accounting for 43.17%, and 17.49% of the total abundance of the identified EcM fungi, respectively (Fig. 2). The relative abundances of the /boletus, /amanita, /sebacina and /pisolithus-scleroderma lineages were 10.38%, 6.58%, 5.65% and 5.28% respectively. The remaining EcM-forming lineages accounted for 11.47% of the total abundance, each contributing less than 4% (Fig. 2).

In addition, the below-ground fungal diversity and abundance differed from those observed above-ground. For instance, the /russula-lactarius, /boletus and /amanita lineages dominated the sporocarps, while the /russula-lactarius, /tomentella-thelephora, and /sebacina lineages dominated the EcMs (data not shown). Other EcM-forming lineages such as the /elaphomyces, /helotiales, and /marcelleina-peziza gerardii were

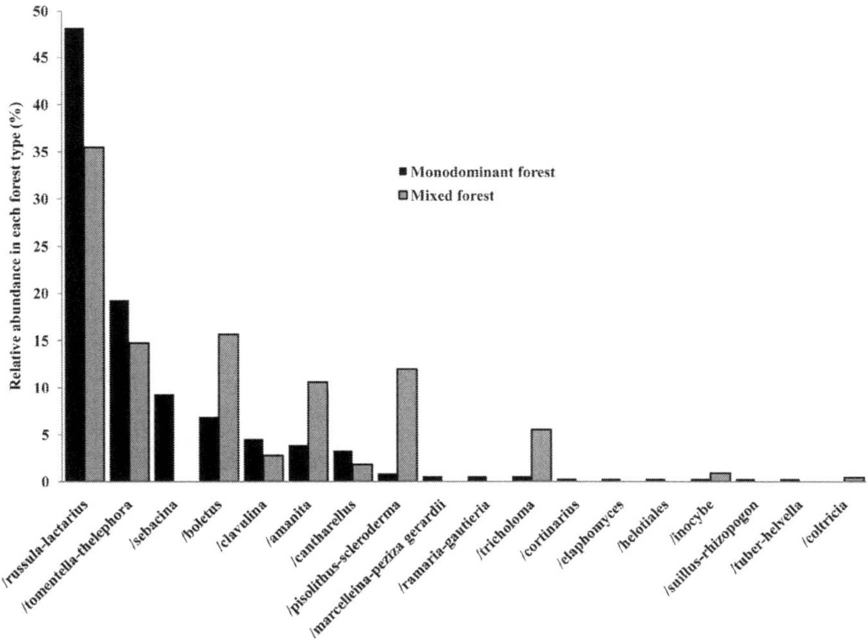

Fig. 2. Relative abundance of the EcM-forming lineages revealed from EcMs and sporocarps collected from each forest type.

found only in EcMs, whereas the/cantharellus,/coltricia, and/inocybe were identified only from sporocarps (Table 1). This result confirms the benefit of combining sporocarp and EcM surveys to predict the EcM association patterns at the root level (Richard et al. 2005, Bâ et al. 2012).

4. Comparisons of the EcM Community between the Mixed and Monodominant Forests

Of the 549 samples representing the EcM-forming lineages, 332 were collected from the monodominant forest of Cameroon and 217 from the mixed forest of Guinea. The samples from the mixed forest were dominated by sporocarps, which represented 63.30%, whereas the samples from the monodominant forest were dominated by EcMs (74.69%). Of the 18 EcM-forming lineages, nine were found in both forest types, eight were found exclusively in the monodominant forest, and one in the mixed forest (Table 1, Fig. 2). Rarefaction analysis performed using the software EstimateS ver. 8.0.0 (Colwell 2006) suggested that, for the minimal samples size $n = 217$, the richness of EcM-forming lineages was lower in the mixed forest (Chao2, Jackknife 2 and ICE were 22.09, 24.57 and 26.27, respectively for the monodominant forest *vs* 10, 11 and 10.49, respectively for the mixed forest). Three non-exclusive reasons can account for the lower richness in terms of EcM-forming lineages in the mixed forest. First, given the seasonal shifts and spatial heterogeneity that exist in EcM fungal communities, the difference in the sampling strategy (e.g., tracing roots vs. extracting soil cores) may provide two different pictures of the EcM fungal communities. Second, the age and history of hosts may influence the EcM communities through competition and selection of more adapted EcM fungal taxa (Selosse et al. 2006). Third, the lower root density of EcM hosts in the mixed forest (our personal observations) may reduce the population size of EcM fungi and subsequently result in lower richness of EcM-forming lineages (Tedersoo and Nara 2010).

Furthermore, with the exception of the /sebacina lineage which accounted for 9.34% of the total abundance of EcM fungi in the monodominant forest, the other EcM-forming lineages found exclusively in a single forest type were represented by either one or two species and displayed relative abundances ≤ 1% (Fig. 2). In this context, we cannot exclude that the members of these latter EcM-forming lineages might have been simply overlooked in one type of forest due to their apparent scarcity. The members of the /sebacina lineage are among the most common EcM fungal species in temperate and Mediterranean forests (Selosse et al. 2002, 2007, Avis et al. 2003, Weiß et al. 2004, Richard et al. 2005) and have been reported from the Neotropics (Moyersoen 2006, Morris et al. 2009, Selosse et al. 2009, Henkel et al. 2012) and the Paleotropics as well (Peay et al. 2010,

Tedersoo et al. 2010a, 2011, Jairus et al. 2011). In regard to this observation and because the members of this lineage produce inconspicuous sporocarps, one could argue that they might have been missed in the mixed forest where relatively few samples were collected. However, some EcM-forming lineages such as /pisolithus-scleroderma and /tricholoma were more abundant in the mixed forest (accounting for 11.98% and 5.53%, respectively) than in the monodominant forest where they accounted for < 1% each (Fig. 2). Thus, although possibly underestimating the number and abundance of EcM-forming lineages in the mixed forest, our results show that the monodominant forest tends to harbor more EcM-forming lineages.

In terms of species richness, the lineages of /russula-lactarius (86 spp.), /tomentella-thelephora (30 spp.), /sebacina (19 spp.), and /boletus (13 spp.) dominated in the monodominant forest, while the /russula-lactarius (42 spp.), /tomentella-thelephora (21 spp.), /boletus (20 spp.), and /amanita (15 spp.) dominated in the mixed forest. Other species-rich lineages included the /pisolithus-scleroderma (10 spp.) and /tricholoma (10 spp.) in the mixed forest and the /clavulina (11 spp.) and /amanita (10 spp.) in the monodominant forest (Fig. 3).

The EcM-forming lineages included 189 and 126 putative EcM species for the monodominant forest and mixed forest, respectively (Table 2). The high EcM species richness observed from these African forests is a

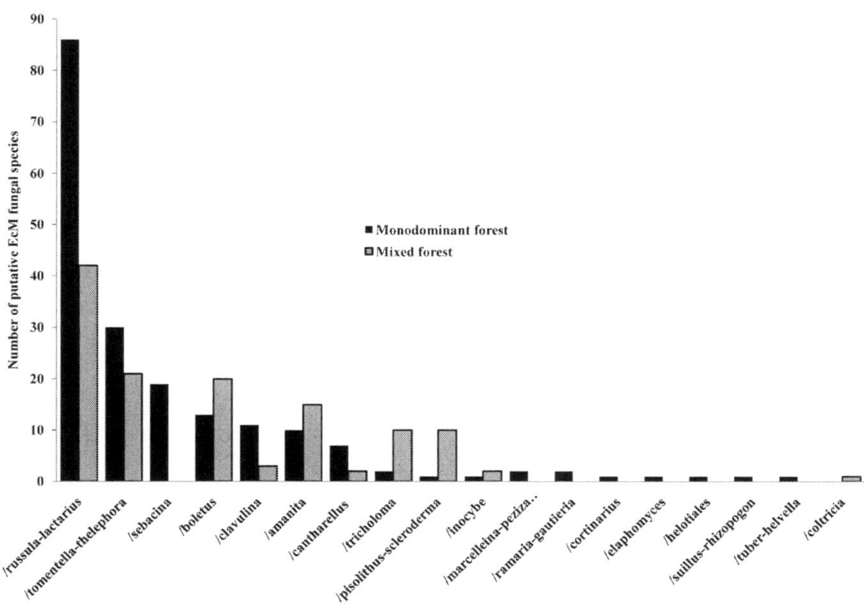

Fig. 3. Number of putative EcM fungal species identified from EcMs and sporocarps collected from each forest type.

Table 2. Number (No.) of EcM samples observed and estimated EcM species richness in each forest type. *The numbers appearing in bold were obtained after rarefaction to 217.

	Monodominant forest	Mixed forest
No. of EcM samples	332	217
No. of putative EcM species	189	126
	137.37*	**126**
Chao2 (95% CI)	575.42 (419.98–835.47)	249.08 (193.87–349.18)
	478.06 (321.39–768.92)	**249.08 (193.87–349.18)**
Jackknife 2	440.97	259.25
	329.82	**259.25**
ICE	675.07	282.2
	523.9	**282.2**
Fisher's alpha ± SD	182.1 ± 17.87	125.51 ± 15.38
	161.82 ± 20.93	**125.51 ± 15.38**
Shannon ± SD	4.86	4.62
	4.64 ± 0.05	**4.62**

common trend in Neotropical (Morris et al. 2009, Smith et al. 2011, 2013, Henkel et al. 2012) and Paleotropical forests (Peay et al. 2010, Tedersoo et al. 2010b, 2011). The calculation of sample-based rarefaction curves and minimal species richness estimates showed that the sampling efforts were insufficient in both types of forests. Indeed, the rarefied accumulation curves of species and minimal species richness estimates (Chao2, Jack2 and ICE) did not reach a clear asymptote with increasing sample size (Fig. 4). The Chao2, Jack2 and ICE richness estimators predicted 249.08, 259.25 and 282.2 putative EcM species, respectively for the mixed forest and 575.42, 440.97 and 675.07 putative EcM species, respectively for the monodominant forest (Table 2).

Surprisingly, when rarefying the samples from the monodominant forest to $n = 217$ to match the sample size of the mix from the mixed forest, we found no significant difference in terms of richness of putative EcM fungal species between the two forests types (Table 2, Fig. 4). Indeed, the 95% CI (confidence intervals) for Chao2 are overlapping and the indices of species diversity, Fisher's alpha and Shannon are not significantly different between the two types of forests (Table 2). This lack of difference in richness of EcM fungal species between the mixed forest and monodominant forest may be in part related to the prevalence of multi-host EcM fungi (Onguene and Kuyper 2002, Diédhiou et al. 2010) which could mask the influence of plant host species on EcM fungal community structure. Interestingly, this rejects the generalization of the idea that host diversity is the sole driver of diversity in EcM community (Dickie 2007, Ishida et al. 2007, Tedersoo

a

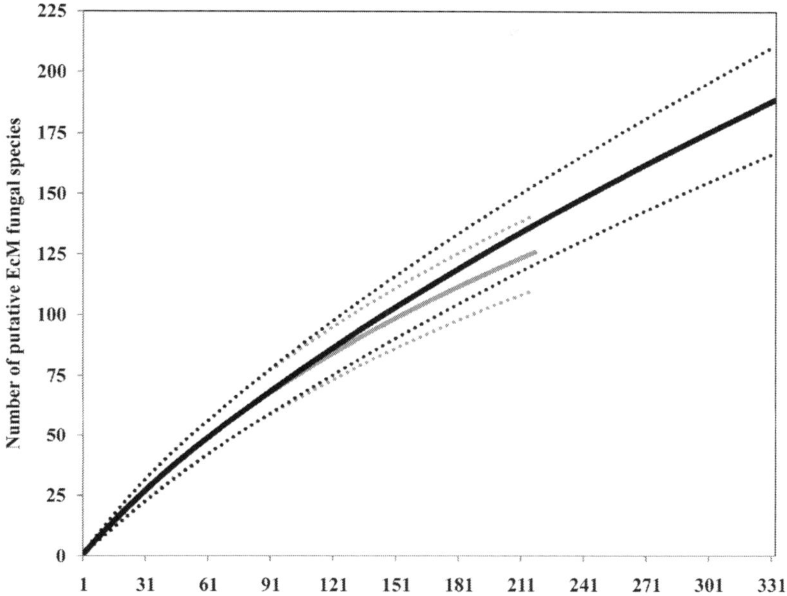

b

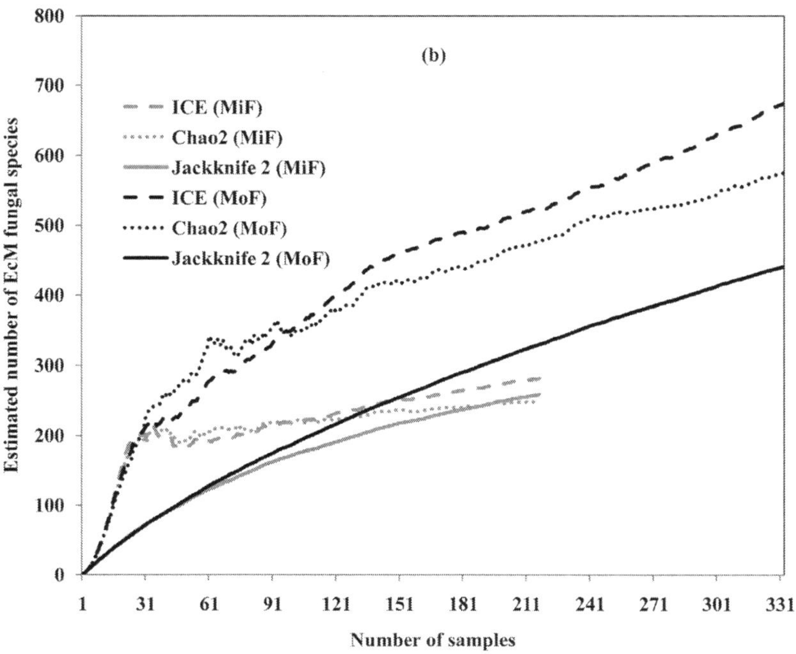

Fig. 4. Rarefied accumulation curves of (a) EcM fungal species and (b) minimal species richness estimates for the monodominant forest (MoF, black) and mixed forest (MiF, grey). Dotted lines in (a) indicate 95% CI.

et al. 2008). This idea may be supported by the finding that African EcM fungal communities are not strongly structured by soil horizon and host at the plant species and family levels (Tedersoo et al. 2011).

5. Conclusion

This study has provided valuable information on the diversity and community structure of EcM fungi from mixed and monodominant forests in tropical Africa. The monodominant forest tends to harbor more EcM-forming lineages than the mixed forest, while there was no significant difference in terms of richness of EcM fungal species between these two forest types. The dominant EcM-forming lineages are similar in the two forest types. On the other hand, a large number of fungi of unknown trophic status were recovered from healthy EcMs; some of them may represent new, overlooked tropical EcM-forming lineages. From this finding, it should be evident that further research is necessary to (i) determine the EcM lifestyle of the latter fungal taxa, (ii) rigorously assess and compare the phylogenetic community composition between the two forest types, and (ii) highlight the role of EcM fungi in the recruitment and establishment of seedlings of EcM tree species in monodominant and mixed forests.

Acknowledegments

We thank the anonymous referees for their valuable comments on this study, and Krista L. McGuire and Caitlyn Gillikin for improving the language.

References

Alexander, I.J. 1989. Mycorrhizas in tropical forests. *In*: J. Proctor (ed.). Mineral Nutrients in Tropical Forest and Savanna Ecosystems. Blackwell Scientific Publications, Oxford, pp. 169–188.

Alexander, I.J. 2006. Ectomycorrhizas-out of Africa? New Phytol. 172: 589–591.

Alexander, I.J. and S.S. Lee. 2005. Mycorrhizas and ecosystem processes in tropical rain forest: implications for diversity. *In*: D. Burslem, M. Pinard and S. Hartley (eds.). Biotic Interactions in the Tropics. Cambridge University Press, Cambridge, pp. 165–203.

Altschul, S.F., W. Gish, W. Miller, E.W. Myers and D.J. Lipman. 1990. Basic local alignment search tool. J. Mol. Biol. 215: 403–410.

Avis, P.G., D.J. McLaughlin, B.C. Dentinger and P.B. Reich. 2003. Long-term increase in nitrogen supply alters above- and below-ground ectomycorrhizal communities and increases the dominance of *Russula* spp. in a temperate oak savanna. New Phytol. 160: 239–253.

Bâ, A.M., J. Garbaye and J. Dexheimer. 1991. Influence of fungal propagules during the early stade of the time sequence of ectomycorrhizal colonization on *Afzelia africana* Sm. Seedlings. Can. J. Bot. 66: 2442–2447.

Bâ, A.M. and D. Thoen. 1990. First syntheses of ectomycorrhizas between *Afzelia africana* Sm. (Caesalpinioideae) and native fungi from West Africa. New Phytol. 103: 441–448.

Bâ, A.M., R. Duponnois, B. Moyersoen and A.G. Diedhiou. 2012. Ectomycorrhizal symbiosis of tropical African trees. Mycorrhiza 22: 1–29.
Bonito, G.M, A.P. Gryganskyi, J.M. Trappe and R. Vilgalys. 2010. A global meta-analysis of Tuber ITS rDNA sequences: species diversity, host associations and long-distance dispersal. Mol. Ecol. 19: 4994–5008.
Brundrett, M.C. 2009. Mycorrhizal associations and other means of nutrition of vascular plants: understanding global diversity of host plants by resolving conflicting information and developing reliable means of diagnosis. Plant Soil 320: 37–77.
Bruns, T.D., M.I. Bidartondo and D.L. Taylor. 2002. Host specificity in ectomycorrhizal communities: what do the exceptions tell us? Integr. Comp. Biol. 42: 352–359.
Burleigh, J.G. and S. Matthews. 2004. Phylogenetic signal in nucleotide data from seed plants: implications for resolving the seed plant tree of life. Am. J. Bot. 91: 1599–1613.
Buyck, B., D. Thoen and R. Walting. 1996. Ectomycorrhizal fungi of the Guinea–Congo region. Proc. R. Soc. Edinb. 104: 313–333.
Colwell, R.K. 2006. Estimates: Statistical Estimation of Species Richness and Shared Species from Samples, Version 8 [WWW document]. URL http://purl.oclc.org/estimates.
Comandini, O., A.C. Rinaldi and T.W. Kuyper. 2012. Measuring and estimating ectomycorrhizal fungal diversity: a continuous challenge. *In*: M.C. Pagano (ed.). Mycorrhiza Occurrence in Natural and Restored Environments. Nova Science Publishers, Hauppauge, NY, pp. 165–200.
de Roman, M., V. Claveria and A.M. de Miguel. 2005. A revision of the descriptions of ectomycorrhizas published since 1961. Mycol. Res. 109: 1063–1104.
Dickie, I.A. 2007. Host preference, niches and fungal diversity. New Phytol. 174: 230–233.
Diédhiou, A.G., A.M. Bâ, S. Sylla, B. Dreyfus, M. Neyra and I. Ndoye. 2004. The early-stage ectomycorrhizal Thelephoroid fungal sp. is competitive and effective on Afzelia africana Sm. in nursery conditions in Senegal. Mycorrhiza 14: 313–322.
Diédhiou, A.G., O. Guèye, M. Diabaté, Y. Prin, R. Duponnois, B. Dreyfus and A.M. Bâ. 2005. Contrasting responses to ectomycorrhizal inoculation in seedlings of six tropical African tree species. Mycorrhiza 16: 11–17.
Diédhiou, A.G., J.-L. Dupouey, M. Buée, E. Dambrine, L. Laüt and J Garbaye. 2009. Response of ectomycorrhizal communities to past Roman occupation in an oak forest. Soil Biol. Bioch. 41: 2206–2213.
Diédhiou, A.G., M.-A. Selosse, A. Galiana, M. Diabaté, B. Dreyfus, A.M. Bâ, S.M. de Faria and G. Béna. 2010. Multi-host ectomycorrhizal fungi are predominant in a Guinean tropical rain forest and shared between canopy and tree seedlings. Environ. Microbiol. 2: 2219–2232.
Hart, T.B. 1995. Seed, seedling and sub-canopy survival in monodominant and mixed forests of the Ituru forest, Africa. J. Trop. Ecol. 11: 443–459.
Hart, T.B., J.A. Hart and P.G. Murphy. 1989. Monodominant and species-rich forests of the humid tropics: causes for their occurrence. Am. Nat. 133: 613–633.
Henkel, T.W., M.C. Aime, M.M.L. Chin, S.L. Miller, R. Vilgalys and M.E. Smith. 2012. Ectomycorrhizal fungal sporocarp diversity and discovery of new taxa in *Dicymbe* monodominant forests of the Guiana Shield. Biodivers. Conserv. 21: 2195–2220.
Högberg, P. and G.D. Piearce. 1986. Mycorrhizas in Zambian trees in relation to host taxonomy, vegetation type and successional patterns. J. Ecol. 74: 775–785.
Hughes, K.W., R.H. Peterson and E.B. Lickey. 2009. Using heterozygosity to estimate a percentage DNA sequence similarity for environmental species' delimitation across basidiomycete fungi. New Phytol. 182: 795–798.
Ishida, T.A., K. Nara and T. Hogetsu. 2007. Host effects on ectomycorrhizal fungal communities: insight from eight host species in mixed conifer-broadleaf forests. New Phytol. 174: 430–440.
Jairus, T., R. Mpumba, S. Chinoya and L. Tedersoo. 2011. Invasion potential and host shifts of Australian and African ectomycorrhizal fungi in mixed eucalypt plantations. New Phytol. 192: 179–187.

Letouzey, R. 1985. Notice de la carte phytogéographique du Cameroun au 1/500.000, Institut de la carte internationale de la végétation, Toulouse, France : 30pp.
Malloch, D.W., K.A. Pirozynski and P.H. Raven. 1980. Ecological and evolutionary significance of mycorrhizal symbioses in vascular plants (a review). Proc. Nad. Acad. Sci. USA 77: 2113–2118.
McGinley, M. 2008. Western Guinean lowland forests. *In:* Cutler J. Cleveland (ed.). Encyclopedia of Earth. World Wildlife Fund, Available: http://www.eoearth.org/article/Western_Guinean_lowland_forests. Accessed 2013 May 6.
McGuire, K.L. 2007. Common ectomycorrhizal networks may maintain monodominace in a tropical rain forest. Ecology 88: 567–574.
Molina, R., H. Massicotte and J.M. Trappe. 1992. Specificity phenomena in mycorrhizal symbioses: community-ecological consequences and practical implications. *In*: M.F. Allen (ed.). Mycorrhizal Functioning. Chapman & Hall, London, pp. 357–423.
Moncalvo, J.-M., R. Vilgalys, S.A. Redhead, J.E. Johnson, T.Y. James, M.C. Aime, V. Hofstetter, S.J.W. Verduin, E. Larsson, T.J. Baroni., R.G. Thorn, S. Jacobsson, H. Clémençon and O.K. Miller Jr. 2002. One hundred and seventeen clades of euagarics. Mol. Phylogenet. Evol. 23: 357–400.
Morris, M.H., M.A. Perez-Perez, M.E. Smith and C.S. Bledsoe. 2009. Influence of host species on ectomycorrhizal communities associated with two co-occurring oaks (*Quercus* spp.) in a tropical cloud forest. FEMS Microbiol. Ecol. 69: 274–287.
Moyersoen, B. 2006. *Pakaraimaea dipterocarpacea* is ectomycorrhizal, indicating an ancient Gondwanaland origin for the ectomycorrhizal habit in Dipterocarpaceae. New Phytol. 172: 753–62.
Newbery, D.M., I.J. Alexander and J.A. Rother. 1997. Phosphorus dynamics in a lowland African rain forest: the influence of ectomycorrhizal trees. Ecol. Monogr. 67: 367–409.
Newbery, D.M., I.J. Alexander and J.A. Rother. 2000. Does proximity to conspecific adults influence the establishment of ectomycorrhizal trees in rain forest? New Phytol. 147: 401–409.
Newbery, D.M., X.M. van der Burgt and M.-A. Moravie. 2004. Structure and inferred dynamics of a large grove of *Microberlinia bisulcata* trees in central African rain forest: the possible role of periods of multiple disturbance events. J. Trop. Ecol. 20: 131–143.
Nicholas, K.B., H.B.J. Nicholas and D.W. Deerfield. 1997. GeneDoc: analysis and visualization of genetic variation. The Europ. Mol. Biol. Network Newsletter 4: 14.
Nilsson, R.H., E. Kristiansson, M. Ryberg, N. Hallenberg and K.H. Larsson. 2008. Intraspecific ITS variability in the Kingdom Fungi as expressed in the international sequence databases and its implications for molecular species identification. Evol. Bioinform. Online 4: 193–201.
Onguene, N.A. and T.W. Kuyper. 2001. Mycorrhizal associations in the rain forest of South Cameroon. For. Ecol. Manag. 140: 277–287.
Onguene, N.A. and T.W. Kuyper. 2002. Importance of the ectomycorrhizal network for seedling survival and ectomycorrhiza formation in rain forests of south Cameroon. Mycorrhiza 12: 13–17.
Peay, K.G., P.G. Kennedy, S.J. Davies, S. Tan and T.D. Bruns. 2010. Potential link between plant and fungal distributions in a dipterocarp rainforest: community and phylogenetic structure of tropical ectomycorrhizal fungi across a plant and soil ecotone. New Phytol. 185: 529–542.
Peh, K.S.-H., B. Sonké, J. Lloyd, C.A. Quesada and S.L. Lewis. 2011. Soil Does Not Explain Monodominance in a Central African Tropical Forest. PLoS ONE 6(2): e16996. doi:10.1371/journal.pone.0016996.
Pringle, A., J.D. Bever, M. Gardes, J.L. Parrent, M.C. Rillig and J.N. Klironomos. 2009. Mycorrhizal Symbioses and Plant Invasions. Annu. Rev. Ecol. Evol. Syst. 40: 699–715.
Richard, F., S. Millot, M. Gardes and M.-A. Selosse. 2005. Diversity and specificity of ectomycorrhizal fungi retrieved from an old-growth Mediterranean forest dominated by *Quercus ilex*. New Phytol. 166: 1011–1023.

Rinaldi, A.C., O. Comadini and T.W. Kuyper. 2008. Ectomycorrhizal fungal diversity: separating the wheat from the chaff. Fungal Divers 33: 1–45.
Rivière, T., A.G. Diédhiou, M. Diabaté, G. Senthilarasu, K. Natarajan, A. Verbeken, B. Buyck, B. Dreyfus, G. Bena and A.M. Bâ. 2007. Genetic diversity of ectomycorrhizal basidiomycetes from African and Indian tropical forests. Mycorrhiza 17: 415–428.
Sanon, K.B., A.M. Bâ and J. Dexheimer. 1997. Mycorrhizal status of some fungi fruiting beneath indigenous trees in Burkina Faso. For. Ecol. Manag. 98: 61–69.
Selosse, M.-A., R. Bauer and B. Moyersoen. 2002. Basal hymenomycetes belonging to the Sebacinaceae are ectomycorrhizal on temperate deciduous trees. New Phytol. 155: 183–195.
Selosse, M.-A., F. Richard, X. He and S.W. Simard. 2006. Mycorrhizal networks: des liaisons dangereuses? Trends Ecol. Evol. 21: 621–628.
Selosse, M.-A., S. Setaro, F. Glatard, F. Richard, C. Urcelay and M. Weiß. 2007. Sebacinales are common mycorrhizal associates of Ericaceae. New Phytol. 174: 864–878.
Selosse, M.-A., M.-P. Dubois and N. Alvarez. 2009. Do Sebacinales commonly associate with plant roots as endophytes? Mycol. Res. 113: 1062–1069.
Smith, M.E., G.W. Douhan and D.M. Rizzo. 2007. Intra-specific and intrasporocarp ITS variation of ectomycorrhizal fungi as assessed by rDNA sequencing of sporocarps and pooled ectomycorrhizal roots from a Quercus woodland. Mycorrhiza 18: 15–22.
Smith, M.E., T.W. Henkel, M.C. Aime, A.K. Fremier and R. Vilgalys. 2011. Ectomycorrhizal fungal diversity and community structure on three co-occurring leguminous canopy tree species in a Neotropical rainforest. New Phytol. 192: 699–712.
Smith, M.E., T.W. Henkel, J.K. Uehling, A.K. Fremier, H.D. Clarke and R. Vilgalys. 2013. The Ectomycorrhizal Fungal Community in a Neotropical Forest Dominated by the Endemic Dipterocarp *Pakaraimaea dipterocarpacea*. PLoS ONE 8: e55160. doi:10.1371/journal.pone.0055160.
Smith, S. and J. Read. 2008. Mycorrhizal Symbiosis. Third edition: 800pp.
Tedersoo, L. and K. Nara. 2010. General latitudinal gradient of biodiversity is reversed in ectomycorrhizal fungi. New Phytol. 185: 351–354.
Tedersoo, L., T. Suvi, K. Beaver and U. Kõljalg. 2007. Ectomycorrhizal fungi of the Seychelles: diversity patterns and hosts shifts from the native *Vateriopsis seychellarum* (Dipterocarpaceae) and *Intsia bijuga* (Caesalpinioideae) to the introduced *Eucalyptus robusta* (Myrtaceae), but not *Pinus caribea* (Pinaceae). New Phytol. 175: 321–333.
Tedersoo, L., J. Teele, B.M. Horton, K. Abarenkov, T. Suvi, I. Saar and U. Kõljalg. 2008. Strong host preference of ectomycorrhizal fungi in a Tasmanian wet sclerophyll forest as revealed by DNA barcoding and taxon-specific primers. New Phytol. 180: 479–490.
Tedersoo, L., R. Nilsson, K. Abarenkov and T. Jairus. 2010a. 454 Pyrosequencing and Sanger sequencing of tropical mycorrhizal fungi provide similar results but reveal substantial methodological biases. New Phytol. 188: 291–301.
Tedersoo, L., T.W. May and M.E. Smith. 2010b. Ectomycorrhizal lifestyle in fungi: global diversity, distribution, and evolution of phylogenetic lineages. Mycorrhiza 20: 217–263.
Tedersoo, L., M. Bahram, T. Jairus, E. Bechem, S. Chinoya, R. Mpumba, M. Leal, E. Randrianjohany, S. Razafimandimbison, A. Sadam, T. Naadel and U. Kõljalg. 2011. Spatial structure and the effects of host and soil environments on communities of ectomycorrhizal fungi in wooded savannas and rain forests of Continental Africa and Madagascar. Mol. Ecol. 20: 3071–3080.
Tedersoo, L., M. Bahram, M. Toots, A.G. Diédhiou, T.W. Henkel, R. Kjoller, M.H. Morris, K. Nara, E. Nouhra, K.G. Peay, S. Põlme, M. Ryberg, M.E. Smith and U. Kõljalg. 2012.

Towards global patterns in the diversity and community structure of ectomycorrhizal fungi. Mol. Ecol. 21: 4160–4170.
Thoen, D. and A.M. Bâ. 1989. Ectomycorrhizas and putative ectomycorrhizal fungi of *Afzelia africana* Sm. and *Uapaca guineensis* Müll. Arg. in southern Senegal. New Phytol. 113: 549–559.
Thoen, D. and M. Ducousso. 1989. Champignons et ectomycorhizes du Fouta Djalon. Bois For. Trop. 221: 45–63.
Thompson, J.D., T.J. Gibson, F. Plewniak, F. Jeanmorgin and D.G. Haggins. 1997. The clustalx windows, interface: flexible, strategies for multiple sequence alignment aided by quality analysis tools. Nucleic Acid Res. 25: 4876–4882.
Weiß, M., M.-A. Selosse, K.-H. Rexer, A. Urban and F. Oberwinkler. 2004. Sebacinales: a hitherto overlooked cosm of heterobasidiomycetes with a broad mycorrhizal potential. Mycol. Res. 108: 1003–1010.

[1]Département de Biologie Végétale, Université Cheikh Anta Diop (UCAD), BP 5005 Dakar, Sénégal.
[2]Laboratoire Commun de Microbiologie IRD/UCAD/ISRA, Centre de Recherche de Bel Air, BP. 1386, Dakar, Sénégal.
[3]Centre d'Ecologie Fonctionnelle et Evolutive (CNRS, UMR 5175), 1919 Route de Mende, 34 293 Montpellier cedex 5, France.
[4]Muséum national d'Histoire naturelle, Département Systématique et Evolution (UMR 7205 OSEB), 45 rue Buffon, 75005 Paris, France.
[5]Institut de Recherche Agronomique pour le Développement, BP. 2123, Yaoundé, Cameroun.
[6]Laboratoire des Symbioses Tropicales et Méditerranéennes (LSTM), UMR113, Campus de Baillarguet, A10/J, 34398 Montpellier, Cedex 5, France.
*Corresponding author: fungalmolecol@gmail.com

CHAPTER 2

Ectomycorrhizas of Three Species of Nyctaginaceae in the Tropical Mountain Rain Forest of South Ecuador

Ingeborg Haug,[1,*] *Ingrid Kottke*[1] and *Juan Pablo Suárez*[2]

1. Introduction

Arbuscular mycorrhizas are the most ancient mycorrhizal association (Taylor 1995) and are found in more than 80% of the plant species (Smith and Read 2008). In a limited number of plant groups, the arbuscular association was replaced by an ectomycorrhizal association (Smith and Read 2008). Certain families such as Pinaceae and Fagaceae exclusively have ectomycorrhizal associations, but there are also several isolated lineages of ectomycorrhization in other families (Brundrett 2002). For instance, in the Nyctaginaceae, only a few ectomycorrhiza forming species are known; these are found in *Neea* Ruiz & Pav. (Harley and Smith 1983), *Guapira* Aubl. (syn *Torrubia*, Harley and Smith 1983) and *Pisonia* L. (Ashford and Allaway 1982). All three genera belong to the tribe Pisonieae (Bittrich and Kühn

[1]Institute of Evolution and Ecology, Plant Evolutionary Ecolology, Eberhard-Karls-University Tübingen, Auf der Morgenstelle 1, D-72076 Tübingen, Germany.
[2]Departamento de Ciencias Naturales, Universidad Técnica Particular de Loja, Loja, Ecuador.
*Corresponding author: ingeborg.haug@uni-tuebingen.de

1993). Species of *Neea* were found to be ectomycorrhizal in the Amazonian rainforest (Singer and Araujo 1979, Janos 1980a,b, Tedersoo et al. 2010) and in Peru (Alexander and Högberg 1986) or ecto- and arbuscular mycorrhizal in an Amazonian lowland rain forest (St. John 1980a,b), in the South of Venezuela (Moyersoen 1993), and in French Guiana (Béreau et al. 1997). Species of *Guapira* were found to be ecto- and arbuscular mycorrhizal in the South of Venezuela (Moyersoen 1993) and ectomycorrhizal in Western Amazonia (Tedersoo et al. 2010). Ectomycorrhizas of *Pisonia grandis* R.Br. were described from coral cays in the Great Barrier Reef (Ashford and Allaway 1982, Chambers et al. 2005) and from the Seychelles (Ashford and Allaway 1985, Suvi et al. 2010). The mycorrhizas associated with *Pisonia grandis* have a unique structure with a hyphal mantle, Hartig net, and transfer cells (Ashford and Allaway 1982, 1985). Arbuscular mycorrhizal infection was detected in other species of Nyctaginaceae, namely *Boerhavia repens, Bougainvillea spectabilis, Mirabilis jalapa, Pisonia umbellifera* (Koske et al. 1992) and *Pisonia seychellarum* (Suvi et al. 2010). A low degree of mycorrhization with only vesicles and mycelium was reported for *Boerhavia diffusa* (Rachel et al. 1989) and no mycorrhization was reported for *Mirabilis jalapa, Boerhavia diffusa* (Muthukumar and Udaiyan 2000) and *Boerhavia coccinea* (Khan 1974). Based on these results, the Nyctaginaceae were considered to be a predominantly non-mycorrhizal family with isolated ectomycorrhizal and arbuscular mycorrhizal species (Tester et al. 1987, Brundrett 2002, Wang and Qiu 2006). However, a very small percentage of the ca. 400 species included in the family (Daly and Roberts 2004) have been investigated so far (Wang and Qiu 2006). In summary, the available data of ectomycorrhizas of Nyctaginaceae show typical and unique structures. Only a few fungal partners are identified with molecular methods. In this study we aim to enhance the knowledge of morphological features of ectomycorrhizas in the neotropics and to identify their fungal partners.

2. Ectomycorrhizas of Nyctaginaceae in the Tropical Mountain Rain Forest of South Ecuador

Two species of *Neea* and one species of *Pisonia* sampled in the species-rich Neotropical mountain rain forest in South Ecuador (Haug et al. 2005) were found to be ectomycorrhizal. While *Neea* sp. 1 has typical ectomycorrhizal structures, *Neea* aff. *floribunda* Poepp. & Endl. and *Pisonia* sp. showed distinctive and unique anatomical and morphological features.

2.1 Ectomycorrhizas of Neea species 1

The root system consists of long and short roots. Ectomycorrhizas are formed on short roots (diameter 0.2 to 0.6 mm). Five different mycorrhizal morphotypes were distinguished (Fig. 1a–e). All morphotypes revealed the typical features of ectomycorrhizas with a hyphal mantle and Hartig net located between rhizodermal cells (Haug et al. 2005).

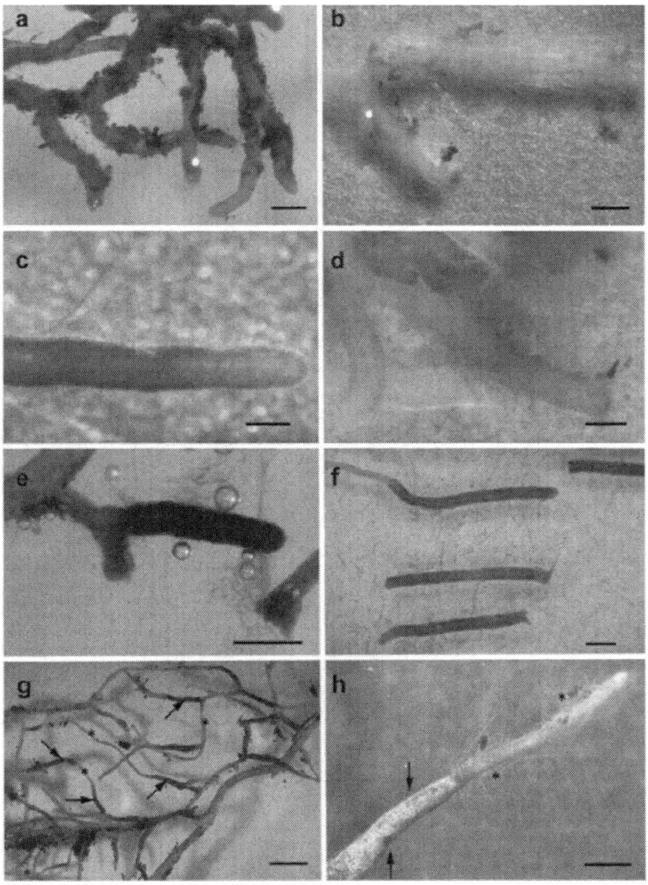

Fig. 1. Micrographs of mycorrhizas of *Neea* species 1 (a-e), *Neea* aff. *floribunda*, *Pisonia* sp. (g-h), a. *Russula puiggarii* – *Neea* species 1, b. *Russula-Lactarius* – *Neea* species 1, **c.** *Tomentella-Thelephora* species 1 – *Neea* species 1, d. *Tomentella-Thelephora* species 2 – *Neea* species 1, e. Ascomycete – *Neea* species 1, f. *Tomentella-Thelephora* species 3 – *Neea* aff. *floribunda*, g.,h. *Tomentella-Thelephora* species 3 – *Pisonia* sp., g. overview: long roots partially with hyphal mantle (arrows), partially with root hairs and no hyphal mantle (*), h. one root shown at higher magnification: distal portion with root hairs (*), proximal portion with hyphal mantle (arrows). Scale bars: a.,b.,f.,h. 1 mm; c.-e. 0,5 mm; g. 5 mm. Data from Haug et al. 2005.

Color image of this figure appears in the color plate section at the end of the book.

Molecular analyses of the ectomycorrhizal partners of *Neea* species 1 revealed six different fungi: two Thelephoraceae belonging to the genera *Tomentella-Thelephora* (AY667418, AY667419), two Russulaceae (*Russula-Lactarius*, AY667426, AY667427), one species of *Clavulina* (Tedersoo et al. 2010), and one ascomycete. Most frequent were the thelephoraceous ectomycorrhizas.

2.2 Ectomycorrhizas of Neea aff. floribunda Poepp. & Endl.

The root system of *Neea* aff. *floribunda* consists of brown, straight long roots which bear bright, unramified fine roots. The fine roots (diameter 0.3 to 0.6 mm) are partially covered with a hyphal mantle (Fig. 1f). Only one morphotype with a brown and smooth hyphal mantle was found. The outer mantle layers are loosely plectenchymatous (Fig. 2a) while the middle

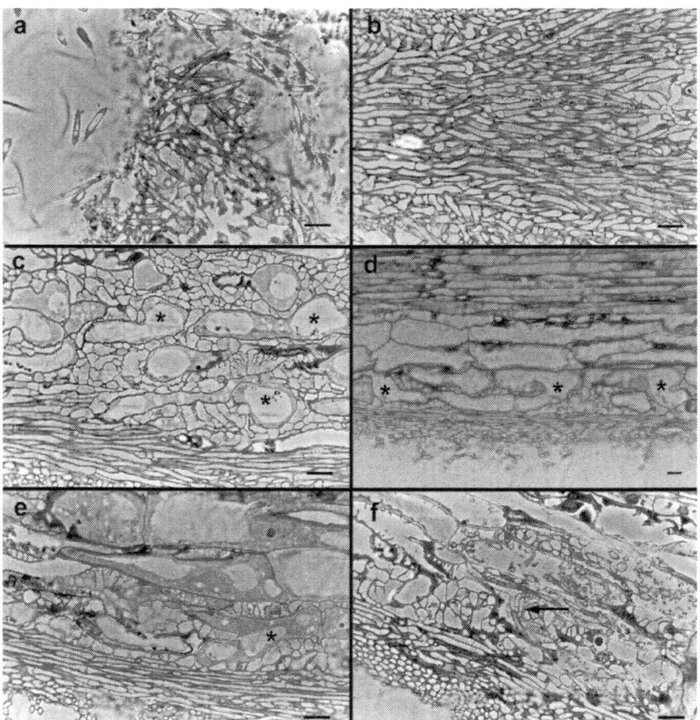

Fig. 2. Longitudinal sections of mycorrhizas of *Neea* aff. *floribunda*. (a) Outer hyphal mantle: loose plectenchyma, (b) middle and inner hyphal mantle: dense plectenchyma, (c) tangential section of the epidermal layer: epidermal outgrowths (*) and Hartig net, d,e. median section through mycorrhiza: hyphal mantle, epidermal cells with outgrowths (*) and Hartig net, f. hyphal mantle, Hartig net and intracellular hyphae (with clamps, →) in epidermal and cortical cells (scale bars: a. - f. 15 µm). Data from Haug et al. 2005.

and inner layers are compactly plectenchymatous (Fig. 2b). A prominent Hartig net develops between root-hair-like outgrowths of the epidermal cells (Figs. 2c–e). The root-hair-like outgrowths are attached to the root surface (Figs. 2d,e) and are elongated cells with a rounded base in tangential section (Fig. 2c). In median sections, connections of the outgrowths to epidermal cells are evident (Fig. 2d,*), which occasionally join up and look like a septum (Fig. 2d). Epidermal cells, cortical cells, and hyphae are plasmatic. There are regions with intracellular hyphae in the epidermal and cortical cells and these hyphae form clamps (Fig. 2f). Molecular analyses of different mycorrhizal tips of *Neea* aff. *floribunda* yielded the same sequence (AY667424). A BLAST search and a NJ-tree clearly showed a membership within the genera *Thelephora/Tomentella* (Haug et al. 2005).

2.3 Ectomycorrhizas of Pisonia sp.

The fine root systems of *Pisonia* consist of only long roots. There are sections of the long roots (diameter 0.3 to 0.6 mm) covered with a hyphal mantle and sections without hyphal mantle but with root hairs (Figs. 1g–h). Root tips covered with a hyphal mantle are rare (Fig. 1h) and many roots have no hyphal mantle at all. Only one morphotype is distinguishable. The hyphal mantle is dark brown with silvery patches on the surface (Figs. 1g, h). Emanating hyphae are colourless and bear clamps. The hyphal mantle is plectenchymatous throughout, where the outer layers are loosely (Fig. 3a) and the middle layers are densely plectenchymatous (Fig. 3b). The epidermal layer develops a Hartig net (Figs. 3c–e). Root segments without hyphal mantle have root hairs. Loose hyphae and intracellular colonization in the root hairs and in the epidermal cells are visible (Figs. 3f–g). Since ultrastructural details are the same as in the hyphae of the Hartig net and hyphal mantle (Figs. 3h-i), we conclude that the intracellular infection is due to the ectomycorrhizal fungus. All mycorrhizas from four individuals of the species of *Pisonia* had identical ITS sequences (AY667420-423). Phylogenetic analyses of the rDNA ITS region revealed a membership within the genera *Thelephora/Tomentella* (Haug et al. 2005).

Morphological and molecular results from ectomycorrhizas of the three Nyctaginacean species are summarized in Table 1.

2.4 Ectomycorrhizas of Nyctaginaceae

The morphology of ectomycorrhizas found on the short roots of *Neea* species 1 was also seen in *Neea obovata* (Moyersoen 1993). Fine root systems of only long roots such as those of *Neea* aff. *floribunda* and the species of *Pisonia* were reported from *Neea robusta* (Moyersoen 1993) and *Pisonia grandis*

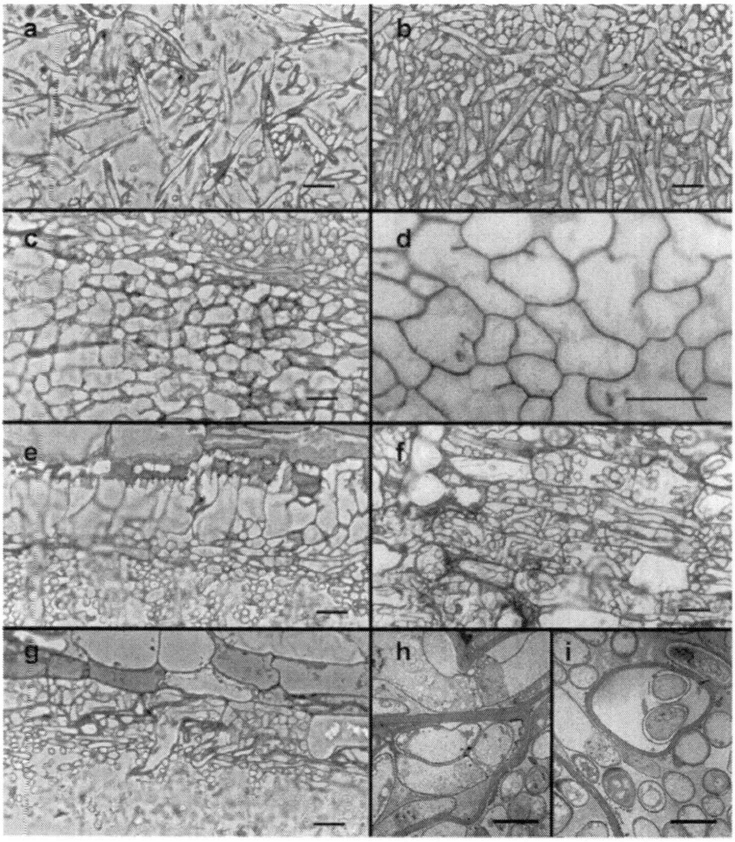

Fig. 3. Longitudinal sections of mycorrhizas of *Pisonia* sp. (a) Outer hyphal mantle: loose plectenchyma, (b) middle hyphal mantle: dense plectenchyma, (c,d). tangential sections of the Hartig net layer adjacent to the outer walls of the epidermal cells, (e) median section through mycorrhiza: hyphal mantle and prominent Hartig net, (f) tangential section of the epidermal layer: intracellular hyphae, (g) median section: intracellular hyphae in epidermal cells and root hairs; (h,i) transmission electron micrographs of hyphal mantle hyphae, intracellular hyphae in cortical cells and root hairs (scale bars: a. - g. 15 µm, h. - i. 2,5 µm). Data from Haug et al. 2005.

(Ashford and Allaway 1985). The differences in the root systems correspond to differing degrees of mycorrhization. While mycorrhization of rootlets was close to 100% in species forming short root systems (*Neea* species 1, *Neea obovata*, *Guapira sancarlosiana*; Moyersoen 1993 and results of this study), species with long root systems showed an incomplete development of the hyphal mantle and no suppression of root hair formation (*Pisonia grandis*, *Neea robusta*, *Neea* species 2, *Pisonia* sp., Ashford and Allaway 1985, Moyersoen 1993, and results presented here). The combination of long root systems that are only partly transformed into ectomycorrhizas, with root

Table 1. Ectomycorrhizal types of three Nyctaginacean species in South Ecuador.

Host plant	Fungal partner	Morphology	Sequences
Neea species 1	*Russula puiggarii*	typical ectomycorrhiza	AY667426
Neea species 1	*Russula-Lactarius*	typical ectomycorrhiza	AY667427
Neea species 1	*Tomentella-Thelephora* 1	typical ectomycorrhiza	AY667411 AY667418
Neea species 1	*Tomentella-Thelephora* 2	typical ectomycorrhiza	AY667412 AY667419
Neea species 1	Ascomycete	typical ectomycorrhiza	no sequence
Neea species 1	*Clavulina* sp.	typical ectomycorrhiza	Tedersoo et al. 2010
Neea aff. *floribunda*	*Tomentella-Thelephora* 3	long roots covered partly with a hyphal mantle, root-hair-like epidermal outgrowths integrated in the hyphal mantle	AY667424
Pisonia sp.	*Tomentella-Thelephora* 4	long roots covered partly with a hyphal mantle	AY667420-423

hair formation that is not suppressed, occasional intracellular penetration of hyphae, and sporadic formation of transfer cell-like structures has not been described from any other plant family so far. We hypothesize that this set of characters represents an early evolutionary step in ectomycorrhiza formation. Subsequently, the typical ectomycorrhizal state evolved as found in the short root-forming Nyctaginacean species. Hence, the Nyctaginaceae may be a model for the evolutionary change from arbuscular mycorrhization to ectomycorrhization.

Our molecular analyses of the Nyctaginacean fungal partners revealed two tree species (*Neea* aff. *floribunda*, *Pisonia* sp.) that were associated with only one thelephoracean taxon. For *Neea* species 1, six different fungal partners were found belonging mainly to Thelephoraceae and Russulaceae. Further investigations of ectomycorrhizal fungi of *Neea* spp. and *Guapira* spp. were done by Tedersoo et al. (2010) in Yasuni National Park in North East Ecuador. *Guapira* harboured three to six fungal partners belonging to five lineages (/cantharellus, /clavulina, /inocybe, /russula-lactarius, /tomentella-thelephora). *Neea* had one to ten fungal partners belonging to four lineages (/amanita, /clavulina, /russula-lactarius, /tomentella-thelephora).

Pisonia grandis is probably only associated with species of Thelephoraceae throughout the area and five species of *Thelephora-Tomentella* are known to form ectomycorrhizas (Chambers et al. 1998, Chambers et al. 2005, Suvi et al. 2010).

All investigations revealed relatively low richness of ectomycorrhizal fungi when compared with the richness of temperate regions (Smith and

Read 2008, Bahram et al. 2012). This may be because of the low amount of ectomycorrhizal trees in these tropical forests (3 out of 115 investigated tree species; Kottke and Haug 2004) and because of the scattered distribution of individual trees. The sampling effort is still low, but species of *Neea, Pisonia,* and *Guapira* have fungal partners that mainly belong to three lineages (*Tomentella-Thelephora, Russula-Lactarius, Clavulina*) and several of the ectomycorrhizal fungi preferred one host genus (Haug et al. 2005, Tedersoo et al. 2010), a feature uncommon in boreal forests.

The high frequency of Thelephoraceae as fungal partners is conspicuous. Many species of Thelephoraceae are saprotrophic (Brundrett et al. 1996) and it is possible that ectomycorrhizal species may also exhibit saprotrophic activity in order to survive periods without a host. Some of the thelephoraceous sequences of the different Nyctaginacean hosts were scattered throughout the phylogram of Thelephoraceae but others were clustered as sister species (Suvi et al. 2010).

3. Conclusion

Although the common ancestor of Nyctaginaceae may have been involved with one thelephoroid taxon, coevolution of Nyctaginaceae and ectomycorrhizal Thelephoraceae in different habitats may have led to the evolution of several different species. It will be a future challenge to study the mycorrhizal status of additional species of Nyctaginaceae in South America and identify their mycobionts to clarify the question of potential co-evolution.

Acknowledgements

This research was generously supported by the Deutsche Forschungsgemeinschaft (DFG project FOR 402, 816). We thank the Fundacíon Científica San Francisco, Ecuador, the NCI for providing research facilities, Jürgen Homeier and Jeremy Hayward for identification of Nyctaginaceae, and New Phytologist for the permission to reprint the figures (copyright 2005). The skilful technical assistance of Jutta Bloschies is much appreciated. We would like to thank Tanja Schuster for editing the manuscript. We also thank Krista L. McGuire and Caitlyn Gillikin for improving the English.

References

Alexander, I.J. and P. Högberg. 1986. Ectomycorrhizas of tropical angiospermous trees. New Phytol. 102: 541–549.

Ashford, A.E. and W.G. Allaway. 1982. A sheathing mycorrhiza on *Pisonia grandis* R. Br. (Nyctaginaceae) with development of transfer cells rather than a Hartig net. New Phytologist 90: 511–519.

Ashford, A.E. and W.G. Allaway. 1985. Transfer cells and Hartig net in the root epidermis of the sheathing mycorrhiza of *Pisonia grandis* R. Br. from seychelles. New Phytol. 100: 595–612.

Bahram, M., S. Polme, U. Koljalg, S. Zarre and L. Tedersoo. 2012. Regional and local patterns of ectomycorrhizal fungal diversity and community structure along an altitudinal gradient in the Hyrcanian forests of northern Iran. New Phytol. 193: 465–473.

Béreau, M., M. Gazel and J. Garbaye. 1997. Les symbioses mycorrhiziennes des arbres de la forêt tropicale humide de Guyane française. Can. J. Bot. 75: 711–716.

Bittrich, V. and U. Kühn. 1993. Nyctaginaceae. *In*: K. Kubitzki, J.G. Rohwer and V. Bittrich (eds.). The families and Genera of Vascular Plants, vol. 2. Springer Verlag, Berlin, pp. 473–486.

Brundrett, M.C. 2002. Coevolution of roots and mycorrhizas of land plants. New Phytol. 154: 275–304.

Brundrett, M., N. Bougher, B. Dell, T. Grove and N. Malajczuk. 1996. Working with mycorrhizas in forestry and agriculture. ACIAR Monograph 32, 374 pp.

Chambers, S.M., J.M. Sharples and J.W.G. Cairney. 1998. Towards a molecular identification of the *Pisonia* mycobiont. Mycorrhiza 7: 319–321.

Chambers, S.M., C.J. Hitchcock and J.W.G. Cairney. 2005. Ectomycorrhizal mycobionts of *Pisonia grandis* on coral cays in the Capricorn-Bunker group, Great Barrier Reef, Australia. Mycol. Research 109: 1105–1111.

Daly, D.C. and A.S. Roberts. 2004. Nyctaginaceae. *In*: N. Smith, S.A. Mori, A. Henderson, D.W. Stevenson and S.V. Heald (eds.). Flowering plants of the Neotropics. Princeton University Press, Princeton, pp. 269–271.

Harley, J.L. and S.E. Smith. 1983. Mycorrhizal Symbiosis. Academic Press, London.

Haug, I., M. Weiß, J. Homeier, F. Oberwinkler and I. Kottke. 2005. Russulaceae and Thelephoraceae form ectomycorrhizas with members of Nyctaginaceae (Caryophyllales) in the tropical mountain rain forest of southern Ecuador. New Phytol. 165: 923–936.

Janos, D. 1980a. Vesicular-arbuscular mycorrhizal infection in an amazonian rainforest. Acta Amazonica 10: 527–533.

Janos, D. 1980b. Mycorrhiza influence tropical succession. Biotropica 12(Suppl.): 56–64.

Khan, A.G. 1974. The occurrence of mycorrhizas in Halophytes, Hydrophytes and Xerophytes and of Endogone spores in adjacent soils. Journal of General Microbiology 81: 7–14.

Koske, R.E., J.N. Gemma and T. Flynn. 1992. Mycorrhizae in Haiwaiian angiosperms: A survey with implications for the origin of the native flora. American J. Bot. 79: 853–862.

Kottke, I. and I. Haug. 2004. The significance of mycorrhizal diversity of trees in the tropical mountain forest of southern Ecuador. Lyonia 7: 49–56.

Moyersoen, B. 1993. Ectomicorrizas y micorrizas vesicula-arbusculares en Caatinga Amazónica del Sur de Venezuela. Scientia Guianae 3, Caracas.

Muthukumar, T. and K. Udaiyan. 2000. Arbuscular mycorrhizas of plants growing in the Western Ghats region, Southern India. Mycorrhiza 9: 297–313.

Rachel, E.K., S.R. Reddy and S.M. Reddy. 1989. VA mycorrhizal colonization of different angiospermic plant species in the semi-arid soils of Andhra Pradesh. Acta Bot. India 17: 225–228.

Singer, R. and I. Araujo. 1979. Litter decomposition and ectomycorrhiza in amazonian forests. Acta Amazonica 9: 25–41.

Smith, S.E. and D.J. Read. 2008. Mycorrhizal Symbiosis. 3rd ed., Academic Press, San Diego, London.
St. John, T.V. 1980a. Una lista de espécies de plantas tropicais brasileiras naturalmente infectadas com micorriza vesicular-arbuscular. Acta Amazonica 10: 229–234.
St. John, T.V. 1980b. A survey of mycorrhizal infection in an amazonian rain forest. Acta Amazonica 10: 527–533.
Suvi, T., L. Tedersoo, K. Abarenkov, K. Beaver, J. Gerlach and U. Koljalg. 2010. Mycorrhizal symbionts of *Pisonia grandis* and *P. sechellarum* in Seychelles: identification of mycorrhizal fungi and description of new *Tomentella* species. Mycologia 102: 522–533.
Taylor, D.W. 1995. Cretaceous to tertiary geology and angiosperm paleogeographic history of the Andes. *In*: S.P. Churchill, H. Blslev, E. Forero and J.L. Luteyn (eds.). Biodiversity and Conservation of Neotropical Montane Forests. New York Botanical Garden, Bronx, New York, pp. 3–9.
Tedersoo, L., A. Sadam, M. Zambrano, R. Valencia and M. Baharam. 2010. Low diversity and high host preference of ectomycorrhizal fungi in Western Amazonia, a neotropical biodiversity hotspot. The ISME Journal 4: 465–471.
Tester, M., S.E. Smith and F.A. Smith. 1987. The phenomenon of "nonmycorrhizal" plants. Can. J. Bot. 65: 419–431.
Wang, B. and Y.-L. Qiu. 2006. Phylogenetic distribution and evolution of mycorrhizas in land plants. Mycorrhiza 16: 299–363.

CHAPTER 3

Diversity and Abundance of Ectomycorrhizal Associations in Rain Forests of Cameroon under Different Disturbance Regimes

Onguene Awana Nérée,[1,*] *Judith Marthiale Tsamo,*[2] *Christelle Michaella Ebenye,*[3] *Amadou Mustapha Bâ*[3] and *Thomas Kuyper*[4]

1. Introduction

In the rainforests of tropical Africa, two kinds of mycorrhizal associations occur: arbuscular mycorrhiza (AM) and ectomycorrhiza (Alexander 1989, Onguene and Kuyper 2001, Bâ et al. 2011). Whereas most timber trees form AM, the ectomycorrhizal (ECM) habit occurs in a very limited number of plant families, viz. Asterpeiaceae, Caesalpiniaceae (a family that is partly arbuscular mycorrhizal), Dipterocarpaceae, Papilionoideae, Phyllanthaceae

[1]Institute of Agricultural Research for Development, Soil, Water and Atmosphere department, Yaoundé, Cameroun.
[2]Department of Plant biology and physiology, University of Yaoundé I, Cameroon.
[3]Laboratoire Commun de Microbiologie (LCM) IRD/UCAD/ISRA, Centre de Recherche de Bel Air, BP. 1386, CP. 18524 Dakar, Senegal.
[4]Department of soil quality, University of Wageningen, the Netherlands.
*Corresponding author: nereeoa678@yahoo.fr

(formerly Uapacaceae), Sarcolaenaceae, Sapotaceae, and Gnetaceae. This may explain the "high latitudinal bias" of the "ecological insignificance" of ECM in the tropics found in earlier literature (Alexander 1989, Tedersoo et al. 2009, Bâ et al. 2011). ECM associations prevail in both the Paleo- and Neotropics (Fassi and Moser 1991). In the Congo basin, they occur in the forest continuum from the Cameroonian Atlantic Ocean coast to the open forest in the DR Congo (Buyck 1994), through the Congo, parts of the Central African Republic and most of the Gabon (De Saint Aubin 1963). These forests could be the hotspots of soil fungal biodiversity in Central Africa. Yet, little information and limited data exists on their occurrence, regeneration requirements, and present functioning under such threats as mining, climate change, agriculture, and logging.

Recent inventories of ectomycorrhizal fungi based on molecular characterization had revealed that native ECM fungal genera in tropical Africa belong, to a large extent, to the same Basidiomycete taxa as those that provide the fungal partners of temperate ECM plants (Boa 2006, Rivière et al. 2007). Common ECM taxa in the tropics belong to the Amanitaceae, Boletaceae, Cantharellaceae, Russulaceae and Sclerodermataceae (De Kessel et al. 2002, Smits 1994). However, on species level, hardly any ECM fungal species (except when introduced) are shared between both regions (Boa 2006, Rivière et al. 2007). Furthermore, many ECM fungal species are poorly documented or have their microscopic details poorly illustrated: a reliable key to all African ECM fungal species is lacking and many species remained unnamed. Various species of Amanitaceae, Boletaceae, Cantharellaceae and Russulaceae have been described from West African rainforests, with virtually no members of the Cortinariaceae (Buyck et al. 1996). Mycoinventories in forests of Korup National Park in SW Cameroon yielded close to 40 species of suspected mycorrhizal fungi. In forests of SW Burkina Faso, 27 ECM fungal species were described beneath indigenous *Afzelia* and *Uapaca* trees (Sanon et al. 1997). Using a fragment of the mitochondrial large subunit rRNA gene, more than 100 ECM sporocarps were typed from Ceasalpiniaceae and Phyllantaceae in a Guinean tropical rain forest (Rivière et al. 2007). It was also concluded that from the same Guinean forest, six taxonomic groups such as boletaceae, sclerodermataceae, russulaceae, telephoraceae, amanitaceae and trichomolotaceae predominate and are shared between canopy trees and seedlings (Diédhiou et al. 2010). The presence of these multi-host ECM fungi in a Guinean rainforest suggest the formation of common ectomycorrhizal networks between differently aged trees just as observed in the rainforest of south Cameroon (Onguene and Kuyper 2002).

Our primary goals were to characterize the below-ground ectomycorrhizal diversity and taxonomic structure of ECM forests in Cameroon, and to test whether ectomycorrhizal fungal community composition changed following land use changes. The objectives of this study were to report on the inventory of ectomycorrhizal associations in humid forests of Cameroon including their habitats (soil and litter characteristics), host tree species, ectomycorrhizas and sporocarps, and assessing the effect of logging and agriculture practices on ECM inoculum potential.

2. Study Sites and Selection of Vegetation Types

The study was carried out in western portions of the Atlantic Biafrean forest of south Cameroon, a humid, tropical climate region with two distinct wet seasons (March-June and August–November) and two dry seasons. Rainfall decreases in an easterly direction from Ebimimbang (1500 mm rainfall) to Nyangong and Bityili (1900 mm rainfall). Along the same gradient, pH and phosphorus availability decrease, as well as land use intensity and consequently forest vegetation (Table 1). In the lowlands, only a few undisturbed, near-primary forest fragments remain while a large part of the area is deforested; in the hilly areas, undisturbed late-secondary forests are more common. Four experimental sites were selected in Ebimimbang (low elevation), Ebom (mid elevation), Nyangong, and Bityili (high elevation).

In each site, field plots were selected in seven vegetation types, viz. ectomycorrhizal forest clumps (EF), late-successional forest stands outside the crown projection of ectomycorrhizal clumps (LS), early-successional forest stands (ES), agricultural fields of food crops with plantain (*Musa* spp.), cocoyam (*Xanthosomas esculenta*), groundnut (*Peanut hypogea*), and cassava (*Manihot esculenta*) as the major crops (FI), *Chromolaena odorata* fallows (FA), *C. odorata* fallows with the liana *Gnetum* (GN), and sites of forestry practices such as skid trails and bare landings. The presence of *Gnetum* spp.,

Table 1. Localization, elevation, rainfall and soil physicochemical characteristics of forest clumps at the four research sites.

Research sites	Ebimimbang	Ebom	Nyangong	Bityili
Localization	3°02.67′N; 10°28.25′E	3°04.73′N; 10°41.24′E	2°58.11′N; 10°45.18′E	2°56.07′N; 10°49.55′E
Altitude (m.a.s.l)	0–350	350–500	500–800	>800
Rainfall (mm)	1556	1987	1677	1800
Soil texture	Sandy	Highly clayey	Highly clayey	Very highly clayey
pH	5–6	4–5	3–4	3–4
Carbon (%)	1.70	2.30	3.28	5.70
Nitrogen (%)	0.11	0.14	0.20	0.36
C/N ratio	15,5	16,4	16,4	15,9
Phosphorus (μm/ml soil)	0.01	0.005	0.002	0.000

an ectomycorrhizal plant, was considered important as this liana might provide ECM inoculum that facilitates establishment of ectomycorrhizal seedlings in former agricultural lands.

3. Sampling and Identification of Sporocarps

Along forest trails leading to permanent sampling plots, sporocarps collections were carried out during both wet seasons with the help of local assistants, one week per month, for three years. Collected specimens were macroscopically described in fresh state before twilight. After description, mushrooms were dried for 2–3 days at about 40°C. Dried mushrooms were temporarily preserved in a local herbarium in Kribi, Cameroon (2°57′N; 9°55′E) before being sent to Wageningen, the Netherlands, for microscopic examination and identification.

Distinction between *Amanita* was based on margin striation, presence of a saccate volva, cap color, hygrophany, and spore amyloidy; Distinction between *Cortinarius* species was based on cap hygrophany, translucence, color and texture of hymenium margins, and presence of a cortina or fibrillose dry ring; Species of *Inocybe* were separated on the basis of pileus texture (radially fibrillose cap), and shape and texture of stipe; Genera of Boletales varied in pileus color and shape, and were distinguished on the basis of poroid hymenium, presence of large cystidia with or without clamp connections and spore ornamentation. *Russula* species varied in pileus size, color and shape, brittleness of gills, presence or absence of lamelullae and an evanescent ring on stipe; they were differentiated from *Lactarius* on the basis of exudation upon flesh injuries and the extension of the lactiferous system into the hymenium. Species of *Cantharellus* differed in pileus texture, hymenium color, and spore ornamentation. Several provisional keys were elaborated based on morphological and anatomical characters of sporocarps (Onguene 2000). Voucher specimens are kept at the National Herbarium of the Netherlands (Leiden Branch) and some duplicates are preserved at the agricultural research station in Kribi, Cameroon. Several manuals were used for species identification (Onguene and Kuyper 2012).

4. Morphological and Anatomical Description of Root Tips and Ectomycorrhizae

After identification and selection of ectomycorrhizal host trees, four root samples were collected around the base of each host tree, washed gently in water, and preserved in 50% alcohol or wrapped in aluminum foil with litter and soil (when alcohol was exhausted in the field). In the laboratory, root systems were cleared of soil debris by careful washing under a water flow while immersed in tap water. Seven to ten selected root tips for each morphotype representative were removed from the root sample and morphologically and anatomically observed. The key morphological and anatomical features examined under a dissecting microscope at 40x included root tip branching, shape of branches, mantle colour and surface texture, presence of rhizomorphs, emanating hyphae. Tips were also observed under a photonic microscope between 25 and 40x to confirm the presence of a mantle and a Hartig net, various layers of hyphal arrangement, and to determine the presence of specialized cells, rhizomorphs and cystidia, using both cross—and longitudinal sections (Agerer 1995).

5. Effects of Agriculture and Logging Practices on ECM Inoculum Potential

To assess whether agriculture or logging practices impact ECM inoculum potential and whether arbuscular mycorrhizal (AM) and ectomycorrhizal fungi share the same niche, three bioassays were carried out using intact soil cores collected from the seven disturbance stages.

In 1-ha forest plots, 100 m^2 quadrats were laid down. In each quadrat, soil cores were dug out of three spots, each 50 m apart. Relatively undisturbed, intact, cylindrical soil monoliths were collected (about 4.2–4.5 kg; wet weight basis) by driving a 15 cm diameter x 45 cm long PVC tube into the ground with a hammer dropped from a constant height (10–20 cm) onto a flat steel plate, placed on top of the PVC tubes. In agricultural food crops, fallow, and forestry plots, plot sizes were 100 m^2 and 100 m long, respectively, and intact cores were removed as previously described. Triplicate core samples were removed intact from each land-use per site. However, due to different fruiting patterns of both trees, the bioassays were carried out at different periods of the year: March to June for *Tetraberlinia* and July to October for *Afzelia*.

Two native timber species, both belonging to the Ceasalpiniaceae, were used for the bio-assay. *Tetraberlinia bifoliolata* is an ectomycorrhizal tree that is valuable as a potential novel timber tree. *Afzelia bipindensis* is a dual mycorrhizal tree that provides a highly priced redwood timber. Hereafter, trees will be designated by their generic names only. *Tetraberlinia* typically occurs in clumps together with other caesalp species, while *Afzelia* usually occurs isolated in matrices of AM trees and has not been observed in clumps with other ECM trees. Both tree species have large pods (10–20 x 5–8 cm) with a small number of large and heavy seeds; average seed size of *Tetraberlinia* was 20–30 x 15–25 x 5–7 mm and that of *Afzelia* 30–40 x 20–30 x 10–20 mm; average seed mass of *Tetraberlinia* was 1.5 g (0.8–2.7 g) and that of *Afzelia* 11.5 g (6.2–17.4 g). Seedlings of both species possess coarsely branched roots with few root hairs (Onguene and Kuyper 2001).

Seeds were germinated for a week in steam-sterilized sand without pregermination treatment. One 1-week old seedling of each tree was placed in a small hole at the center of the soil core. Cores were placed on benches and plants grown for four months under natural light conditions in a greenhouse in Kribi in a randomized complete block design, watered every three days to maintain soils at field capacity, without nutrient amendments.

To determine whether ECM and EM fungal propagules share the same niche, a local variety of cowpea, *Vigna unguiculata* (Fabaceae), was grown in soil cores from ECM clumps. In addition, four intact soil cores were collected around the stem base of ECM tree species, at 5 m and 10 m distance away from the stem base of *A. bipindensis* and *Brachystegia cynometroides* at the Ebom site, *T. bifoliolata* and *Paraberlinia bifoliolata* at the Bityili site. Cowpea plants were raised for a month in the greenhouse under the same conditions as previously described. At harvest, ECM fractional colonization was assessed in water without staining while portions of *Afzelia* root samples stained with acid fuchsine were assessed by the gridline intersect method (Onguene and Kuyper 2001).

6. Experimental Design and Statistical Analysis

For *Tetraberlinia*, the experiment was a full factorial with two factors, site (3 levels) and disturbance stage (7 levels). For *Afzelia*, a partial factorial experiment was executed with soils from three sites and three forest types (ECM clumps, late-successional stands, early-successional stands) owing to seed shortage. The full set of disturbance stages was only investigated for the Ebom soil, where the tree was fairly common and widespread.

The SAS package (SAS Inc. 2004) was used for statistical analysis. Data were first tested for normality and homogeneity of variances using the Levene test in the one-way analysis of variance (ANOVA). Data on ECM fractional root colonization by *Tetraberlinia* contained many zeroes and did not meet the requirements of normal distribution and homogeneous variances. Therefore, the non-parametric Kruskall-Wallis test was applied. When data analysis was restricted to the three forest stands (ECM clumps, late- and early-successional stands), data of fractional ECM root colonization, after arcsin square root transformation did meet the requirements for ANOVA. Data on ECM and AM root colonization by *Afzelia* were also arcsin square-root transformed.

7. Habitat Types of Ectomycorrhizal Associations in Cameroon

Ectomycorrhizal forest clumps occurred on sandy, highly clayey and very highly clayey soils, throughout the entire landscape, even on swamps like *Gilbertiodendron* clumps. They were acidic to very strongly acidic. The mineral contents of N and available P in ECM forest clumps were very low, sometimes undetectable such as in the Bityli soils, irrespective of sites and soil texture. C/N ratio was always larger than 10 (Table 1). The layers of litter were thin (one to two), tough, and varied from one forest clump type to the other.

In the humid forests of Cameroon, ectomycorrhizal associations occurred as small to large forest clumps of five types: *Gilbertiodendron* monodominant clumps, *Uapaca* monodominant clumps, monodominant *Microberlinia* clumps in the Korup National Park (Newbery et al. 1988), oligo-dominant ceasalp clumps locally called "Ekop," and mixed clumps of "Ekop" with *Uapaca* spp.

Twenty-four tree species were found to be symbiotically associated with ectomycorrhizal fungal species on the basis of root tip colonization. They belonged to two botanical families: Caesalpiniaceae and Phyllantaceae. Whereas most ECM host tree species occurred in one of the five ECM clumps, six host tree species including *Afzelia*, *Anthonotha* and *Berlinia* were always isolated in matrices of arbuscular mycorrhizal host plants (Table 2).

The contribution of ECM host tree species to basal area of forest clumps varied from 20 to 80%, and 100% in *Gilbertiodendron* clumps. The basal area diameter of ECM trees was low in young and high in adult ECM trees, respectively (Table 3).

8. Cameroonian Ectomycorrhizae and Sporocarp Diversity

Nineteen ectomycorrhizal morphotypes were described and found differing by mantle color (Black, reddish, whitish, brown black, chestnut), external textures (smooth, strongly rhizomorphic), thickness, and presence or absence of cystidia (Table 4).

Found throughout all land types, including swampland, ECM forest clumps were the only habitats for most Cameroonian ECM fungal diversity: more than 1000 sporocarps were collected over three years and nearly 200 ECM fungal species were identified, belonging to eight families and 27 genera, mostly new to science. Only *Lactarius gymnocarpus* and all *Cantharella* species were locally edible (Table 5).

Table 2. List of ectomycorrhizal tree species by types of ectomycorrhizal forest habitats in Cameroon.

Habitat types	Family	Species	Pilot name
Not observed in clumps	Caesalpiniaceae (Tribe Detareae)	*Afzelia bipindensis* *A. pachyloba*	Doussie rouge Doussie blanc
	Caesalpiniaceae (Tribe Amherstieae)	*Anthonotha fragans* *A.macrophylla* *Berlinia bracteosa* *B. confusa*	Enak Enak Ebiara Ebiara
Gilbertiodendron monodominant clumps	Caesalpiniaceae	*Gilbertiodendron dewevrei* *G. brachystegioides*	Abem Abem
Uapaca monodominant or mixed clumps	Phyllantaceae	*Uapaca acuminata* *U. buchholizianum* *U. guineensis* *U. vanhoutei*	Rikio Rikio Rikio Rikio
Microberlinia monodominant clumps	Caesalpiniaceae	*Microberlinia bisulcuta**	Zingana
Mixed oligodominant ceasalp clumps	Caesalpiniaceae (Tribe Amherstieae)	*Brachystegia cynometrioides* *B. euricoma* *B. zenkeri* *Didelotia africana* *Didelotia letouzeyi* *Julbernadia seretii* *Monopethalanthus letestui* *M. microphyllus* *Paraberlinia bifoliolata* *Tetraberlinia bifoliolata* *Touabouate brevipaniculata*	Ekop naga Ekop evene Ekop gombe Ekop rouge Ekop Ekop Ekop blanc Ekop mayo Ekop mayo ngang Ekop beli Ekop ribi Ekop zing
Mixed *Uapaca* and ceasalp clumps	Phyllantaceae and Caesalpiniaceae (Tribe Amherstieae)	Various *Uapaca* and Ekop species	Rikio and Ekop

Notes: Pilot names of "Ekop" were taken in Letouzey and Mouranche (1952). *Data from Newbery et al. (1988).

Table 3. Basal stem diameter in different forest clump types in 1000 m^2 area (Average observations of three years during the dry season).

Basal stem diameter size (cm)	Mixed oligodominant ceasalp clumps	*Gilbertiodendron* monodominant clumps	*Uapaca* monodominant	*Microberlinia* monodominant
<10	7	0	10	0
10–49	15	0	57	0
50–99	45	50	157	176
>99	164	250	98	>350

Table 4. Morphological and anatomical characteristics of some Cameroonian ectomycorrhizae.

Ectomycorrhizal morphotype collection number	Color	Shape	Texture	Texture of mantle		Thickness (μm)	Presence of rhizomorphs	Presence of cystidia
				External mantle	Internal mantle			
CM 400 *Gnetum* spp.	Yellow gold	monopennate	Thread like	Felty prosenchyma	Synenchyma	100.8–201.5	Present	Absent
CM 401 *Paraberlinia bifoliolata*	Milky white	monopennate	Smooth	Felty prosenchyma	Synenchyma	255–408	Absent	Absent
CM 402 *Uapaca guineensis*	Orange brown	monopyramidal	Smooth	Felty prosenchyma	Synenchyma	100.5	Absent	Present
CM 403 *Gilbertiodendron dewevrei*	Brown	monopyramidal	Smooth	Regular Synenchyma	Net prosenchyma Irregular Synenchyma	61.2	Absent	Absent
CM 404 *Uapaca guineensis*	Brown	monopyramidal	Spiny	Felty prosenchyma	Synenchyma	100.2	Absent	Present
CM 405 *Didelotia letouzeyi*	White	monopyramidal	Felty or velvety	Felty prosenchyma	Synenchyma	>533	Absent	Absent
CM 406 *Julbernadia seretii*	Brown	monopyramidal	Cotton like	Felty prosenchyma	Synenchyma	408–510	Absent	Absent
CM 407 *Uapaca acuminata*	Brown	monopyramidal	Smooth	Felty prosenchyma	Synenchyma	25–30	Absent	Present
CM 408 *Monopethalanthus letestui*	White	monopyramidal	Felty	Felty prosenchyma	Synenchyma	100.8–183.9	Absent	Present
CM 409	White	monopyramidal	Felty	Felty prosenchyma	Net prosenchyma	162–200.4	Present	Absent

Table 4. contd....

Table 4. contd.

Ectomycorrhizal morphotype collection number	**Color**	**Shape**	**Texture**	**Texture of mantle**		**Thickness (μm)**	**Presence of rhizomorphs**	**Presence of cystidia**
				External mantle	**Internal mantle**			
CM 410 *Touabouate brevipaniculata*	White	monopyramidal	Cotton like	Felty prosenchyma	Synenchyma	255	Present	Absent
CM 411 *Gilbertiodendron dewevrei*	White	monopennate	Cotton like	Felty prosenchyma	Synenchyma	390–650	Present	Absent
CM 412 *Paraberlinia bifoliolata*	Brown	monopennate	woolly	Felty prosenchyma	Synenchyma	110–119.2	Present	Absent
CM 413 *Monopethalanthus letestui*	Brown	monopyramidal	woolly	Felty prosenchyma	Synenchyma	306–561	Present	Absent
CM 414 *Didelotia letouzeyi*	Brown	monopyramidal	woolly	Felty prosenchyma	Synenchyma	540–615	Present	Absent
CM 415 *Tetraberlinia bifoliolata*	Brown or maroon or brown maroon	monopennate	woolly	Felty prosenchyma	Irregular Synenchyma	510–765	Present	Absent
CM 416 *Uapaca guineensis*	Brown	monopyramidal	woolly	Felty prosenchyma	Net prosenchyma	306–459	Present	Absent
CM 417 *Uapaca acuminata*	Brown maroon à maroon	monopennate	Cotton like to woolly	Felty prosenchyma	Net prosenchyma	104	Absent	Absent
CM 418 *Touabouate brevipaniculata*	Brown	monopyramidal	Cotton like	Felty prosenchyma	Synenchyma	403–502.3	Absent	Present

Table 5. List of putative ectomycorrhizal mushroom species by family, genus and species in humid forests of South Cameroon.

Amanitaceae	**Boletaceae**	**Russulaceae**
Amanita afzeliae sp.	*Afroboletus luteolus* (Heinem)	*Russula afronigricans* Buyck
A. albidodisca sp.	P. & Y.	*R. albospissa* Buyck
A. albopulverulenta sp.	*Boletus boletiformis* sp.	*R. annulata* Heim
A. afrobispora sp.	*B. macrocystis* sp.	*R. annulatobadia* Beeli
A. afroflavescens sp.	*B. nyangongensis* sp.	*R. apsila* Buyck
A. afrorubescens sp.	*B. pustulatus* Beeli	*R. intrica* Lizoii
A. afrovaginata sp.	*B. suspinulosus* sp.	*R. areolata* Buyck
A. annulatovaginata Beeli	*Boletellus sulcatipes* Hein. & Goo.	*R. aurantiofloccosa* Buyck
A. aureofloccosa Bas	*Chalciporus clypeatus* sp.	*R. camerunensis* sp.
A. bingensis Beeli	*Gyrodon aberrans* sp.	*R. cellulata* Buyck
A. brachystegiae sp.	*Gyroporus microsporus* (Sing. &	*R. chrysotricha* sp.
A. calopus Beeli	Grinl.) Heinem & Rammeloo	*R. declinata* Buyck
A. crassiconus Bas	*Leccinum excedens* sp.	*R. diffusa* var. *fissurans*
A. elegans Beeli	*Paxillus brunneotomentosus* sp.	*R. discopus* Heim
A. flavovirens sp.	*Paxillus camerunensis* sp.	*R. echnosperma* R. Heim &
A. fulvosquamulosa Beeli	*Phlebopus braunii* (Bres.) Heinem	Gilles
A. gigavolvata sp.	*P. silvaticus* Heinem	*R. fulvoochrascens* Buyck
A. griseofarinosa Hongo	*Phlebopus* sp.	*R. heliochroma* Heim
A. lanosa Beeli	*Phylloporus depressus* Heinem	*R. intricate* Buyck
A. leucoagaricioides sp.	*Pulveroboletus aberrans* Heinem	*R. kivuensis* Buyck
A. luteoflava Beeli	*P. viridis* Heinem	*R. lamprocystidia* (Nakasone)
A. monopetalanthi sp.	*Rubinoboletus luteopurpureus*	*R. liberiensis* Buyck
A. nigropyramis sp.	(Beeli)	*R. macrocystis* sp.
A. pseudoafroalba sp.	*Strobilomyces echinatus* Beeli	*R. mimetic* sp.
A. pseudolanosa sp.	*S. luteolus* Heinem	*R. pausiaca* Buyck
A. pustulata sp.	*S. strobilaceus* (Berk.)	*R. pseudocarmesina* Buyck
A. roseocinnamomea nov.	*S. velutipes* Corner	*R. pseudopurpurea* Buyck
A. rubescens (Pers.:Fr) G	*Tubosaeta alveolata* Heinem	*R. speudostriatoviridis* Buyck
A. strobilaceovolvata Beeli	*T. brunneosetosa* (Singer) E. Horak	*R. striatoviridis* Buyck
A. subviscosa Beeli	*T. goosseniae* sp.	*R. testaceoaurantiaca* Beeli
A. sulphurea sp.	*Tylopus violaceus* sp.	*R. velutina* (DC per Pers.:Fr)P
A. virella Beeli		*R. yaeneroensis* Buyck
	Clavulinaceae	
	Clavulina vanderstii sp.	
	Sclerodermataceae	
	Scleroderma sinnamariense Mont	
	S. roseacarneum sp.	
	Coltriciaceae	
	Coltricia spathulata (Hooker)	
	Murill	
	C. pyrophila (Wakef) Ryvarden	

Table 5. contd....

Table 5. contd.

Inocybe	**Cantharellaceae**	***Lactarius***
Inocybe afronodulosa sp.	*Gomphus brunneus** (Heinem)	*Lactarius acutus* Heim
I. afrostellata sp.	*G. clavatus* (Pers.) Gray	*L. annalatoangustifolius* sp.
I. korupensis sp.	*Craterullus crispus* Fr	*L. claricolor* sp.
I. bipindensis sp.	*C. cornucopioides* Persoon	*L. denigricans* Ver. & Kar.
I. perpusilla Velen	*Cantharellus camerunensis* sp.	*L. densifolius* Ver. & Kar.
I. zingii sp.	*C. cibarius var.roseocanus* Arora & Dunham	*L. gymnocarpus* Heim
		L. kivuensis deWitte
Cortinarius	*C. congolensis* Beeli	*L. medusa* Verbeken
Cortinarius afroconicus sp.	*C. dichrous* (Fr.) Bres	*L. pumilus* Verbeken
	C. densifolius Heinem	*L. pulcrispermus* sp.
C. diobensis sp.	*C. floridulus* Heinem	*L. sesemotani* Beeli
C. ionopygmaeus sp.	*C. isabellinus* Heinem	*L. undulates* Verbeken
	C. luteopunctatus (Beeli) Heinem	
	C. miniatescens Heinem	
	C. microcibarius Heinem	
	C. pseudocibarius Hennings	
	C. rufopunctatus (Beeli) Heinem	

Note: *Gomphus brunneus* is a *cantharella* species restricted to Cameroon, RD Congo and Uganda.

Ten provisionary keys to Cameroonian species were developed for Amanitaceae, Boletaceae, Cantharellaceae, Clavulinaceae, Coltriciaceae, *Cortinariaceae, Inocybaceae, Russulaceae* (*Russula* and *Lactarius*) and Sclerodermaceae (Onguene 2000).

9. Effect of Disturbance Regimes on Ectomycorrhizal Inoculum Potential

Non-parametric analysis of variance indicated that fractional ectomycorrhizal colonization of *Tetraberlinia* was significantly affected by disturbance stage ($p < 0.001$), but not by site ($p = 0.125$). Four-month seedlings grown in soils from forestry practices, agricultural fields and fallow without *Gnetum* remained free of ECM colonization. Seedlings in soils from fallow with *Gnetum* from all three sites were colonized to some extent. A two-way analysis of variance for ECM colonization of the three forest stands (EC, LS, and ES) indicated that both site and disturbance stages were statistically significant, whereas their interaction was not (Table 6). A rank by Mann-Whitney U-test indicated that colonization was highest in soils from ECM clumps and late-successional forests (Fig. 1A). ECM inoculum increased during succession, with forest clumps showing a significantly higher colonization than late-successional stands, but being low in early-successional stands. ECM colonization was the highest in soils from Nyangong and lowest in soils from Ebom (Table 7).

Table 6. Two-way analysis of variance of site and disturbance stages on ectomycorrhizal colonization of four-month old seedlings of *Tetraberlinia bifoliolata*.

Sources of variation	Df	P	F
Site	2	52.7	0.000
Disturbance stage	2	88.9	0.000
Site x disturbance stage	4	0.4	0.836

Table 7. Ectomycorrhizal colonization (percent root length colonized) of seedlings of *Tetraberlinia bifoliolata* (T) and *Afzelia bipindensis* (A) in various forest stands. Letters indicate significant differences according to Duncan's Multiple Range Test at $p < 0.05$. Late-SF: late successional forest; Early-SF: Early successional forest.

Forest types	Ebimimbang		Ebom		Nyangong		Bityili	
	T	A	T	A	T	A	T	A
Forest clump	48^{bc}	22^{a}	32^{cd}	5^{c}	81^{a}	34^{a}	89^{a}	53^{a}
Late-SF	28^{d}	26^{a}	12^{e}	0^{c}	52^{b}	13^{b}	47^{b}	28^{b}
Early-SF	4^{f}	1^{c}	0^{f}	11^{b}	20^{dc}	0^{c}	15^{c}	31^{c}

A non-parametric analysis of variance indicated that ectomycorrhizal colonization of *Afzelia* in soils from the three forest stands (EC, LS, and ES) was neither significantly influenced by site nor by disturbance stage ($p > 0.1$). No or very little colonization by ectomycorrhizal fungi was observed in soils from a *Gilbertiodendron* clump in Ebom and in soils from early-successional forest from Ebimimbang and Nyangong. For the Ebom soils only, cores taken under a mature *Afzelia* (CO) resulted in the highest fractional ectomycorrhizal root colonization. Colonization was high in sites of agricultural practices (fields, fallow) and declined in soils from later-successional stages. In soils from forestry practices and ectomycorrhizal clumps, no ectomycorrhizal colonization was observed (Fig. 1B). In the three forested disturbance stages, there was no correlation between ectomycorrhizal inoculum potential as assessed by *Tetraberlinia* and *Afzelia* ($r = 0.50$, $n = 9$; $p > 0.1$).

Analysis of variance for ectomycorrhizal colonization of *Afzelia* in soils of forest stands (EC, LS, and ES) indicated only the interaction of disturbance stage and site was statistically significant (Table 6). No or very little colonization by native ECM fungi was observed in soils from a *Gilbertiodendron* monodominant clump in Ebom and in soils from early-successional forest from Ebimimbang and Nyangong. For the Ebom soils only, ECM colonization was highest in sites of agricultural practices (fields,

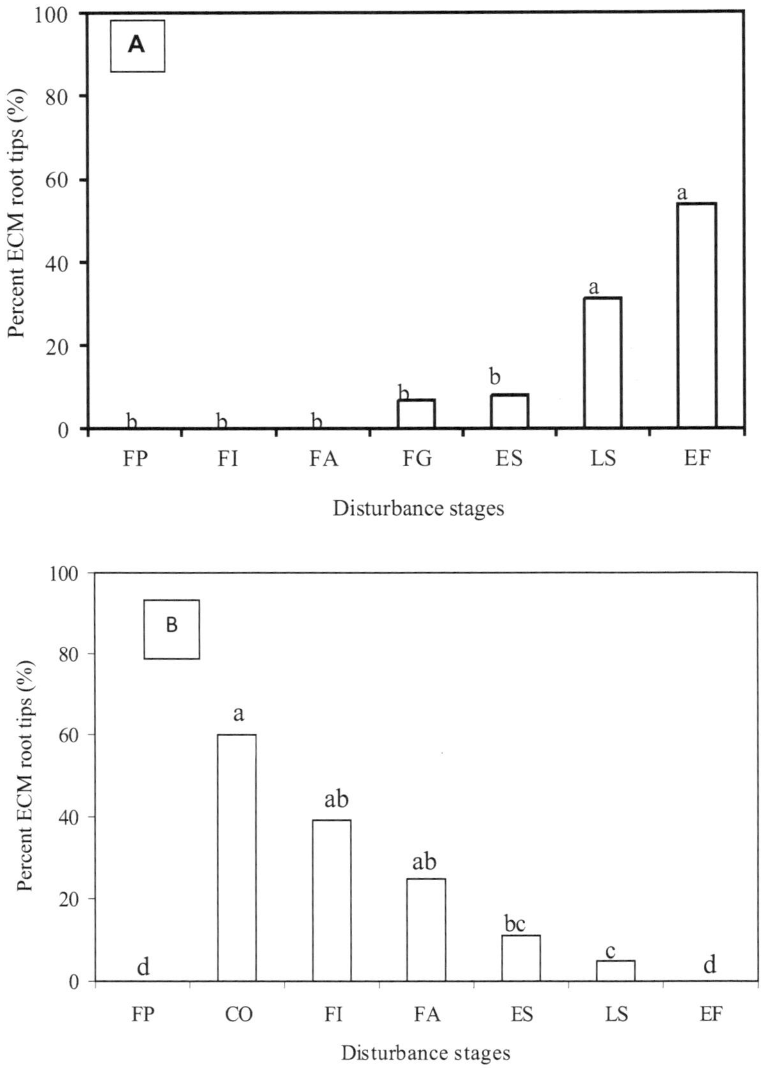

Fig. 1. Ectomycorrhizal fractional colonization of seedlings of Tetraberlinia (A) and Afzelia (B) grown in soils from different disturbance stages (average from three sites). Significant differences between disturbance stages (Mann-Whitney U-test; $p < 0.05$) are indicated by different letters. Abbreviations are as follows: EF = ectomycorrhizal forest clumps; LS = late-successional forest stands outside the crown projection of ectomycorrhizal clumps; ES = early-successional forest stands; FI = agricultural fields of food crops with plantain (*Musa* spp.), cocoyam (*Xanthosomas esculenta*), groundnut (*Peanut hypogea*), and cassava (*Manihot esculenta*) as the major crops; FA = *Chromolaena odorata* fallows; FG = *C. odorata* fallows with the liana Gnetum; FP = sites of forestry practices such as skid trails and bare landings.

fallow) and decreased in soils from late-successional stages. Soil cores taken under mature *Afzelia* trees resulted in even higher fractional ECM root colonization. In soils from forestry practices and ECM clumps, no ECM colonization was observed (Fig. 1). Fractional AM colonization was always lower than 5%; no colonization by indigenous AM fungi was observed in soil cores from forestry practices and from ECM clumps.

Most soil cores did not produce abundant arbuscular mycorrhizal fungal colonization on roots of *Vigna unguiculata*. AM colonization was detected in 56% (54 out of 96) soil cores from forest clumps. The sparse AM colonization varied with sites: very low to low in clumps in Ebimimbang and Ebom, and completely absent in Nyangong. No AM colonization was observed in soil cores around the stem base of *Afzelia*, *Brachystegia*, and *Paraberlinia*, but AM colonization varied from 2 to 22% in the vicinity of *Tetraberlinia* trees.

In humid forests of south Cameroon, five types of ectomycorrhizal forest clumps exist but no longer regenerate, as observed by the dominance of only the adult size class of individuals: *Gilbertiodendron* monodominant, *Microberlinia* monodominant in the Korup National Park (Newbery et al. 1988), *Uapaca* monodominant, *Uapaca* oligo-dominant, mixed oligo-dominant ceasalps, mixed ceasalp and *Uapaca* trees. If six ECM timber species out of 24 occur isolated in the midst of arbuscular mycorrhizal plant species, most ECM host species formed clumps, either canopy dominant or oligo-dominant. They varied in size from small to medium in the Bipindi-Lolodorf-AkomII zone, large in the Korup National Park and very large in the forests of south-east Cameroon, like the 10 km long stretch of monodominant *Gilbertiondendron* clumps along the road from Dja River to Ngoïla (Onguene, Amadou and Ebenye, pers.obs.). Botanical inventories in Gabon, Congo, Central African Republic, and the Democratic Republic of Congo also depicted such clumps of ceasalp and *Uapaca* tree species (De Saint Aubin 1963, Ndong et al. 2011). On the other hand, ECM associations forming clumps are limited elsewhere in the Neotropics. Only two genera, *Dycimbe* and *Aldinia* have consistently been shown to form such associations (Singer et al. 1983, Smith et al. 2011). ECM fungi have also been described in humid forest relics of West Africa (Diédhiou et al. 2010) and in the open forest of the Zambesian region (Buyck 1994). Such a variety of habitat conditions cannot be explained by environmental context and floristics alone (Smith et al. 2011).

9.1 Ectomycorrhizal forest clumps grow only in specific regeneration niches

In tropical Africa, most timber trees form arbuscular mycorrhiza (AM) but the ectomycorrhizal habit occurs in a very limited number of plant families.

Co-occurrence of both mycorrhizal types has raised the question of whether trees with different kinds of mycorrhiza show niche partitioning. Niche partitioning could occur along various axes. ECM and AM trees could show edaphic specialization. This niche differentiation has been suggested for forests in the Amazonian region, with forests on podzolic or white sands being dominated by ECM trees and forests on brown sand being dominated by AM trees (Singer and Araujo 1986). Some supporting evidence for this kind of niche differentiation has also been put forward for the rainforest of Korup National Park, in south-west Cameroon (Newbery et al. 1988).

However, spatial separation of ECM and AM trees does not necessarily provide support for a hypothesis on edaphic niche differentiation. Tropical ECM trees often show conspicuous gregarious behavior in monospecific or at least plant species-poor stands (Connell and Lowman 1989) and this habit is well-known from ECM trees in the African rainforest (Letouzey 1968, Newbery et al. 1997, Onguene and Kuyper 2001). In such clumps, taxa from the Caesalpiniaceae, tribe Amherstieae, constitute the largest contribution to basal area. The occurrence of related tree taxa in these clumps has been previously noted by tree prospectors in Cameroon and Gabon, where these unnamed caesalps were collectively known as "ekop" and "andoung", respectively (Letouzey and Mouranche 1952, De Saint Aubin 1963).

Richards (1996) stated that the gregarious behaviour might be a consequence of the limited dispersal ability of either partner forming the ECM symbiosis. As both ECM tree and ECM fungus do not possess the capacity to grow and reproduce independently of the other symbiotic partner, mycorrhizal establishment on new sites might be a rare chance event (Janos 1996). After the ECM symbiosis is initiated, such trees could serve as focal points for the establishment of other trees that are compatible with the ECM fungus, thereby setting up the regeneration niche of ECM trees. Our data provide better support to this hypothesis. All over the area, ECM clumps did not show strong regeneration of intermediate plant size (Table 3).

Our data provide support for the hypothesis that ectomycorrhizal and arbuscular mycorrhizal trees differ in their regeneration niche (Grubb 1977). In a separate study, it was found that inoculum potential of AM fungi was about twice as high in early-successional as in late-successional forest (Onguene 2000). In that study, the effects of ECM clumps on inoculum potential of native AM fungi had not been addressed. Mycorrhizal colonization of *Vigna unguiculata* indicated that in ECM clumps the inoculum potential of AM fungi is very low. Apparently, the buildup of ECM inoculum together with AM inoculum decline. Direct competitive interactions between ECM and AM fungi have also been postulated by Moyersoen et al. (1998). Inoculum of ECM fungi in forested sites was highest in Bityili and lowest in Ebom. These data are consistent with the relatively large contribution of ECM trees to basal area (Onguene and Kuyper 2001) and to inventories of ECM fungi (Onguene and Kuyper 2012), where Bityili was highest and Ebom lowest.

9.2 Ectomycorrhizal regeneration niches are fungal biodiversity hotspots

Nineteen ECM anatomo-morphotypes were described for the first time, though not from *Microberlinia* forests. They differed in external texture, mantle thickness, and presence or absence of rhizomorphs and/or cystidia. They also differed markedly from the conspicuous yellow gold ECM of *Gnetum* species (Eneke and Alexander 2012). Therefore, it appears that *Gnetum* plants found in fallows of *Chromolaena odorata* could not serve as focal points for the regeneration of ECM trees owing to differences of ECM specificity. Though most ECM from different host tree species could not be specific, thereby confirming the multi-host fungi observed in the Guinean forest (Diédhiou et al. 2010), the only ECM observed in *Gilbertiodendron* clumps were whitish and large rhizomorphs. Apparently, rhizomorphs could play a role in water relations of host trees (Agerer 1995). Hence, ECM clumps and isolated ECM tree species depict a particular biodiversity that needs to be preserved.

Ectomycorrhizal forest habitats were the only sites where more than 200 ECM sporocarp species in 27 genera and eight families were collected and identified. It is the first time in Africa that such a highly diverse and abundant putative ECM sporocarp collection has been reported from the same area. In the Pakaraima Mountains of Guyana, 75 morphotypes of putatively ECM fungi were identified from discrete groves formed by *Dycimbe* and *Aldina* species (Henkel et al. 2002). In the Miombo savanna and humid forest of West Africa, only five families of ECM sporocarps were recorded (Rivière et al. 2007, Sanon et al. 1997). The same high number of members of the Russulaceae (44 species) from this study supports existing data (Buyck et al. 1996). Species-rich genera included *Amanita*, *Russula*, and members of Boletales and Cantheralles. Less frequent ECM fungal species, *Clavulina*, *Coltricia* and *Scleroderma* were also collected for the first time as well as rare ECM sporocarp species, viz. two, three and six species of *Paxillus*, *Cortinarius* and *Inocybe*, respectively (Table 3), confirming earlier observations made elsewhere in West Africa, the Neotropics and the Laojun Mountain region in southwestern China (Rivière et al. 2007, Henkel et al. 2002). Lack of frequent inventories and experienced mycologists in Africa could explain the poor accounts for tropical ECM sporocarp diversity. Only 546 records of wild macrofungi have been made from South Saharan countries (Boa 2006). In Benin, one species of *Craterellus*, *Hebeloma*, *Inocybe*, *Russula* and eight *Lactarius* species were recorded (De Kessel et al. 2002). Five out of six macrofungi in Burkina Faso were identified as *Amanita hemibapha*, *Cantharellus pseudofriessi*, *Lactarius gymnocarpus*, *Phaeogyroporous sudanicus* and *Tubosaeta brunneosetosa* (Sanon et al. 1997, Rammeloo and Walleyn 1993) as well as an unidentified *Calvatia* (Walleyn and Rammeloo 1994). Recently, more than two dozen ECM fungi have been recorded (Ndong et al. 2011), with most of them also found in the rain forest of Cameroon (Onguene and Kuyper 2012). No less than 40 ECM fungi have been recorded from the Republic of Congo including 15 *Cantharellus* species and dozens of *Russula* and *Lactarius*. The highest number of *Lactarius* species (19 species) has been thus recorded in Tanzania. Other African countries are less prolific in ECM fungi (Boa 2006).

Ectomycorrhizal forest communities could contribute not only to the preservation of forest refuges in South Cameroon with endemic species, but also to the protection of the biodiversity hot spots of Africa's rainforests. Maintenance of such ECM forest communities with their associated edible mushrooms such as chanterelles also helps to preserve a valuable source of alternative protein-rich food for local communities who depend on non-timber forest products (Malaisse 1967, Van Dijk et al. 2003), in addition to creating new ecological jobs for mushroom collectors. Thus, ECM forest clumps should be protected as biodiversity sanctuaries, owing to the present lack of knowledge on their regeneration requirements.

Fractional ectomycorrhizal colonization of *Tetraberlinia* was significantly reduced by both shifting agriculture and selective logging. ECM colonization increased during succession, with forest clumps showing a significantly higher colonization than late-successional stands, but being low in early-successional stands. ECM colonization was the highest in soils from Bityili and lowest in soils from Ebom (Table 7). No or very little colonization by native ECM fungi was observed in soils from a *Gilbertiodendron* monodominant clump in Ebom and in soils from early-successional forest from Ebimimbang and Nyangong. Apparently, each ECM clump may have different ECM fungal consortia, confirming the observation of multi-species fungal forests (Diédhiou et al. 2010). The question remains whether the initial soil and climatic conditions regulate ECM symbiosis.

Various types of propagules, such as basidiospores, hyphal fragments, or rhizomorphs and dying roots, can contribute to the mycorrhizal inoculum potential. We were not able to evaluate the relative importance of various propagule sources in our soil cores. Baiting techniques to assess ECM inoculum might yield different results if plants are baited in the field within reach of live mature ECM plants or are baited in the greenhouse (Diédhiou et al. 2010, Onguene and Kuyper 2002, 2005). Studies where seedlings were baited in intact vegetation in the field supported the view that the species colonizing naturally regenerating seedlings in natural vegetation were similar to that of the ectomycorrhizas of that surrounding vegetation

(Jonsson et al. 1999), whereas in the absence of surrounding vegetation of ECM plants, a different suite of ECM fungi will be encountered (Taylor and Bruns 1999).

In soils from forestry practices, both seedlings of *Tetraberlinia* and *Afzelia* remained devoid of ECM. Although selective logging at present concentrates on AM trees (most ECM trees are considered potential timber species), lack of ECM inoculum on skid trails and landings suggests that dispersal and survival of ECM propagules is limited. Lack of ECM inoculum after severe disturbances is consistent with the results of earlier studies (Boerner et al. 1996, Janos 1996).

Agricultural practices also affected ectomycorrhizal inoculum potential of *Tetraberlinia* but not of *Afzelia*. The most likely explanation of the differential behavior of the two ceasalps is that both taxa differ in their specificity towards ECM fungi, thus confirming the above remark. The issue of host plant specificity of ECM fungi has been repeatedly discussed. Smits (1994) emphasized specificity of ECM fungi in dipterocarp forests in Kalimantan (Indonesia) and implied that tropical ECM fungi were different in that respect from ECM fungi in temperate areas. Kuyper (unpublished observations) revised the taxonomy of ECM fungi from these forests and concluded that ECM specificity was not different between tropical and temperate forests.

Our results demonstrate that if most ectomycorrhizal tree species behave similarly to *Tetraberlinia*, conservation of forest patches and clumps where these trees occur is urgently needed. In the framework of sustainable management of tropical rainforests, it would be important to preserve seed-bearing trees and to assess whether addition of soil with ECM inoculum to sites where selective logging has occurred would increase chances for seedlings of ECM ceasalps to become ectomycorrhizal and hence contribute to the maintenance of the diversity of ECM trees and fungi.

10. Conclusion

Cameroonian ectomycorrhizal fungi abundantly fruited mostly in five types of clumps including *Gilbertiodendron* monodominant, *Uapaca* monodominant, monodominant *Microbelinia* in Korup National Park, oligo-dominant ceasalp clumps locally called "Ekop" and mixed clumps of "Ekop" with *Uapaca* spp., independent of elevation, rainfall, topography and soil texture. Though poor in plant diversity and no longer or barely regenerating, these habitats recruited abundant and various ECM fungal species on which native ECM tree species depend for survival, and some others such as chanterelles which serve as a ready source of protein for local people during harsh periods and could be a cash flow. Therefore, forest clumps should be conserved as biodiversity sanctuaries. The presence of ECM fungi could serve as indicators for sustainable management of African humid forests and climate change. Hence, the presence of such tree species and stands should be made known to all forest stakeholders for their ecological specificity and relevance in the management of humid forests of the Congo basin.

Acknowledgements

We thank the anonymous referees for their valuable comments on this study, and Krista L. McGuire and Caitlyn Gillikin for improving the English. Financial support from NWO (Priority programme—Biodiversity in disturbed ecosystems) and logistic support by the Tropenbos Cameroon Programme and the Government of Cameroon through financial support to the Institute of Agricultural Research for Development (IRAD) are gratefully acknowledged. Field assistance was provided by Jean-Baptiste Mva, Edouard Nsomoto, Serge Aba'a Aba's, Roger Eyene, and Thomas Mba. The technical assistance by Veronique O. Anaba during the project is very much appreciated. Thanks to the reviewer.

References

Agerer, R. 1995. Anatomical characteristics of identified ectomycorrhizas: an attempt towards a natural classification. *In*: A.K. Varma and B. Hock (eds.). Mycorrhiza: Structure, Function, Molecular Biology and Biotechnology. Springer Verlag, Heidelberg, pp. 685–734.

Alexander, I.J. 1989. Mycorrhizas in tropical forests. *In*: J. Proctor (ed.). Mineral Nutrients in Tropical Forest and Savanna Ecosystems. Blackwell Scientific Publications, Oxford, pp. 169–188.

Bâ, A.M., R. Duponnois, M. Diabaté and B. Dreyfus. 2011. Les champignons ectomycorhiziens des arbres forestiers en Afrique de l'Ouest. Méthodes d'étude, diversité, écologie, utilisation en foresterie et comestibilité. Collections Actiques IRD, IRD Editions 268pp.

Boerner, R.E.J., B.G. Demars and P.N. Leicht. 1996. Spatial patterns of mycorrhizal infectiveness of soils along a successional chronosequence. Mycorrhiza 6: 79–90.

Buyck, B., D. Thoen and R. Watling. 1996. Ectomycorrhizal fungi of the Guineo-Congo Region. Proc. R. Soc. Edinb. Sect. B 104: 313–333.

Boa, E. 2006. Champignons comestibles sauvages—Vue d'ensemble sur leurs utilisations et leur importance pour les populations. FAO, Rome.

Buyck, B. 1994. Ubwoba: les champignons comestibles de l'Ouest du Burundi. Adm. Gén. Coop. Dév. Publi. Agric. Bruxelles 34: 123.

Connell, J.H. and M. Lowman. 1989. Low-diversity tropical rainforests: some possible mechanisms for their existence. Am. Nat. 134: 88–119.

De Kessel, A., J.T.C. Codjia and N.S. Yorou. 2002. Guide des champignons comestibles du Bénin. Cotonou, République du Bénin, Jardin Botanique National de Belgique et Centre International d'Ecodéveloppement Intégré (CECODI). Impr. Coco-Mutimedia 275pp.

De Saint Aubin, G. 1963. La forêt du Gabon—Centre Techn. For. Trop., Nogent-sur-Marne.

Diédhiou, A.G., M.-A. Selosse, A. Galiana, M. Diabaté, B. Dreyfus, A.M. Bâ, S. de Faria and G. Béna. 2010. Multi-host ectomycorrhizal fungi are predominant in a Guinean tropical rainforest and shared between canopy and seedlings. Environ. Microbiol. 12: 2219–2232.

Eneke, E.T.B and I.J. Alexander. 2012. Mycorrhiza status of *Gnetum* spp. in Cameroon: evaluating diversity with a view to ameliorating domestication efforts. Mycorrhiza 22: 99–108.

Fassi, B. and M. Moser. 1991. Mycorrhizae in the natural forests of tropical Africa and the Neotropics. *In*: A. Fontana (ed.). Funghi, Piante e Suolo, Centro di studio Micologia del Consiglio nazionalle delle Ricerche, Torino, Italy, pp. 183–202.

Grubb, P.J. 1977. The maintenance of species richness in plant communities: the importance of the regeneration niche. Biol. Reviews 52: 107–145.

Henkel, T.W., J. Terborgh and R.J. Vilgalys. 2002. Ectomycorrhizal fungi and their leguminous hosts in the Pakaraima Mountains of Guyana. Mycol. Res. 106: 515–531.

Janos, D.P. 1996. Mycorrhizas, succession and rehabilitation of deforested lands in the humid tropics. *In*: J.C. Frankland, N. Magan and G.M. Gadd (eds.). Fungi and Environmental Change. Cambridge University Press, Cambridge, pp. 129–161.

Jonsson, L., A. Dahlberg, M.-C. Nilsson, O. Karén and O. Zackrisson. 1999. Continuity of ectomycorrhizal fungi in self-regenerating boreal *Pinus sylvestris* forests studied by comparing mycobiont diversity on seedlings and mature trees. New Phytol. 142: 151–162.

Letouzey, R. 1968. Etude phytogéographique du Cameroon. Paris, Editions P. le Chevalier.

Letouzey, R. and R. Mouranche. 1952. Ekop du Cameroun. Centre Technique Forestier Tropical. Nogent-sur-Marne.

Malaisse F. 1967. Se nourrir en forêt claire africaine—Approche écologique et nutritionnelle. Les Presses Agronomiques de Gembloux, Gembloux, Belgique.

Moyersoen, B., A.H. Fitter and I.J. Alexander. 1998. Spatial distribution of ectomycorrhizas and arbuscular mycorrhizas in Korup National Park, Cameroon, in relation to edaphic parameters. New Phytol. 139: 311–320.

Ndong, E.H., J. Degreef and A. De Kesel. 2011. Champignons comestibles des forêts denses d'Afrique centrale. Taxonomie et identification. Vol. 10. ABC Taxa. ISSN 1784–1291: 262pp.

Newbery, D.M., I.J. Alexander, D.W. Thomas and J.S. Gartlan. 1988. Ectomycorrhizal rain forest legumes and soil phosphorus in Korup National Park, Cameroon. New Phytol. 109: 433–450.

Newbery, D.M., I.J. Alexander and J.A. Rother. 1997. Phosphorus dynamics in lowland African rainforest: the influence of ectomycorrhizal trees. Ecol. Monographs 67: 367–409.

Onguene, N.A. 2000. Diversity and dynamics of mycorrhizal associations in tropical rain forests with different disturbance regimes in south Cameroon. Tropenbos Cameroon Ser. 3: 1–167.

Onguene, N.A. and T.W. Kuyper. 2012. Habitat and diversity of ectomycorrhizal mushrooms in humid forests of South Cameroon. Cameroon J. Exp. Biol. 8: 26–34.

Onguene, N.A. and T.W. Kuyper. 2001. Mycorrhizal associations in the rain forest of South Cameroon. For. Ecol. Manag. 140: 277–287.

Onguene, N.A. and T.W. Kuyper. 2002. Importance of the ectomycorrhizal network for the seedling survival and ectomycorrhizal formation in rain forests of south Cameroon. Mycorrhiza 12: 13–17.

Onguene, N.A. and T.W. Kuyper. 2005. Growth response of three timber species to soils with different arbuscular mycorrhizal inoculum potentials in South Cameroon. Indigenous inoculum and effect of addition of grass inoculum. For. Ecol. Manag. 210: 283–290.

Rammeloo, J. and R. Walleyn. 1993. The edible fungi of Africa south of Biafra: a literaturesurvey. Scripta Bot. Belg. 5: 1–62.

Richards, P.W. 1996. The tropical rain forest—an ecological study, 2nd edition. Cambridge, Cambridge University Press.

Rivière, T., A.G. Diédhiou, M. Diabaté, G. Senthilarasu, K. Natarajan, A. Verbeken, B. Buyck, B. Dreyfus, G. Béna and A.M. Bâ. 2007. Genetic diversity of ectomycorrhizal basidiomycetes from African and Indian tropical forests. Mycorrhiza 17: 415–428.

Sanon, K.B., A.M. Bâ and J. Dexheimer. 1997. Mycorrhizal status of some fungi beneath indigenous trees in Burkina Faso. For. Ecol. Manag. 98: 61–69.

Singer, R. and I. Araujo. 1986. Litter decomposition and ectomycorrhizal Basidiomycetes in an igapo forest. Plant Syst. Evol. 153: 107–117.

Singer, R., I. Araujo and M.H. Ivory. 1983. The ectotrophically mycorrhizal fungi of the neotropical lowlands, especially Central Amazonia. Nova Hedw. 77: 1–352.

SAS Inc. 2004. SAS—X user's guide, 4th edition. Gorinchem.

Smith, M.E., T.W. Henkel, M.C. Aime, A.K. Fremier and R. Vilgalys. 2011. Ectomycorrhizal fungal diversity and community structure on three co-occurring leguminous canopy tree species in a Neotropical rainforest. New Phytol. 192: 699–712.

Smits, W. 1994. Dipterocarpaceae: mycorrhizae and regeneration. Tropenbos Series 9: 1–243.

Taylor, D.L. and T.D. Bruns. 1999. Community structure of ectomycorrhizal fungi in a *Pinus muricata* forest: minimal overlap between the mature forest and resistant propagule communities. Mol. Ecol. 8: 1837–1850.

Tedersoo, L., A. Sadam, M. Zambrano, R. Valencia and M. Bahram. 2009. Low diversity and high host preference of ectomycorrhizal fungi in Western Amazonia, a neotropical biodiversity hotspot. Int. Soc. Micro. Ecol. 1–7.

Van Dijk, H., N.A. Onguene and T.W. Kuyper. 2003. Knowledge and utilization of edible mushrooms by local populations of the rain forest of South Cameroon. Ambio. 32: 19–23.

Walleyn, R. and J. Rammeloo. 1994. The poisonous and useful fungi of Africa south of the Sahara. Scripta Bot. Belg. 10: 1–56.

CHAPTER 4

Mycorrhizal Fungi Diversity and their Importance on the Establishment of Native Species Seedlings within Madagascarian Degraded Sclerophyllous Forest

Rondro Harinisainana Baohanta,[1,*] *Herizo Andrianantoandro Randriambanona,*[1] *Marc Ducousso,*[2] *Christophe Nirina Rakotoarimanga,*[1] *Yves Prin,*[2] *Heriniaina Ramanankierana*[1] and *Robin Duponnois*[1]

1. Introduction

The impact of human activities on tropical ecosystems has increased dramatically in recent decades leading to a global reduction of primary forests (Laurance 1999, Morris 2010). For tree species, fragmentation of forests into patches has led to degradation of both their habitat and

[1]Laboratoire de Microbiologie de l'Environnement, Centre National de Recherches sur l'Environnement BP 1739 Fiadanana Antananarivo 101, Madagascar.
[2]Laboratoire des Symbioses Tropicales et Méditerranéennes (LSTM), UMR 113 CIRAD/INRA/IRD/SupAgro/UM2, Campus International de Baillarguet, TA A-82/J, Montpellier, France.
*Corresponding author: ninish.rondro@yahoo.fr

ecological processes such as biogeochemical cycling and community dynamics (Weaver and Kellman 1981). The forest ecosystems of Madagascar are recognized as being among the most species-rich around the world (Myers et al. 2000). Approximately 80% of the plant species of Madagascar are endemic and the island contains a wealth of fauna and flora (Mittermeier et al. 2004), but they have been seriously impacted by a combination of agricultural practices such as slash and burn culture, livestock production and logging, and ancient climate changes. More recently, during the last two centuries, the extent of the Malagasy forest has decreased dramatically. Today, primary vegetation probably covers only about 10% of the original area and dense forest has been reduced to a fragmented landscape (Myers et al. 2000). Similarly, in the highland of Madagascar, sclerophylous woodland dominated by *Uapaca bojeri* (Tapia forest) is the last remnant of primary forest. The main components of this formation are small endemic trees of *Uapaca bojeri* (family Euphorbiaceae), in some patches endemic shrubs such as *Sarcolaena* sp. (family Sarcolaenaceae), or *Asteropeia* sp. (family Asteropeiaceae) which also grow as subdominants. The Tapia forest plays an important role in local community livelihood as it provides fuel, wood and timber, medicinal plants, non-timber forest products, and many other sources of food and income (Kull et al. 2002). Additionally, Tapia forest provides important environmental services: e.g., protection against erosion, protection of water sources, and carbon sequestration. However, Tapia trees are threatened by human destruction through bush fires, firewood collection, and charcoal production, and restoration efforts using native trees of this forest formation should be undertaken to preserve the entire ecosystem.

For several years, the trend for ecological restoration has been the use of mycorrhizal fungi, considered an important component that stabilizes the soil and enhances plant growth by alleviating nutrient and drought stress (Mikola 1980, Högberg 1982, Williams et al. 2012). Mycorrhizal inoculation has been used to promote native plant growth and reestablishment in various ecosystems (Smith et al. 1998). Likewise, in restoration programs of vulnerable or endangered native species, it is important to know the mycorrhizal status of the plants. Cooke and Lefor (1998) reported that the presence or absence of mycorrhizae in individuals of plant species used in wetland restorations might be an important factor in the reestablishment of wetland plant associations. Classified as biological tools in ecological restoration (Ruiz-Jaen and Aide 2005), mycorrhizae colonization can significantly affect plant growth and patterns of succession after a disturbance (Haselwandter 1997).

During the last decade, an increasing number of studies have also emphasized the importance of mycorrhizal facilitation in driving plant community succession and vegetation restoration (Callaway 1995, 1997,

Brooker et al. 2008). In Madagascar, the description of mycorrhizal status of native species and the role of facilitative interactions in plant communities have received attention over the last few years, enhancing the knowledge of the ecology of those threatened ecosystems (Ramanankierana et al. 2007, Baohanta et al. 2012). For example, in the Tapia forest, it has been reported that *Leptolaena bojeriana* (Bail.) Cavaco. (family Sarcolaenaceae), an endemic shrub, can facilitate the ectomycorrhizal (ECM) colonization of *Uapaca. bojeri* and mitigates the negative effects of the introduction of exotic tree species (*Pinus patula, Eucalyptus camaldulensis*) on the early growth of the native tree (Baohanta et al. 2012). The current tendency is to utilize the native plants to restore degraded ecosystems (Castro et al. 2006). According to Ren et al. (2008), the best nurse plants are the native species that offer microhabitats for target plant establishment or recruitment on degraded environment. In this chapter, we aim to review some of the recent advances in the understanding of the implication of ECM symbioses in the development and regeneration of native shrub and tree species in Madagascarian highland sclerophyllous forest. We will focus on (i) the sporocarp survey within Madagascarian sclerophyllous forest, (ii) the importance of mycorrhizal inoculation on early development of *U. bojeri*, an endemic tree species, and (iii) the nurse plant phenomenon between endemic shrub and tree species within sclerophyllous forest in the Malagasy high plateau.

2. Sporocarp Survey within Sclerophyllous Forests in Central Highland of Madagascar

The sclerophyllous forest dominated by *U. bojeri* constitutes one of the few remaining patches of forest and woodland on the central highland of Madagascar more than 1000 m above sea level. The sclerophyllous woodlands are valuable examples of a very restricted vegetation type that has mostly been replaced by artificial, fire-maintained grassland (Dupuy and Moat 1996). It is estimated that *U. bojeri* forest size has been reduced by approximately 43% since the 1970s (Moat and Smith 2007). This forest formation is characterized by a low specific richness of plants, but dominated by a few ECM endemic trees of the Euphorbiaceae family and shrub species of Sarcolaenaceae and Asteropeiaceae. Three sites of this sclerophyllous forest were visited by Ramanankierana et al. (2007) at 2 to 3 week intervals during the sampling seasons, mid-November 1992 to early February 1993, July–August 1994, and July to mid-September 1995 in order to collect ECM fungi fruiting under ectomycorrhizal tree and shrub species. The forests were located at 50 km to the west of Antananarivo (site A), 20 km to the south of Antsirabe (Ambositra site, site B), and 100 km to the east of Toliara (Isalo site, site C). The vegetation sampled included savannas (sites

A and B) and deciduous forest (site C). The main chemical characteristics of the upper soil layer (2–20 cm) of these sites are shown in Table 1.

These investigations showed that sclerophyllous woodlands dominated by *U. bojeri* contain a wide range of sporocarps belonging to at least four different fungal families: Russulaceae, Cantharellaceae, Boletaceae, and Amanitaceae. Overall, 94 sporocarps of putative epigeous ECM fungi were collected in three survey sites. They were identified as belonging to the ECM genera *Afroboletus, Amanita, Boletus, Cantharellus, Leccinum, Gyroporus, Rubinoboletus, Russula, Scleroderma, Suillus, Tricholoma*, and *Xerocomus* (Table 2). *Russula* was the most frequent ECM genus recorded (32.9% of the above-ground sporocarp diversity) followed by the genera *Amanita* (17.1%) and *Cantharellus* (Fig. 1a). At species level, 21 species were recorded for *Russula* followed by 14 *Amanita* species,10 *Cantharellus* species, and 10 *Boletus* species (Fig. 1a, b). The fungal richness of the above-ground sporocarps decreased from site A (40 species) to site C and B with 29 and 27 species, respectively.

The results of Ramanankierana et al. (2007) showed a large diversity of sporophores recorded under ectomycorrhizal tree and shrub species in the sclerophyllous forests of Madagascarian highlands. Different fungal families (Russulaceae, Cantharellaceae, Boletaceae, and Amanitaceae) observed at this forest have been recorded in other tropical forests under *Afzelia africana, Monotes kerstingii, Uapaca guineensis*, and *U. somon* in Africa (Thoen and Bâ 1989, Sanon et al. 1997) and under dipterocarps in Asia (Lee 1998). It has been already demonstrated that Russulaceae are often dominant in the tropical rainforests of Africa, Asia, and Madagascar (Buyck et al. 1996, Lee et al. 1997, Walting and Lee 1998, Rivière et al. 2006).

In the northern part of this sclerophyllous forest (Site A), three ectotrophic plant species (*U. bojeri, Leptolaena pauciflora* Baker and *Leptolaena bojeriana* (Baill.) Cavaco. were identified by Baohanta et al. (2012). The highest fungal diversity of the above-ground sporocarpswas recorded at

Table 1. Main chemical characteristics of the upper soil layer (2–20 cm). Data from Ramanankierana et al. (2007).

Site	Site A	Site B	Site C
pH (H_2O)	4.96	5.37	4.54
pH (KCl)	4.75	5.23	4.45
Total C (%)	1.12	3.09	1.33
Total N (%)	0.07	0.15	0.91
Total organic matter (%)	1.92	5.31	2.28
C/N	16.0	21.0	14.6
Total P (mg g^{-1})	15.2	15.2	17.3
Available P (mg g^{-1})	3.42	7.01	5.24

Table 2. Description of putative ectomycorrhizal fungi collected from the three studied sites beneath ectomycorrhizal tree and shrub species within sclerophyllous forest in Madagascarian highlands. Data from Ramanankierana et al. (2007).

Species	Prominent features	Habitat	Sites		
Amanitaceae			Site A	Site B	Site C
Amanita rubescens Gray	White pinkish cap (8 cm diameter) covered with white powdered and flat scales, remnant veil visible at the margin, white stem reddening by wound, often eaten by insect larvae	Solitary, scarce		x	
Amanita virosa (Fr.) Bertillon	White yellowish fruiting body (7 to 12 cm diameter), white and chinated stem (1.2 cm diameter) with ring and cup at the base	Patch of 5 to 6 individuals	x	x	x
Amanita phalloides var. *verna* Bull	White fruiting body (5.5 to 11 cm diameter), stem (0.6 diameter by 9.5 cm high) with a large pendant ring and a bulbous cup at the base	Patch of 5 to 7 individuals	x	x	x
Amanita strobiliformis Bertillon	White and big fruiting body (10 to 12 cm diameter), fleecy remnant veil on the cap, club-shaped stem (2.2 cm diameter) with a ring	Solitary, scarce	x		
Amanita cf. *Baccata* (Fr.) Gillet	Big white fruiting body similar features than previous species but with no ring, stem (2 cm diameter by 7 cm high)	Solitary, scarce		x	
Amanita sp1	White finely scaled fruiting body (4 to 6 cm diameter) turning yellowish when aging or by wound, concolored gills and flesh	Solitary, scarce	x		
Amanita cf. *strobiloceovolvata* Beeli	White fruiting body (8.5 to 11cm diameter), stem (1.2 cm diameter by 10.5 cm high) without ring, well-developed bulbous cup at the base	Patch of 3 to 4 individuals	x	x	x
Amanita sp2	White and big species with a convex scaly cap (10 to 13 cm diameter by 9 to 10 cm high), strong bulbous stem (3 to 4cm diameter) with a pendant ring	Solitary, scarce		x	
Amanita sp3	Pale grey cap (4.5 cm diameter) with few veil remanences on surface, bulbous stem (0.7 to 6cm) with grey chinates	Solitary, scarce	x		
Amanita sp4	Yellow conical and mucronated cap (2.5 to 3 cm diameter), paler to whitish gills and stem (0.5 cm diameter by 12 cm high), white scaly basal cup	Solitary, scarce	x		

Table 2. contd....

Table 2 .contd.

Species	Prominent features	Habitat	Sites		
Amanitaceae			Site A	Site B	Site C
Amanita cf. *cecilia* (Berk. et Broome) Bas	Yellow grey cap (4 to 5 cm diameter) with rised scales, white gills and concoloured stem (0.7 cm diameter to 6 cm high), bulbous base covered by grey chinates and veil remanences	Solitary, scarce	x		
Amanita sp5	Convex and grey purplish-blue cap (4 to 4.5 cm diameter) with grey flat scales at the centre and hairy ones at the margin, white flesh and gills, white bulbous stem (0.9 cm diameter by 6 cm high) turning to grey by touch with a pendant ring	Solitary, scarce	x		
Amanita sp6	Small white species (2 to 3 cm diameter) with yellowish scales, bulbous based stem with pendant ring	Solitary, scarce		x	
Amanita sp7	Big white flat cap species (9 to 13 cm diameter) with veil remanences at the margin, strong bulbous stem (3 to 4 cm diameter) with a ring	Patch of 2 to 3 individuals	x	x	x
Boletaceae					
Rubinoboletus griseus	Big red-pink and grey-brownish dry and smooth cap (10 to 12 cm diameter by 8 to 9 cm high), white flesh (1.8 cm thick) partially burnishing after sectioning, pale reticulated hairy scaled stem, burnishing like pores by touch	Patch of 5 to 6 individuals	x	x	x
Gyroporus cf. *cyanescens* (Bulliard Fr.) Quélet	Big white yellowish smooth cap (10 to 12 diameter by 8-9 cm high), concolored tubes and stem turning to blue by wound	Patch of 3 to 4 individuals	x	x	x
Boletus sp1	Brownish to brown cap, with large darker flat scales, cylindrical and dark stem, red-reticulated becoming yellow at the base like rhizomorph, flesh and pores turning blue by wound	Patch of 3 to 4 individuals	x		
Leccinum sp1	Small grey boletus (1.8 to 3 cm diameter by 3 to 4cm high), yellow pores, red hairy scales on the stem, base of the stem yellow like the rhizomophs	Patch of 3 to 4 individuals	x	x	x
Boletus sp2	Big brownish-brown wet cap (7 to 8 cm diameter to 12 to 15 cm high), white and smooth flesh	Patch of 5 to 6 individuals	x		

Xerocomus sp1	Brown scaly cap (8.5 cm diameter) showing white flesh between scales, white stem (1.4 cm diameter by 5 to 6 cm high) with some red zone	Solitary, scarce	x		
Leccinum sp2	Yellow grey scaled boletus (4.5 to 6 cm diameter by 6 to 7 cm high), stem yellow at the base and red in its upper part, yellow blueishing pores	Solitary, scarce	x		
Boletus sp3	Brown cap (7.5 cm diameter) with red-pink pigments, yellow and red pores, greenishing and blueishing tubes, yellowish stem with some red pigments	Solitary, scarce	x		
Leccinum sp3	Red purplish-blue wet cap (7 cm diameter), yellow burnishing stem (0.8 cm diameter by 6 cm high), concolored yellow flesh and pores, blueishing after air exposure	Solitary, scarce	x		
Boletus sp4	Big smooth and shiny red boletus (8 to 12 cm diameter by 7 to 8 cm high), yellowish stem with some pink pigments, concolored flesh (1.6 cm thick)	Patch of 2 to 3 individuals	x	x	x
Xerocomus sp2	Pale to dark brown scaly dry cap (5 cm diameter), white dirty stem (0.8 cm diameter by 4 cm high) with a white-yellowish flesh, yellow greenish and pink pores	Solitary, scarce	x		
Boletus sp5	Yellowish brown cap (8 cm diameter) with flat partially pink scales, yellow pores and stem (1.2 cm diameter by 6 cm high), white flesh (1.6 cm thick)	Solitary, scarce		x	
Boletus sp6	Dark brown scaly cap showing yellow flesh, pale concolored pores and stem	Solitary, scarce			x
Boletus sp7	Brown boletus with dry and silky cap (4.5 cm diameter), concolored dark stem (2.2 cm diameter by 5.2 cm high), white flesh (1.6 cm thick) rapidly turning to red, then black after air exposure	Solitary, scarce			x
Boletus sp8	Pale brown boletus with silky cap (5 cm diameter), white stem (1.5 cm diameter by 5.2 cm high) and flesh (1.3 cm thick) turning purplish-blue after air exposure	Solitary, scarce			x

Table 2 .contd....

Table 2 .contd.

Species	Prominent features	Habitat	Sites		
Amanitaceae			Site A	Site B	Site C
Leccinum sp4	Yellow and wet cap (3.5 cm) with hairy grey scales, yellow pores, yellow and red stem (0.5 cm diameter) with dark scales and a narrow base	Solitary, scarce			x
Leccinum sp5	Yellowish-brown dry cap (3.5 cm diameter), red pores and lighter stem (0.6 cm diameter by 4 cm high) turning to dark-brownish in section, white flesh turning burnish after air exposure	Solitary, scarce			x
Suillus sp2	Yellow and grey scaly cap (5 cm diameter), yellow pores covered by a yellow partial veil when young, yellow stem (1.4 cm diameter by 4.5 cm high) with greenish grey scales, becoming very slimy	Patch of 2 to 3 individuals		x	
Boletus sp9	Yellow brownish boletus (7 to 8 cm diameter) with a sticky surface, yellow pores and stem, yellowish flesh (1.7 cm thick)	Solitary, scarce		x	
Leccinum sp6	Pale brown cap (5 to 4 cm diameter) with red brownish scales at the centre, white pores and white flesh turning rapidly to red, then black by wound	Solitary, scarce		x	
Boletus sp10	Yellow brown boletus (4.5 to 5.7 cm diameter) with wet and smooth surface, yellow pores, yellow stem (1.2 diameter by 3 cm high), white flesh (1 cm thick)	Solitary, scarce		x	
Cantharellaceae					
Cantharellus sp1	Tall thick and lobed fasciculate bright yellow caps (4 to 6 cm diameter) forming patches of 4 to 5 individuals (12 cm), grained gills, pale yellow stem (1.8 cm), white flesh	Patch of 8 to 10 individuals	x	x	x
Cantharellus sp2	Small orange-brownish cap (2 to 2.2 cm diameter), white pinkish gills, pink stem and white flesh	Solitary, scarce	x		
Cantharellus sp3	Yellowish to pale brown cap (3.5 to 3.2 cm diameter), yellow grained gills, pale yellow stem (0.6 to 2.5 cm)	Solitary, scarce	x	x	x

Cantharellus sp4	Red orange cap (3.2 to 3.5 cm diameter), largely spaced yellowish grained gills, pale yellow to reddish stem (0.9 cm)	Patch of 8 to 10 individuals	x	x	x
Cantharellus sp5	Pale brown cap (3.2 to 3.5 cm diameter), pale pink grained gills, white stem and flesh, turning to yellow by touch or sectioning	Solitary, scarce	x	x	x
Cantharellus sp6	Red pinkish fasciculate caps (2.5 cm diameter) forming small patch (3.5 to 4 cm), yellowish grained gills, pink orange stem and white fibrous flesh	Patchy			x
Cantharellus sp7	Small and fragile bright yellow cap (2 to 3.2 cm diameter), pale yellow gills, concolored short stem (0.3 cm)	Solitary, scarce			x
Cantharellus cf *decolorans* Eyss. et Buyck	Small pink orange cap (0.7 to 1.5 cm diameter, 2.5 to 3.5 cm high), concolored gills and short stem (0.2 cm)	Patch of 5 to 6 individuals		x	
Cantharellus cf. *cyanoxanthus* R. Heim	Yellow and purple cap (4 cm diameter), pale pink grained gills, pale yellow stem (1.8 cm), fibrous flesh	Patch of 2 to 3 individuals			x
Cantharellus rubber R. Heim	Pale pink cap (3.5 to 4 cm diameter), concoloured stem and gills	Patch of 2 to 3 individuals			x
Russulaceae					
Russula subfistulosa Buyck	White-greyish (darker at the centre) umbilicated cap (3 to 12 cm diameter).	Solitary to patch of 3 individuals	x	x	x
Russula ochraceorivulosa	Greyish to purplish-blue grey cap (7 to 8 cm diameter). Convex cap with an undulating margin	Solitary	x	x	x
Russula patouiillardi	Pale yellow and purple (darker at the centre) dry scaly cap, white and purple stem	Solitary to patch of 5 individuals	x	x	x
Russula liberiensis Buyck	White-greyish fibrillose cap (3 to 12 cm diameter) turning brown when ageing, closely spaced decurrent gills	Solitary to patch of 3 individuals	x	x	x
Russula cf. *cyanoxantha*	Pink to purple-red cap (5 to 15 cm diameter), white stem	Patch of 2 to 3 individuals	x		
Russula cellulata Buyck	Brown scaly cap (3 to 9 cm diameter), closely spaced decurrent gills	Patch of 2 to 3 individuals	x		

Table 2 .contd....

Table 2 .contd.

Species	Prominent features	Habitat	Sites		
Amanitaceae			Site A	Site B	Site C
Russula cf. *archae R. Heim*	White smooth and flat cap (4.5 to 6 cm diameter)	Solitary	x		
Russula cf. *nigricans*	White-greyish cap turning to brown when ageing, white flesh turn rapidly pink to red by air exposure	Solitary	x		
Russula cf. *subfistulosa*	White-greyish convex cap (3 to 8 cm diameter)	Solitary to patch of 4 individuals	x		
Russula sp3	White to pale yellow gluey and convex cap (3 to 13 cm diameter)	Solitary to patch of 3 individuals	x		
Russula sp5	Yellow smooth umbilicated cap (6 to 12 cm diameter), with a very regular margin	Patch of 2 to 3 individuals	x	x	x
Russula sp6	White-yellowish flat or slightly umbilicated cap (4 to 10 cm diameter), white flesh turning reddish after air exposure	Patch of 3 to 5 individuals	x		
Russula sp7	Dark grey to brown convex cap (3.5 to 8 cm diameter), involucrated margin, wet surface covered by orange to yellow layers, white-yellowish flesh	Solitary, rarely patchy	x		
Russula sp8	White convex to slightly umbilicate cap (4 to 13 cm diameter) turning brown when ageing, smooth surface with involucrated margin, white flesh turning reddish after air exposure	Solitary, scarce	x		
Russula sp10	Dark grey to brown when fully mature convex to flat cap (4 to 9 cm diameter), white flesh	Patch of 2 to 4 individuals	x		
Russula sp11	Small purple to purple-reddish umbilicate when young to flat when ageing cap (2 to 7 cm diameter), sticky surface, regular margin, adnate white to yellowish closely spaced gills, white flesh	Solitary to patch of 3 individuals	x		
Russula sp13	Brown-reddish convex and smooth glutinous cap (6 to 15 cm diameter), decurrent gills, white flesh turning greyish by air exposure	Solitary		x	

Russula sp14	Dark yellow to brown convex to flat sticky cap (4 to 10 cm diameter), adnate closely spaced gills	Patch of 3 to 5 individuals			x
Russula sp15	Yellow to orange-yellow flat slightly umbilicated with an involucrated yellow margin cap (2 to 8 cm diameter) with a smooth surface with small strias	Patch of 2 to 5 individuals			x
Russula sp16	Pink to reddish (darker at the centre) fragile convex glutinous cap, (2 to 6 cm diameter) with a smooth or dusty surface, white flesh	Patch of 2 to 4 individuals			x
Russula sp17	Slightly umbilicated convex and glutinous cap (4 to 10 cm diameter), dark yellow tending to brown, yellow to pale orange closely spaced gills	Solitary to patch of 4 individuals			x
Strobilomycetacea					
Afroboletus sp1	Brown-purple scaly cap (3 to12 cm diameter), fibrous stem, pale yellow flesh turning purplish by air exposure	Patch of 3 to 5 individuals	x		
Afroboletus sp2	Flat-convex dusty cap (3 to 10 cm diameter) with dark-brown to black scales, fibrous stem inflated at the base, greyish-yellow flesh	Patch of 2 to 3 individuals			x
Sclerodermataceae					
Scleroderma sp1	Whitish to yellowish small pyriformic fruit bodies, size below 3 cm in diameter, dark grey gleba	Solitary to patch of 5 individuals			x
Scleroderma sp2	Whitish to yellowish 3 to 7cm diameter fruit bodies with grey spots at the top, dark grey gleba	Solitary, rarely patchy	x		
Tricholoma sp2	Yellow cap (3 to 9 cm diameter), dry surface, involucrated margin, thick widely spaced gills, yellow flesh keeping yellow even after exposure to air	Solitary to patch of 4 individuals	x		
Tricholoma sp3	Yellow-greyish cap (3 to 12 cm diameter), dry surface, white yellowish stalk, white flesh	Solitary to patch of 4 individuals		x	
Tricholoma sp4	Dark-grey cap (3 to 15 cm diameter), smooth dry surface, thick gills, white flesh	Solitary		x	

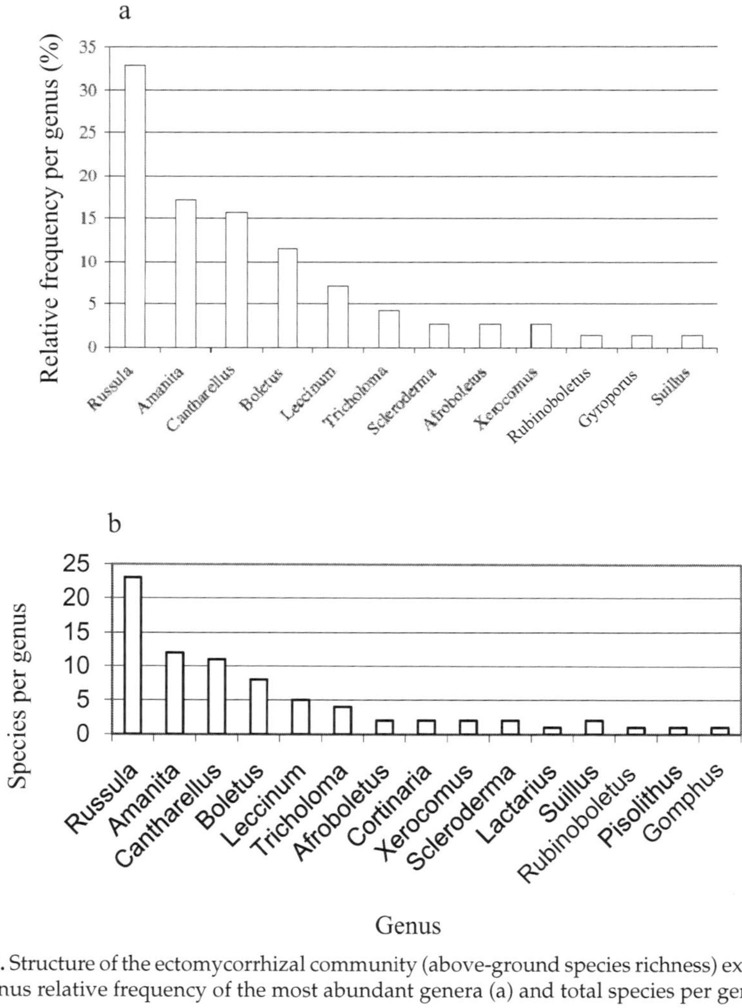

Fig. 1. Structure of the ectomycorrhizal community (above-ground species richness) expressed as genus relative frequency of the most abundant genera (a) and total species per genus (b).

this site (40 species). With other tropical ectomycorrhizal tree species, Lee et al. (1997) recorded only 28 fungal species under *Shorea leprosula*, while Sanon et al. (1997) have identified 14 fungal species under *U. guineensis* and 11 species under *U. somon* in Burkina Faso. However, numerous studies in temperate areas indicate little correlation between above-ground (sporocarps) and below-ground (morphotypes of ectomycorrhizas) fungal diversity (Buscot et al. 2000, Horton and Burns 2001). Further investigations using molecular approaches are needed to entirely describe the fungal community associated with plants.

3. Time Sequence of Mycorrhizal Colonization on *Uapaca bojeri*

The arbuscular mycorrhizal (AM)-ectomycorrhizal succession may be linked to spatial competition for colonization sites and differential colonization rates by the two types of fungi (Last et al. 1983, Chilvers et al. 1987, Last et al. 1987, Duchesne et al. 1988).

Uapaca bojeri constitutes a fundamental species in the sclerophyllous forests of the Madagascarian highlands. It has been illustrated that this endemic Euphorbiaceae of Madagascar is associated both with arbuscular mycorrhizas and ectomycorrhizas (Ramanankierana et al. 2007). Using soils collected beneath an adult tree of *U. bojeri*, Ramanankierana et al. (2007) demonstrated that ECM and AM structures were recorded on root systems of this plant after two months growing time (Fig. 2). Three native ECM fungi colonized approximately 50% of the lateral roots sampled from this tree species after 5 months growing time in pot culture.

In Madagascar, the result of Ramanankierana et al. (2007) illustrated, for the first time, the occurrence of two types of mycorrhizal fungi and their importance on seedling development of *U. bojeri*, an endemic tree species. These investigations highlighted the succession of three ECM morphotypes on root system during the early development of this tree (Fig. 2). However, the dominance of mycorrhizal structures on Malagasy tree and shrub species has been already shown within five coastal humid forests located along the eastern coast of Madagascar (Ducousso et al. 2008). These authors demonstrated that all of tree and shrub species belonging to Asteropeiacea

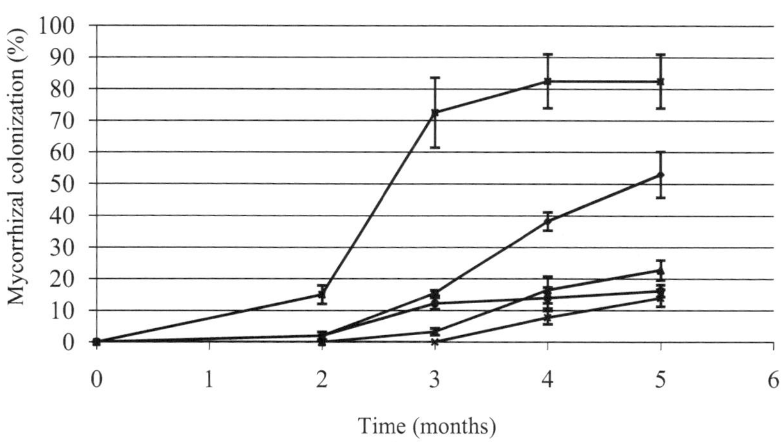

Fig. 2. Mycorrhizal colonization and sequence of ectomycorrhizal morphotypes on *U. bojeri* seedlings during 5 months growing time on soil collected from the native stand of this tree, ■: AM colonization, ◆: Total ectomycorrhizal colonization, ▲: morphotype 1, ●: morphotype 2, x : morphotype 3. Data from Ramanankierana et al. (2007).

and Sarcolaenaceae, two botanical families entirely endemic of Madagascar, were found to have both ectomycorrhizal and arbuscular mycorrhizal structures. However, the succession of each mycorrhizal structure on the root system of these plants was not described. In other tropical ecosystems, more attention has been focused on mycorrhizal colonization patterns of commercial forest tree species (Oliveira et al. 1997, Dos Santos et al. 2001, Founoune et al. 2002) and of forest-dominating tropical trees (McGuire et al. 2008). Because of the many desirable characteristics of these trees, such as rapid growth, high cellulose production, and resistance to disease and adverse environmental conditions, their use in reforestation programs in the tropics has been increasing steadily by using advanced mycorrhizal inoculation technologies such as controlled mycorrhization in nursery. In Madagascar, ecological restoration and reforestation programs using native plants have been undertaken in recent years in order to preserve, valorize, and sustainably manage the high biodiversity of this island. In this way, *U. bojeri* constitutes a potential candidate for reforestation programs in the highland region of Madagascar by managing the associated mycorrhizal fungi of this tree. Furthermore, research on mycorrhizal inoculation technology has been expanded to other valuable endemic trees in recent years, such as *Dalbergia trichocarpa* Jum. *Adansonia za* Jum & H. Perrier.

4. Mycorrhizal Dependency of *U. bojeri*

New opportunities to stimulate the growth of these plants or to reduce their mortality rate after transplantation to natural sites have arisen by using seedlings inoculated with both arbuscular mycorrhizas and ectomycorrhizas. This dual symbiotic association is well documented for *Populus* (Lodge and Wentworth 1990), *Salix* (Dhillion 1994), *Eucalyptus* (Lapeyrie and Chilvers 1985), *Alnus* (Molina et al. 1994), Pinaceae (Cazares and Trappe 1993), *Quercus* (Egerton-Warburton and Allen 2001), and Casuarinaceae (Duponnois et al. 2003). Using two types of mycorhizas (arbuscular endomycorrhizas and ectomycorrhizas) to inoculate *Acacia holosericea* (A. Cunn. Ex G. Don), Founoune et al. (2002) showed that inoculating with both forms of mycorrhiza resulted in greater plant development than if only one inoculant was used. However, the possibility of negative interactions among mycorrhizal fungi has been suggested (Lodge and Wentworth 1990). In Madagascar, single or dual mycorrhization of *U. bojeri* by *Scleroderma* sp. SC1 and/or *Glomus intraradices* (Schenk and Smith DAOM 181602, Ottawa Agricultural Herbarium) was conducted on sterilized sandy soil in greenhouse conditions (daylight of approximately 12 h, average daily temperature of 25°C) (Ramanankierana et al. 2007). *Scleroderma* sp. SC1 was a native ectomycorrhizal strain isolated from sporocarps collected under an adult tree of *U. bojeri*. The shoot dry weight of plants inoculated with *G. intraradices* or *Scleroderma* sp. SC1 was significantly higher than in the control treatment (non-inoculated plants) (Table 3). The shoot growth of inoculated plants was stimulated 1.9 times with *Scleroderma* sp. SC1, whereas it was 1.7 times higher for plants inoculated with *G. intraradices* compared to the control treatment. Inoculation with both fungal symbionts significantly increased the shoot dry weight of plants over the single inoculation treatments.

Table 3. Shoot growth, mycorrhizal development, and relative mycorrhizal dependency of *U. bojeri* seedlings 5 months after *G. intraradices* and/or *Scleroderma* sp. SC1 inoculation in pot culture. Data from Ramanankierana et al. (2007).

Treatments	Shoot biomass (mg per plant)	Ectomycorrhizal colonization (%)	AM colonization (%)	RMD (%)[a]
Control	91.1a[b]	0a	0a	-
Scleroderma sp. SC1	180.2b	8.7b	0a	47.6a
G. intraradices	160.0b	0a	77.5b	42.7a
Scleroderma sp. SC1 + *G. intraradices*	360.3c	11.5b	85.5b	70.7b

5. Dominance of Ectotrophic Endemic Shrub Species in Degraded Area of Sclerophyllous Forest in Malagasy Highland

Human activities such as land use change, slash and burn agriculture, or overexploitation of forest trees lead to soil degradation and loss of biodiversity in tropical areas. These anthropogenic impacts not only degrade natural plant communities (vegetation structure and species diversity) but also physicochemical and soil biological properties such as nutrient availability, microbial activity, and soil structure (Agarwal et al. 2005). It is well known that changing the vegetation structure can modify the dynamics of mycorrhizal fungi communities in soil (Dickie et al. 2002, Dickie and Reich 2005). Studies were carried out within two disturbed sclerophyllous forest ecosystems of *Uapaca bojeri*, located at Arivonimamo (S 19° 00′ 15″; E 47° 07′ 00″) and Ambatofinandrahana (S 20° 39′ 41.5″; E 047° 06′ 53.1″) in the Central part of Madagascar. The relative importance of mycorrhizal associations on early established shrub species in the degraded areas of these forests ecosystem was studied by using the transect method at three sampling points which were situated at the forest edge (0 m), passed through the area colonized mainly by shrubs species (25 m), and ended at the most degraded area without vegetation influences (50 m). It has been observed that a large part of these two respective natural forest ecosystems was replaced by exotic tree plantations, shrub formations, and grassland,

and/or by degraded surface areas where the regeneration of native species is quite limited or non-existent. Within the two study sites, the results showed that most of the shrub species identified are mycotrophic. Indeed, among all species inventoried, eight shrub species had AM infections and three, represented by *Leptolaena bojeriana, Leptolaena pauciflora, Sarcolaena oblongifolia*, were found with both AM and ECM colonization (Table 4).

In the Arivonimamo forest ecosystem, ten shrub species distributed among three sampling points were recorded. They were identified as belonging to the botanical families of Sarcolaenaceae, Ulmaceae, Ericaceae, Flacourtiaceae, Anacardiaceae, Asteraceae, and Rosaceae. The shrub species *Leptolaena bojeriana* was the most dominant at 25m from the *U. bojeri* adult tree, followed by *L. pauciflora* (Figs. 3A, C) (Baohanta 2011). Most of those shrub species were also found in the Ambatofinandrahana forest, which was characterized by the dominance of the shrub species *Sarcolaena oblongifolia*. These results corroborate those found by Pidwirny (2006) illustrating that the degradation of natural forests often leads to the formation of bare areas devoid of vegetation or of zones colonized by early colonizers such as grasses or shrubs. Particularly, ectomycorrhizal shrub species (*L. bojeriana* and/or *S. oblongifoila*) were widespread throughout degraded areas of the Madagascarian highland sclerophyllous forest. Some authors such as Schatz et al. (2001) and Ducousso et al. (2004, 2008) have already reported that most of the genera belonging to the Sarcolaenaceae family are ectotrophic. These properties could be the basis of their great capacity to tolerate environmental stresses found at degraded areas.

Table 4. Mycorrhizal status of early established shrub species in the degraded areas of the Arivonimamo and Ambatofinandrahana forests. Data from Baohanta (2011).

Shrub and tree species	**Family**	**Mycorrhizal status**	**Study site**
Leptolaena pauciflora Baker.	Sarcolaenaceae	ECM & AM	Arv & Amb
Leptolaena bojeriana (Baill.) Cavaco.	Sarcolaenaceae	ECM & AM	Arv & Amb
Trema sp.	Cannabaceae	AM	Arv & Amb
Vaccinium emirnense Hook.	Ericaceae	AM	Arv & Amb
Aphloia theaeformis (Vahl.) Benn.	Aphloiaceae	AM	Arv & Amb
Rhus taratana (Baker.) *H. Perrier*	Anacardiaceae	AM	Arv
Helychrysum russillonii Hochr.	Asteraceae	AM	Arv & Amb
Psiadia altissima (D.C.) Drake	Asteraceae	AM	Arv
Rubus apetalus Poir.	Rosaceae	AM	Arv & Amb
Erica sp.	Ericaceae	AM	Arv & Amb
Sarcolaena oblongifolia (Baill.) Cavaco.	Sarcolaenaceae	ECM & AM	Amb

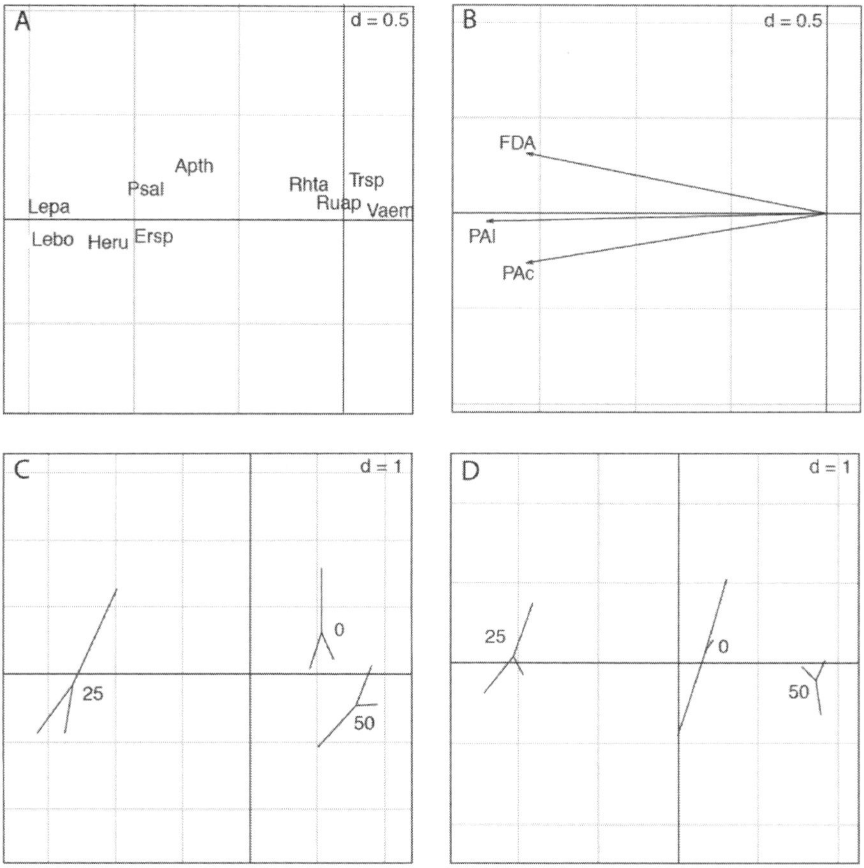

Fig. 3. Vegetation structure (A and C), global microbial activity and phosphatase activities of soil (B and D) at degraded forest edge (from an individual adult tree of *U. bojeri*) of sclerophyllous forest in high plateau of Madagascar.

6. Early Established Ectotrophic Shrub Species Affect Soil Chemical and Microbial Properties

In some cases, the propagation of pioneer species in degraded areas could affect soil restoration processes such as chemical and/or microbial properties (Holl 2002, Gómez-Aparicio et al. 2005). Within the two study sites in the high plateau of Madagascar, Baohanta (2011) observed the same results by analyzing total microbial activities [by fluorescein diacetate (FDA) hydrolysis assay] and soil phosphatase activities. Values of these two parameters were significantly high on soil samples mainly colonized by shrub species (25 m) compared to those recorded on bare soil (50 m) and on *U. bojeri* soil (0 m) (Fig. 3B, D). Correlation between these two soil parameters and the distribution of shrub species was observed in the Arivonimamo forest. It has been shown that the abundance of *L. pauciflora* and *L. bojeriana* was associated with a high value of global microbial and phosphatase activities of soil (Fig. 3).

Some authors (Gómez-Aparicio et al. 2005, Montané et al. 2010) have already reported that the propagation of some shrub species enhance litter quality and the amount of organic matter in soil. Such situations could lead to a definite increase in soil fertility. However, more experiments must be undertaken in order to confirm that modified soil conditions by pioneer shrub species facilitate both the establishment of tree species as well as the development of late successional species in Madagascarian highland sclerophyllous forest. On degraded soil, mycorrhizal fungi constitute one of the well-known biological components that can improve host performance by enhancing nutrient and water uptake from the soil and protecting host roots from pathogens and toxic compound (Smith and Read 2008).

7. Importance of Ectotrophic Shrub Species on the Establishment of Native Tree Species: The Case of *S. oblongifolia-L. bojeriana / U. bojeri*

Seedling establishment may be both inhibited and facilitated by pre-established plants. In this part of the chapter, we aim to review the recent advances in the capacity of two pioneer shrub species (*S. oblongifolia* and *L. bojeriana*) to facilitate seedling development of *U. bojeri* by sharing ECM symbionts in Ambatofinandrahana and Arivonimamo sclerophyllous forests respectively. By providing compatible fungal symbionts or by sharing interspecific hyphal links, it has been demonstrated that mycorrhizal fungi can mediate plant coexistence and succession (Dickie et al. 2004, Simard and Durall 2004, Baohanta et al. 2012).

Numerous studies have reported that the introduction of exotic tree species has an environmental impact on soil characteristics (i.e., soil nutrient contents, water dynamics) (Smith et al. 2000, Sicardi et al. 2004). By monitoring the development of *U. bojeri* seedlings in culture or not with *L. bojeriana* under glasshouse bioassays during five months growing time, Baohanta et al. (2012) illustrated that the occurrence of the shrub species significantly enhances the development of tree seedlings in both bare and *U. bojeri* origin soils or soil invaded by exotic plant species (*Eucalyptus camaldulensis* or *Pinus patula*) from Arivonimamo forests (Table 5). This form of facilitation was reported by several authors as the ability of a plant species to promote the development of other plant species by enhancing soil fertility or by stimulating soil beneficial microorganisms (Gómez-Aparicio et al. 2005).

Table 5. Development and ectomycorrhizal colonization of *U. bojeri* seedlings in culture or not with *L. bojeriana* on degraded and bare soils or in soil invaded by exotic plant species (*Eucalyptus camaldulensis* or *Pinus patula*).

Treatments	SB[4]	RB[5]	RB/SB[6]	N[7]	P[8]	ECM[9]
Bulk soil						
Control[1]	131 (11)[10] a[11]	113 (12)a	0.88 '0.13)b	0.89 (0.06)a	71.1 (7.3)a	36 (2.1)a
L. bojeriana[2]	277 (11)b	140 (10)ab	0.51 (0.04)a	3.02 (0.12)b	253.4 (10.9)b	42 (6)a
L. bojeriana WA[3]	309 (26)b	166 (3)b	0.55 (0.04)ab	3.08 (0.27)b	332.1 (29.1)b	90. 3 (3.2)b
U. bojeri soil						
Control	125 (15)a	295 (35)a	2.37 (0.16)b	0.85 (0.1)a	94.1 (9.9)a	73.7 (3.2)a
L. bojeriana	222 (38)ab	242 (28)a	1.21 (0.33)a	2.14 (0.32)b	197.7 (34.1)b	78 (2.1)a
L. bojeriana WA	332 (19)b	219(39)a	0.67 (0.14)a	3.58 (0.19)c	303.9 (14.1)c	90.7 (2.4)b
E. camaldulensis soil						
Control[1]	83 (0.9)a	27 (4)a	0.34 (0.08)a	0.65 (0.07)a	62.3 (7.3)a	16.3 (2.4)a
L. bojeriana[2]	233 (41)b	99 (6)b	0.45 (0.09)a	2.30 (0.41)b	194 (35.5)b	65.3 (3.3)b
*L. bojeriana*WA[3]	250 (42)b	129 (12)b	0.57 (0.17)a	3.17 (0.57)b	268 (44.9)b	79.3 (4.1)b
P. patula soil						
Control	85 (12)a	119 (10)a	1.42 (0.12)b	0.65 (0.09)a	58.9 (8.7)a	29.3 (5.5)a
L. bojeriana	233 (9)b	146 (27)a	0.62 (0.11)a	2.28 (0.10)b	181.3 (5.7) b	30.3 (2.4)a
L. bojeriana WA	333 (66)b	127 (7)a	0.41 (0.08)a	3.90 (0.78)b	278.1 (53.9)b	65.3 (1.5)b

[1]*U. bojeri* without pre- and dual cultivation with *L. bojeriana*. [2]Pre-cultvation with *L. bojeriana* and dual cultivation with *L. bojeriana* seedlings with aerial parts.[3]Pre-cultivation with *L. bojeriana* and dual cultivation with *L. bojeriana* seedlings without aerial parts.[4]Shoot biomass (mgdry weight).[5]Root biomass (mg dry weight).[6]Root:shoot ratio. [7]N leaf mineral content (mg per plant). [8]P leaf mineral content (mg per plant). [9]Ectomycorrhizal colonization (%). [10]Standard error of the mean. [11]Data in the same column and for each soil origin followed by the same letter are not significantly different according to the Newman-Keuls test ($p<0.05$).

In another experiment without the shrub species, these authors showed that seedling development of *U. bojeri* was less stimulated on soil invaded by exotic plant species and bare soil than on *U. bojeri* origin soil of this tree. The development of *U. bojeri* seedlings was significantly stimulated on pre-colonized soil by *L. bojeriana* and dual cultivation with this shrub species without aerial parts. Seeing that experiments were conducted in pot culture, the nutrient resources of soil may constitute a limiting factor for the development of the two plant species after 5 months growing time. However, it has been demonstrated by the results of Baohanta et al. (2012) that for each soil origin, the occurrence of *L. bojeriana* affected soil characteristics, *U. bojeri* growth, and ECM communities. Co-cultured with the shrub species (with or without aerial parts), the development of *U. bojeri* seedlings was accompanied by higher ECM colonization rate and high activity levels of total soil microbial [by fluorescein diacetate (FDA) hydrolysis assay] and soil phosphatase enzymes (Fig. 4A, B). These results suggest that ectotrophic early-successional shrub species such as *L. bojeriana* could lower the negative effects provided by *E. camaldulensis* and *P. patula* by facilitating ECM establishment and consequently improving *U. bojeri* early growth.

As mentioned above, *S. oblongifolia* constitutes the most dominant shrub species in degraded areas of sclerophyllous forest in the Ambatofinandrahana region(20-39′-S, 47-06′-E). The morphological and molecular structure of ECM communities associated with this shrub species and *U. bojeri* was described in this part of the Malagasy highland in December 2006 (Ramanankierana H., unpublished data). Study sites were selected among the surface areas where little to no anthropogenic disturbances had occurred during at least the last five years. Ectomycorrhizas were collected with soil cores from three microhabitats on the basis of ECM host plant composition: (i) a homogenous population of *U. bojeri* trees devoid of *S. oblongifolia* or other ECM host plants, (ii) a clumped population of *S. oblongifolia* saplings devoid of other ECM host plants and, (iii) a mixed formation of *U. bojeri* and *S. oblongifolia*. The main objective was to collect soil cores likely to contain ectomycorrhizas from exclusively *U. bojeri* or *S. oblongifolia* and ectomycorrhizas from both *S. oblongifolia* and *U. bojeri* saplings. To this goal, a sampling scheme was developed in which the nearest undesirable ECM host plant was located at least 20–25 m from the sampling point. Ectomycorrhizas from each core were treated separately until RFLP type identification and three ectomycorrhizas randomly selected from each morphotype group were screened by RFLP analysis using a part of fungal rDNA ITS region with 2 units of endonuclease, *Hae*III and *Hin*fI. Three groups of RFLP types were then created following the host plants species: *U. bojeri*, *S. oblongifolia* and mixture host plants. All RFLP types were sequenced to specify their taxonomic position by using the same primers as

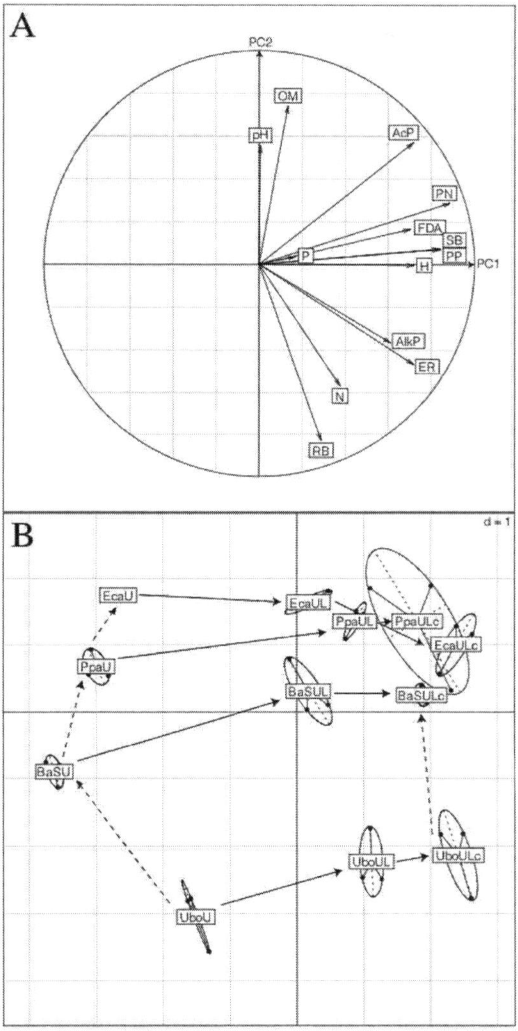

Fig. 4. Results of the PCA on the data table of soil, plant, and microbial activity parameters. (A) Correlation circle of all the parameters. The 12 variables are: pH=pH, P=total phosphorus (mg Kg^{-1}), N=total nitrogen (%), OM=total organic matter (%), FDA= global microbial activity; AcP=acid phosphatase, AlkP=alkaline phosphatase, SB=shoot biomass (g), RB= root biomass (g), ER=ectomycorrhizal rate (%), PN=leaf nitrogen (%), PP=leaf phosphorus (mg Kg^{-1}), H= Shannon diversity index of ectomycorrhizal fungi. (B) Map of sample scores on the first two principal components. Samples are coded as follows. The first three characters correspond to the soil origin; Eca=soil collected under *E. camaldulensis*, Ppa= soil collected under *P. patula*, Ubo=soil collected under *U. bojeri*, BaS=bare soil. The treatment applied to the *U. bojeri* seedlings is coded as follows. U=*Uapaca* plant alone, UL=*Uapaca* plant + *L. bojeriana*, Ulc=*Uapaca* plant + *L. bojeriana* cut after 4 months cultivation. For example, a sample coded "EcaULc" is a *U. bojeri* seedling grown in soil collected under *E. camaldulensis* in which a plant of *L. bojeriana* was grown and cut after 4 months of cultivation.

in fungal rDNA amplification. A total of 2025 ECM root tips were collected from 15 soil cores and classified into 165 ECM morphotypes without mixing root samples from different soil cores. DNA extraction and RFLP analysis were carried out on 495 ectomycorrhizas and revealed 137 RFLP types (Table 6). Considering that each RFLP type corresponds to a single ECM fungal species, the highest diversity of ectomycorrhizas expressed by Simpson's diversity index and by Shannon-Wiener information index was found on mixed vegetation. These diversity indexes were similar between *S. oblongifolia* and mixed host plant vegetation. However, diversity indexes were lower for *U. bojeri*, suggesting that in homogeneous vegetation this tree species harbors more rare species than it does with the multiple assemblage community.

Overall, 52.3% of identified RFLP types were found in both *U. bojeri* and *S. oblongifolia* root systems. The remaining RFLP types were found exclusively on either *U. bojeri* (19%) or *S. oblongifolia* (28.5%) (Fig. 5). Previous studies have reported that ECM fungi are not uniformly distributed in forest ecosystems but rather with variability in terms of presence, abundance, and community composition (Baylis 1980, Perry et al. 1989, Dickie et al. 2004, Dickie and Reich 2005). Environmental factors can influence patterns of fungal distribution such as soil characteristics and soil microbiota (Smith and Read 2008), but it is largely admitted that mycorrhized plant communities are the main factor influencing the availability of ECM fungal propagules in soil (Dickie and Reich 2005). These biological processes provide large implications for the establishment of ECM plant cover and for fungal community ecology.

Table 6. Sampling design, host plant colonization by ectomycorrhizal fungal and fungal diversity.

	Origins		
	U. bojeri	***S. oblongifolia***	**Mixture vegetation**
Number of soil cores	5	5	5
Number of ectomycorrhizal root tips	665	580	780
Number of RFLP types (without soil core comparison)	38	46	53
Percentage of shared identified taxa* between the two host plant species (%)	60	58,33	76,92
Simpson's diversity index	6.75	10.58	11.22
Shannon-Wiener information index	0.87	1.05	1.07

*Taxa are identified as RFLP types based on similarity of fragmented size in GenBank.

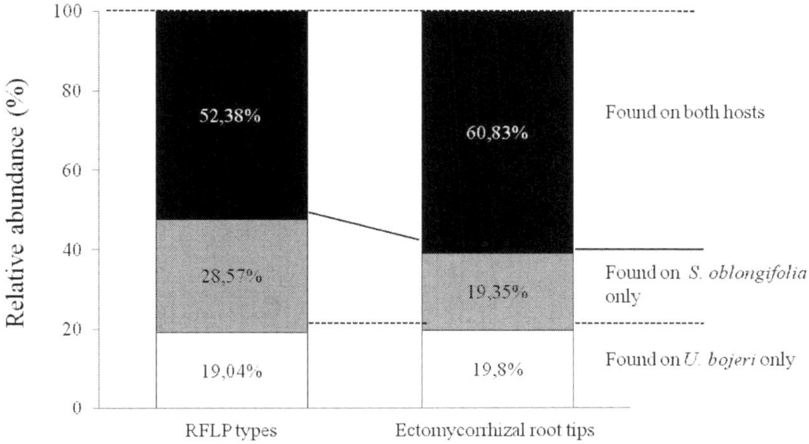

Fig. 5. Distribution of ectomycorrhizal fungi between *Uapaca bojeri* and *Sarcolaenaoblongifolia* expressed by percentage of RFLP types and percentage of total number of sampled ectomycorrhizal root tips.

In the present study, two ECM plant species (*U. bojeri* and *S. oblongifolia*) can support patches of ECM fungi. The same study also illustrated that the growth of *U. bojeri* seedlings was significantly higher in soils sampled from areas with established adult trees of *U. bojeri* than in other soils (Table 7). Well-developed seedlings on these soils were accompanied by higher ECM colonization rate and by significant taxa richness of ECM communities. The stimulation effect of external fungal mycelium radiating from mature trees on young seedling ECM formation was evidenced in a Guinean tropical forest (Diédhiou et al. 2010), in Mediterranean holm oak (Richard et al. 2005), and in temperate coniferous forests (Jonsson et al. 1999, Kranabetter and Friesen 2002). However, in the experimental site, the ECM shrub *S. oblongifolia* displayed an important effect on ECM fungal infection of young seedlings. In this part of the Madagascarian sclerophyllous forest, *S. oblongifolia* was naturally established especially at the forest edge or in degraded surface areas free of *U. bojeri* adult trees. This investigation demonstrates that more than half of collected ECM taxa were shared between *S. oblongifolia* and *U. bojeri* grown separately or in mixed formation. It suggests that *S. oblongifolia*, as a pioneering species, may persist on disturbed sites and facilitate the survival of ECM fungi that could potentially infect *U. bojeri* seedling roots. This nurse effect was particularly efficient by enhancing the ECM fungal diversity associated with young *U. bojeri* seedlings and the rate of ECM infection. These results are similar to reports on facilitation of ectomycorrhizal infection of *Pseudotsuga*

Table 7. Growth and ectomycorrhizal development of *U. bojeri* seedlings after 5 months of culture in soil samples collected at different distances from established *U. bojeri* adult tree.

		Distances from *U. bojeri* adult tree				
	Control	**20 m**	**10 m**	**5 m**	**1 m**	**0 m**
Seedling development						
Shoot biomass (mg per plant)	92 a [(1)]	120 a	313 c	335 c	199 b	183 b
Root biomass (mg per plant)	61 a	75 a	168 bc	211 c	128 b	137 b
Total biomass (mg per plant)	154 a	195 a	481 c	546 c	328 b	320 b
Shoot / Root mass ratio	1.53 ab	1.65 ab	1.86 b	1.59 ab	1.64 ab	1.42 a
Number of root tips per mg of root biomass	0.83 b	0.73 b	0.51 ab	0.39 a	0.63 ab	0.49 a
Ectomycorrhizal colonization (%)	3.19 a	5.88 ab	21.06 c	22.05 c	11.70 bc	10.49 bc
Ectomycorrhizal diversity						
Taxa richness (expressed as RFLP type)	0.5 a	0.8 a	3 c	3.5 c	1.6 ab	2 b

[(1)] Data in the same line followed by the same letter are not significantly different according to the Newman-Keul test ($p< 0.05$).

menziesii seedlings by pioneering hardwoods (Borchers and Perry 1987) or the positive effect of *Helianthemum bicknellii* on ECM infection of *Quercus* spp. (Dickie et al. 2004).

8. Conclusion

This chapter shows that the development of *U. bojeri*, an endemic tree of the Malagasysclerophyllous forest, strongly requires the occurrence of compatible ECM fungi. These ECM symbionts could be from an adult tree of this plant where various putative ECM fungi were observed, or from pioneer shrub species that persist on disturbed sites and facilitate the survival of ECM fungi propagules that could potentially infect roots of *U. bojeri* seedling. In this way, two species of endemic shrub species (*L. bojeriana* and *S. oblongifolia*) could be of great interest. These data suggest that these nurse plant species should be considered to enhance the performance of reforestation efforts in Madagascar by providing ECM inoculum potential.

Acknowledgments

We thank the anonymous referees for their valuable comments on this study, and Krista L. McGuire and Caitlyn Gillikin for improving the language.

References

Agarwal, D.K., J.A. Silander, A.E. Gelfand, R.E. Dewar and J.G. Mickelson. 2005. Tropical deforestation in Madagascar: analyses using hierarchical, spatially explicit, Bayesian regression models. Ecol. Modelling 185 : 105–131.

Baohanta, R.H. 2011. Facilitation de la régénération d'*Uapaca bojeri* L. par la gestion des communautés de champignons mycorhiziens associés aux espèces pionnières de la zone dégradée de la forêt sclérophylle d'Arivonimamo. Thèse de doctorat en Science de la vie. Université d'Antananarivo, p. 114.

Baohanta Rondro, H., J. Thioulouse, H. Ramanankierana, Y. Prin, R. Rasolomampianina, E. Baudoin, N. Rakotoarimanga, A. Galiana, H. Randriambanona, M. Lebrun and R. Duponnois. 2012. Restoring native forest ecosystems after exotic trees plantation in Madagascar: combination of the local ectotrophic species *Leptolena bojeriana* and *Uapaca bojeri* mitigates the negative influence of the exotic species *Eucalyptus camaldulensis* and *Pinus patula*. Biol. Invasion. DOI 10.1007/s10530-012-0238-5.

Baylis, G.T.S. 1980. Mycorrhizas and the spread of beech. New Zeal J. Ecol. 3: 151–153.

Borchers, S. and D. Perry. 1987. Early successional hardwoods as refugia for ectomycorrhizal fungi in clearcut douglas-fir forests of southwestern Oregon. *In*: D.M. Sylvia, L.L. Hung and J.H. Graham (eds.). Mycorrhizae in the Next Decade: Practical Applications and Research Priorities. University of Florida, Gainesville, Florida, p. 84.

Brooker, R.W., F.T. Maestre, R.M. Callaway, C.L. Lortie, L.A. Cavieres, G. Kunstler, P. Liancourt, K. Tielbörger, J.M.J. Travis, F. Anthelme, C. Armas, L. Coll, E. Corcket, S. Delzon, E. Estelle Forey, Z. Kikvidze, J. Olofsson, F. Pugnaire, C.L. Quiroz, P. Saccone, K. Schiffers, M. Seifan, B. Touzard and R. Michalet. 2008. Facilitation in plant communities: the past, the present and the future. J. Ecol. 96: 18–34.

Buscot, F., J.C. Munch, J.Y. Charcosset, M. Gardes, U. Nehls and R. Hampp. 2000. Recent advances in exploring physiology and biodiversity of ectomycorrhizas highlight the functioning of these symbioses in ecosystems. FEMS Microbiol. Rev. 24: 601–614.

Buyck, B., D. Thoen and R. Walting. 1996. Ectomycorrhizal fungi of the Guinea-Congo region. Proc. R. Soc. Edinb. 104: 313–333.

Callaway, R.M. 1995. Positive interactions among plants. Bot. Rev. 61: 306–349.

Callaway, R.M. 1997. Positive interactions in plant communities and the individualistic-continuum concept. Oecologia1 12: 143–149.

Castro, J., R. Zamora and J.A. Hódar. 2006. Restoring *Quercus pyrenaica* forests using pioneer shrubs as nurse plants. Applied Vegetation Sc. 9(1): 137–142.

Cazares, E. and J.M. Trappe. 1993. Vesicular endotrophytes in roots of the Pinaceae. Mycorrhiza 2: 153–156.

Chilvers, G.A., F.F. Lapeyrie and D.P. Horan. 1987. Ectomycorrhizal vs endomycorrhizal Fungi within the same root system. New Phytologist 107: 441–448.

Cooke, J.C. and M.W. Lefor. 1998. The mycorrhizal status of selected plant speciesfrom Connecticut wetlands and transition zones. Restor. Ecol. 6(2): 214–222.

Dhillion, S.S. 1994. Ectomycorrhizae, arbuscular mycorrhizae, and *Rhizoctonia* sp. of alpine and boreal *Salix* spp. in Norway. Arct. Alp. Res. 26: 304–307.

Dickie, I.A., R.C. Guza, S.E. Krazewski and P.B. Reich. 2004. Shared ectomycorrhizal fungi between a herbaceous perennial (*Helianthemum bicknellii*) and oak (*Quercus*) seedlings. New Phytol. 164: 375–382.

Dickie, I.A., R.T. Koide and K.C. Steiner. 2002. Influence of established trees on mycorrhizas, nutrition, and growth of *Quercus rubra* seedlings. Ecol. Monographs 72: 505–521.

Dickie, I.A. and P.B. Reich. 2005. Ectomycorrhizal fungal communities at forest edges. J. Ecol. 93: 244–255.

Diédhiou, A.G., M.A. Selosse, A. Galiana, M. Diabaté, B. Dreyfus, A.M. Bâ and G. Béna. 2010. Multi-host ectomycorrhizal fungi are predominant in a Guinean tropical rainforest and shared between canopy trees and seedlings. Environ. Microbiol. 12: 2219–2232.

Dos Santos, V.L., R.M. Muchovej, A.C. Borges, J.C.L. Neves and M.C.M. Kasuya. 2001. Vesicular-arbuscular-/ecto-mycorrhiza succession in seedlings of *Eucalyptus* spp. Brazilian J. Microbiol. 32: 81–86.

Duchesne, L.C., R.L. Peterson and B.E. Ellis. 1988. Pine root exudates stimulate the synthesis of antifungal compounds by the ectomycorrhizal fungus *Paxillus involutus*. New Phytol. 108: 471–476.

Ducousso, M., C. Bourgeois, B. Buyck, G. Eyssartier, M. Vincelette, R. Rabevohitra, G. Béna, L. Randrihasipara, B. Dreyfus and Y. Prin. 2004. The last common ancestor of Sarcolaenaceae and Asian Dipterocarp trees was ectomycorrhizal before the India—Madagascar separation, about 88 million years ago. Mol. Ecol. 13(1): 231–236.

Ducousso, M., H. Ramanankierana, R. Duponnois, R. Rabévohitra, L. Randrihasipara, M. Vincelette, B. Dreyfus and Y. Prin. 2008. Mycorrhizal status of native trees and shrubs from eastern Madagascar littoral forests with special emphasis on one new ectomycorrhizal endemic family, the Asteropeiaceae. New Phytol. 178: 233–238.

Duponnois, R., S. Diédhiou, J.L. Chotte and M.O. Sy. 2003. Relative importance of the endomycorrhizal and/or ectomycorrhizal associations in *Allocasuarina* and *Casuarina* genera. Can. J. Microbiol. 49(4): 281–287.

DuPuy, D. and J. Moat. 1996. A refined classification of the primary vegetation of Madagascar based on the underlying geology: using GIS to map it distribution and to assess its conservation status. Proceedings of the International Symposium on the "Biogeography de Madagascar", Lourenço WR., 205–218. Paris.

Egerton-Warburton, L. and M.F. Allen. 2001. Endo- and ectomycorrhizas in *Quercus agrifolia* Nee. (Fagaceae): patterns of root colonization and effect on seedling growth. Mycorrhiza 11: 283–290.

Founoune, H., R. Duponnois and A.M. Bâ. 2002. Influence of dual arbuscular endomycorrhizal/ectomycorrhizal symbiosis on the growth of *Acacia holosericea* in glasshouse conditions (A. Cunn. Ex. G. Don). Ann. For. Sci. 59: 93–98.

Gómez-Aparicio, L., J.M. Gómez, R. Zamora and J.L. Boettinger. 2005. Canopy *vs* soil effects of shrubs facilitating tree seedlings in Mediterranean montane ecosystems. J. Vegetation Sc. 16(2): 191–198.

Haselwandter, K. 1997. Soil micro-organisms, mycorrhiza, and restoration ecology. *In*: K.M. Urbanska, N.R. Webb and P.J. Edwards (eds.). Restoration Ecology and Sustainable Development. University Press, Cambridge, pp. 65–80.

Högberg, P. 1982. Mycorrhizal associations in some woodland and forest trees and shrubs in Tanzania. New Phytol. 92: 407–415.

Holl, K.M. 2002. Effects of shrubs on tree seedling establishment in an abandoned tropical pasture. J. Ecol. 90: 179–187.

Horton, T.R. and T.D. Burns. 2001. The molecular revolution in ectomycorrhizal ecology: peeking into the black-box. Mol. Ecol. 10: 1855–1871.

Jonsson, L., A. Dahlberg, M.C. Nilsson, O. Zackrisson and O. Kren. 1999. Ectomycorhizal fungal communities in late-successional Swedish boreal forests, and the composition following wildfire. Mol. Ecol. 8: 205–215.

Kranabetter, J.M. and J. Friesen. 2002. Ectomycorrhizal community structure on western hemlock seedlings transplanted from forests into openings. Can. J. Bot. 80: 861–868.

Kull, C.A., J. Ratsirarson and G. Randriamboavonjy. 2002. Les forêts de tapia des Hautes Terres malgaches. J. Cult. Geog. 19(2): 22–58.

Lapeyrie, F.F. and G.A. Chilvers. 1985. An endomycorrhiza-ectomycorrhiza succession associated with enhanced growth of *Eucalyptus dumosa* seedlings planted in a calcareous soil. New Phytol. 100: 93–104.

Last, F.T., J. Dighton and P.A. Mason. 1987. Successions of sheating mycorrhizal fungi. Trends Ecol. Evol. 2: 157–160.

Last, F.T., P.A. Mason, J. Wilson and J.W. Deacon. 1983. Fine roots and sheating mycorrhizas: their formation, function and dynamics. Plant Soil 71: 9–21.

Laurance, W.F. 1999. Reflections on the tropical deforestation crisis. Biol. Conserv. 91: 109–117.

Lee, S.S. 1998. Root symbiosis and nutrition. *In*: S. Appanah and J.M.A. Turnbull (eds.). Review of Dipterocarp: Taxonomy, Ecology and Sylviculture. CIFOR, Bogor, Indonesia, pp. 99–114.

Lee, S.S., I.J. Alexander and R. Walting. 1997. Ectomycorrhizas and putative ectomycorrhizal fungi of *Shorea loprosula* Miq (Dipterocarpaceae). Mycorrhiza 7: 63–81.

Lodge, D.F.J. and T.R. Wentworth. 1990. Negative associations among VA-mycorrhizal fungi and some ectomycorrhizal fungi inhabiting the some root system. OIKOS 57: 347–356.

McGuire, K.L., T.W. Henkel, I. Granzow de la Cerda, G. Villa, F. Edmund and C. Andrew. 2008. Dual mycorrhizal colonization of forest-dominating tropical trees and the mycorrhizal status of non-dominant tree and liana species. Mycorrhiza 18: 217–222.

Mikola, P. 1980. Tropical mycorrhiza research. Clarendon Press, Oxford, UK.

Mittermeier, R.A., P.R. Gil, M. Hoffman, J. Pilgrim, T. Brooks, C.G. Mittermeier, J.D.A. Lamoureux and G.A.B. Fonsesca. 2004. Hotspots revisited: earth's biologically richest and most endangered ecoregions. CEMEX, Mexico City.

Moat, J. and P. Smith. 2007. Atlas of the vegetation of Madagascar. CI, FTM, MBG, ESRI. RBG KEW, 124p.

Molina, R., D. Myrold and C.Y. Li. 1994. Root symbiosis of red alder: technological opportunities for enhanced regeneration and soil improvement. *In*: D.E. Hibbs, D.S. de Bell and R.F. Tarrant (eds.). The Biology and Management of red alder. Oregon State University Press, Corvallis, pp. 23–46.

Montané, F., J. Romanyà, P. Rovira and P. Casals. 2010. Aboveground litter quality changes may drive soil organic carbon increase after shrub encroachment into mountain grasslands. Plant Soil 337: 151–165.

Morris, R.J. 2010. Anthropogenic impacts on tropical forest biodiversity: a network structure and ecosystem functioning perspective. Philo. Transac. Royal Soc. Biol. 365: 3709–3718.

Myers, N., R.A. Mittermeier, C. Goettsh Mittermeier, G.A. Da Fonseca and J. Kent. 2000. Biodiversity hotspots for conservation priorities. Nature 403: 853–858.

Oliveira, V.L., V.D.B. Schmidt and M.M. Bellei. 1997. Patterns of arbuscular-and ecto-mycorrhizal colonization of *Eucalyptus dunnii* in southern Brazil. Ann. Sci. For. 54: 473–481.

Perry, D.A., H. Margolis, C. Choquette, R. Molina and J.M. Trappe. 1989. Ectomycorrhizal mediation of competition between coniferous tree species. New Phytol. 112: 501–512.

Pidwirny, M. 2006. Plant Succession. Fundamentals of Physical Geography, 2nd Edition.

Ramanankierana, N., M. Ducousso, N. Rakotoarimanga, Y. Prin, J. Thioulouse, E. Randrianjohany, L. Ramaroson, M. Kisa, A. Galiana, and R. Duponnois. 2007. Arbuscular mycorrhizas and ectomycorrhizas of *Uapaca bojeri* L. (Euphorbiaceae): sporophore diversity, patterns of root colonization, and effects on seedling growth and soil microbial catabolic diversity. Mycorrhiza 17(3): 195–208.

Ren, H., L. Yang and N. Liu. 2008. Nurse plant theory and its application in ecological restorationin lower subtropics of China. Prog. Nat. Sci. 18: 137–142.

Richard, F., S. Millot, M. Gardes and M.A. Selosse. 2005. Diversity and specificity of ectomycorrhizal fungal retrieved from an old-growth Mediterranean forest dominated by *Quercus* ilex. New Phytol. 166: 1011–1023.

Rivière, T., K. Natarajan and B. Dreyfus. 2006. Spatial distribution of ectomycorrhizal basidiomycete *Russula* subsect. Foetentinae populations in a primary dipterocarp rainforest. Mycorrhiza 16: 143–148.

Ruiz-Jaen, M.C. and T.M. Aide. 2005. Restoration success: how is it being measured? Restor. Ecol. 13: 569–577.

Sanon, K., A.M. Bâ and J. Dexheimer. 1997. Mycorrhizal status of some fungi fruiting beneath indigenous trees in Burkina Faso. For. Ecol. Manag. 98: 61–69.

Schatz, G.E., P.P. LowryII and A.E. Wolf. 2001. Endemic families of Madagascar.VII. A synoptic revision of *Leptolaena* thouars sensu strict (Sarcolaenaceae).Adansonia série 3. 23(2): 171–189.

Sicardi, M., F. Garcia-Prechac and L. Frioni. 2004. Soil microbial indicators sensitive to land use conversion from pastures to commercial *Eucalyptus grandis* (Hill ex maide) plantations in Uruguay. Appl. Soil Ecol. 27: 125–133.

Simard, S.W. and D.M. Durall. 2004. Mycorrhizal networks: a review of their extent, function, and importance. Can. J. Bot. 82: 1140–1165.

Smith, O.H., G.W. Peterson and B.A. Needelman. 2000. Environmental indicators of agroecosystems. Adv. Agron. 69: 75–97.

Smith, S.E. and D.J. Read. 2008. Mycorrhizal symbiosis. 3rd edition. Academic Press Ltd., Cambridge, UK, p. 787.

Smith, M.R., I. Charvat and R.L. Jacobson. 1998. Arbuscular mycorrhizae promote establishment of prairie species in a tallgrass prairie restoration. Can. J. Bot. 76: 1947–1954.

Thoen, D. and A.M. Bâ. 1989. Ectomycorrhizae and putative ectomycorrhizal fungi of *Afzelia Africana* and *Uapaca guineensis* in Southern Senegal. New Phytol. 113: 549–559.

Walting, R. and S.S. Lee. 1998. Ectomycorrhizal fungi associated with members of the Dipterocarpaceae in Peninsular Malaysia-II. J. Trop. For. Sci. 10: 421–430.

Weaver, M. and M. Kellman. 1981. The effect of fragmentation on tree woodlot biota in Southern Ontario. J. Biogeography 8: 199–210.

Williams, A., H.J. Ridway and D.A. Norton. 2012. Different arbuscular mycorrhizae and competition with an exotic grass affect the growth of *Podocarpus cunninghamii* Colenso cuttings. New For. DOI: 10.10007/s11056-012-9309-9.

CHAPTER 5

Morpho-anatomical Characterization of Three Sebacinales Ectomycorrhizal Species from a *Pakaraimaea dipterocarpacea* ssp. *nitida* (Dipterocarpaceae) Forest in Southern Venezuela

Bernard Moyersoen

1. Introduction

Sebacinales are a genetically diverse group of basidiomycetes with a broad range of lifestyles including a diversity of mycorrhizal and endophytic associations (Weiß et al. 2011). These fungi are widespread and the same species can colonize a wide range of host plants and be involved in different symbiotic associations (Selosse et al. 2002a, Weiß et al. 2011). Anatomically, Sebacinales are characterized by longitudinally septate basidia, imperforate parenthesomes, and a lack of both clamp connections and hymenial cystidia (Weiß et al. 2004). A very high diversity within the Sebacinales has been

School of Biological Sciences, University of Aberdeen, AB243UU, United Kingdom.
Email: bmoyersoen@hotmail.com

described from host plant and soil DNA samples and the observed global diversity was estimated to be 365 ITS taxonomic units (Setaro et al. 2012). Molecular-phylogenetically, Sebacinales are divided into two groups called A and B (Weiß et al. 2004). Up to date, most ectomycorrhizal (EcM) associations have been documented from group A (Weiß et al. 2004, 2011) and only one clade in group B included EcM strains (Hynson et al. 2013). Only a few EcM fruitbody species belonging to *Sebacina*, *Tremellodendron* and *Tremelloscypha* have been described to date (Weiß et al. 2004). These morphospecies include the cosmopolitan cryptic species *S. incrustans* (Pers.) Tul. & C. Tul. and *S. epigea* (Berk. & Broome) Bourdot & Gazin and the polymorphic *Tremelloscypha gelatinosa* (Murrill) Oberw. & K. Wells (Weiß et al. 2004, Wells and Oberwinkler 1982). In addition to the fruitbodies, EcM anatomotypes can be used to hypothesize phylogenetic relationships (Agerer 2006). Amongst Sebacinales, only seven EcM anatomotypes have been described to date and all of these EcMs were from temperate areas (Selosse et al. 2002b, Urban et al. 2003, Azul et al. 2006, Twieg and Durall 2009, Wei and Agerer 2011). A precise taxonomy should rely on a combination of characters (Dayrat 2005). Morphological and anatomical descriptions, in combination with other data, are a prerequisite for accurate EcM biodiversity surveys. A greater knowledge of Sebacinales EcM morpho-anatomy could be particularly useful for the taxonomy and biodiversity surveys of this fungal group with cryptic or unknown fruitbody species.

This study is part of a phylogeographic and taxonomic analysis of *Pakaraimaea dipterocarpacea* Maguire & Ashton ssp. *nitida* Maguire & Steyerm. EcM fungi in Southern Venezuela (Moyersoen 2006, 2012a,b,c,d). *P. dipterocarpacea*, an endemic tree species from Guayana region (Maguire et al. 1977, Maguire and Ashton 1980, Moyersoen 2006), is one of the few known locally dominant EcM tree species in the Neotropical lowland forests. This tree is phylogenetically related to the most important EcM tropical tree family in SE Asia: the Dipterocarpaceae. The phylogeny of *P. dipterocarpacea* EcM fungi might reflect the disjunct distribution of this endemic tree species. A diverse EcM fungal community was observed in a 400 m^2 *P. dipterocarpacea* forest plot (Moyersoen 2012a). In total, 40 EcM fungal species were found and up to 12 different EcM morphotypes could colonize a single 125cm^3 soil core. Floristic comparisons and a molecular-phylogenetic analysis of two newly reported *Inocybe* species (Moyersoen 2012a,d) indicated that host jump and fungal dispersion (Pirozynski 1983) were probably important radiation mechanisms for *P. dipterocarpacea* fungi (Moyersoen 2012a). Possible EcM endemic fungi in the forest indicated that *P. dipterocarpacea* fungal communities might also include old relictual species (Moyersoen 2012a). Similar trends regarding the great EcM fungal diversity, the dominance of host-generalist fungi and the presence of possible unique Agaricales were observed in a *P. dipterocarpacea* ssp. *dipterocarpacea* forest in Guyana (Smith et al. 2013).

Sebacinales were selected to extend the *P. dipterocarpacea* EcM fungi phylogeographic study. *Tremellodendron ocreatum* (Berk.) P. Roberts and *S. incrustans* fruitbodies were first described in a *Dicymbe* rain forest in Guayana region (Henkel et al. 2004). The first report of Sebacinales EcM in the tropics was made on *P. dipterocarpacea* (Moyersoen 2006). In total, three Sebacinales species (specimens BM03M3, PD10 and 6MM2) were found in a *P. dipterocarpacea* forest plot (Moyersoen 2012a). Of the 97 sequences from a total of 113 EcM DNA samples, 10 percent (11 sequences) belonged to the three Sebacinales species. This fungal group contributed to 11% of total below ground fungal richness; it was frequent and was collected in five out of nine soil cores scattered in the plot (B.M., unpublished data).

Tropical Sebacinales biodiversity was recorded mostly from DNA analysis in a wide range of forests across the tropics (Moyersoen 2006, 2012a, Rivière et al. 2007, Morris et al. 2008, Morris et al. 2009, Peay et al. 2009, Roy et al. 2009, Tedersoo et al. 2010a,b, Jairus et al. 2011, Kennedy et al. 2011, Smith et al. 2011, Tedersoo et al. 2011) and only one Sebacinales EcM has been briefly described to date (Moyersoen 2006). The aims of this study were to confirm that the three Sebacinales identified by DNA analysis are EcM, to describe in detail the morpho-anatomy of these three EcM anatomotypes and to compare them with published descriptions. The molecular-phylogeny of the three Sebacinales species will be published separately.

2. Fine Root Sampling, Morpho-anatomical and Ultrastructural Descriptions

The specimens were sampled at 4°20′N, 61°48′W, altitude 500 m, near Icabarú Village, in Gran Sabana, Estado Bolivar, Venezuela. The precise location of the 400 m^2 plot was selected based on a previous report of a stand of *Pakaraimaea dipterocarpacea* Maguire & Ashton ssp. *nitida* Maguire & Steyerm described by Maguire and Steyermark (1981) in the same area. Four separate sampling expeditions were conducted, in November 2003 (one day), March 2006 (one day), July 2007 (five days) and July 2008 (nine days).

Fine roots from the top organic soil layer were either traced from *P. dipterocarpacea* trees (collections BM03M3, PD10) or collected from soil cores (sample 6MM2). The sampling procedures are described in Moyersoen (2006, 2012a). EcM were washed from the soil and clusters were assigned to morphotypes using a field dissecting microscope (Novex AP-7) or a 10X magnifying glass. EcM tips were stored in 2% cetyltrimethylammonium bromide (CTAB). BM03M3 subsamples from the same cluster were stored in water at 5°C (subsample 1) or fixed in glycerol: 100% ethanol: water (1:1:1) (subsample 2). Within three days of collection, for BM03M3 subsample 1, the habit was recorded with a dissecting microscope (Leica MZ6) fitted with a camera (Leica MPS60). The anatomical features of BM03M3 subsample 2, PD10 and 6MM2 were described from mantle peelings using Agerer's (1991) method of description. Drawings were performed using a Normarski interference contrast microscope (Leica DM2500) equipped with a drawing tube. Reference specimens of BM03M3 and PD10, and DNA extract of 6MM2 are stored in the University of Liège Herbarium (LG).

To examine specimen BM03M3 at the ultrastructural level, EcM tips fixed in glycerol:ethanol:water were transferred in 2% gluteraldehyde in 0.1 M sodium cacodylate buffer (pH 7.2) at room temperature for 1 h. These tip samples were then submitted to six transfers in 0.1 M sodium cacodylate buffer and post fixed in 1% osmium tetroxide in the same buffer for 1 h in the dark. They were then washed in distilled water and stained in 1% aqueous uranyl acetate for 1h in the dark. After five washes in distilled water, samples were dehydrated in acetone, using 10 min changes at 10, 25, 50, 70, 95% and three times in 100% acetone. Samples were imbedded in Spurr's plastic (Spurr 1969) and sectioned with a glass knife for semi-thin sections and a diamond knife for ultra-thin sections. Semi-thin sections were stained with new fuchsin and crystal violet, mounted in Entellan and used for light microscopy. Ultra-thin serial sections were mounted on Formvar coated, single-slot copper grids, stained with lead citrate at room temperature for 5 min, and washed with distilled water. They were examined using a Zeiss transmission electron microscope operating at 80 kV.

Sebacinales BM03M3 + *Pakaraimaea dipterocarpacea*

Short description: The mycorrhizae are characterized by a simple or irregularly monopodially-pinnate system with a yellowish colour and a woolly surface. Emanating hyphae and rhizomorphs are frequent and occur throughout. The mantle outer layer is a transition between a dense plectenchyma with ramified hyphae and a pseudoparenchyma with a matrix, covered by an hyphal net; it is a plectenchyma in deeper layers. Septa are simple. Septal pores are dolipore parenthesome type and the parenthesomes are imperforate with electron-dense middle lamella. Walls are smooth and light orange; they are thick in outer mantle layer, rhizomorphs and emanating hyphae. Triangular inflations and short protrusions are present in the outer mantle layer and emanating hyphae. Polytomies with three branches are present in the net. Rhizomorphs are not differentiated and are uniformly compact. Emanating hyphae are smooth, and some are similar to awl-shaped cystidia. The mantle preparation is slightly dextrinoid. The mycorrhizae are hydrophilic and belong to a transition between short distance and smooth subtype of medium distance exploration type.

Morphological characters (Fig. 1a): *Mycorrhizal systems* simple or irregularly monopodially-pinnate. *Main axes* 0.2–0.3 mm thick. *Unramified ends* straight, 0.5–1 mm long and 0.2–0.3 mm diam., yellowish (G in Colour Identification Chart, Royal Botanic Garden Edinburgh, Flora of British Fungi, 1969) in older parts, white at apex (Fig. 1B in Moyersoen 2006). *Surface of unramified ends* woolly; emanating hyphae frequent, occurring throughout, colourless. *Rhizomorphs* colourless to yellowish, approximately 10 µm wide, frequent, circular in cross-section, occurring throughout, unramified. Soil particles often sticking to the mantle surface and rhizomorphs. Hydrophilic. *Exploration type*, a transition between the short distance exploration type and the smooth subtype of the medium distance exploration type.

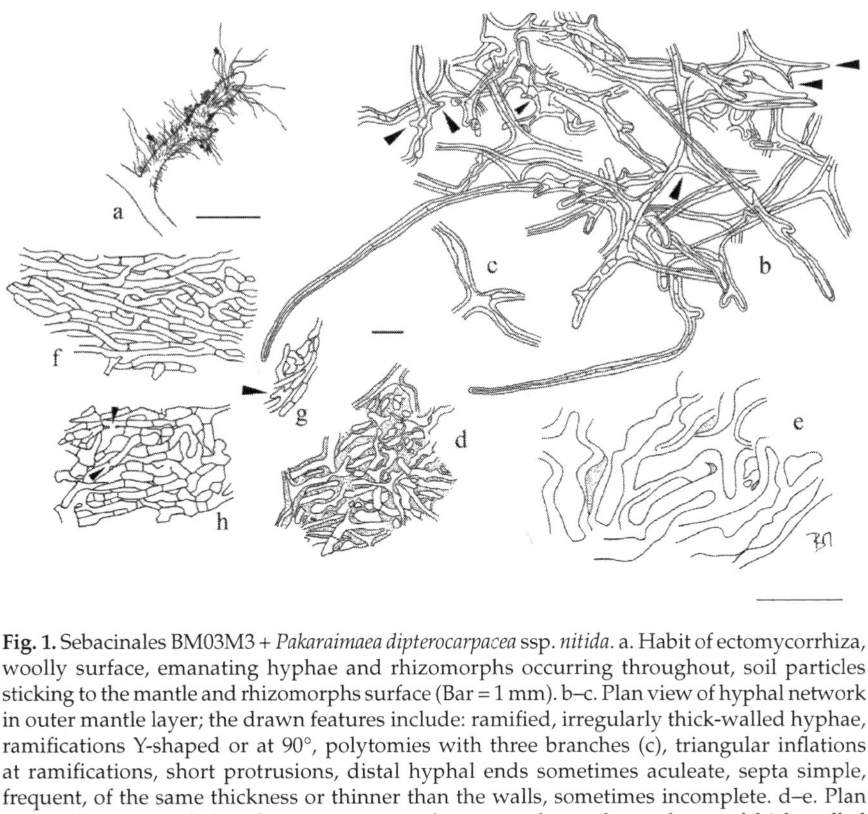

Fig. 1. Sebacinales BM03M3 + *Pakaraimaea dipterocarpacea* ssp. *nitida*. a. Habit of ectomycorrhiza, woolly surface, emanating hyphae and rhizomorphs occurring throughout, soil particles sticking to the mantle and rhizomorphs surface (Bar = 1 mm). b–c. Plan view of hyphal network in outer mantle layer; the drawn features include: ramified, irregularly thick-walled hyphae, ramifications Y-shaped or at 90°, polytomies with three branches (c), triangular inflations at ramifications, short protrusions, distal hyphal ends sometimes aculeate, septa simple, frequent, of the same thickness or thinner than the walls, sometimes incomplete. d–e. Plan views of outer layer below the net, a transition between a dense plectenchyma of thick-walled ramified hyphae and a pseudoparenchyma, with a matrix. f–g. Plan view of middle mantle layer, dense plectenchyma of ramified hyphae, anastomoses short, open (g). h. Plan view of inner mantle layer, dense plectenchyma of short, ramified hyphae, anastomosis short, open, some incomplete septa. (Bars = 10 µm).

Anatomical characters of mantle in plan views (Figs. 1b–h): *Outer mantle layers* (Figs. 1b–e), a transition between a dense plectenchyma of thick-walled ramified hyphae and a pseudoparenchyma (type E/C/H, Agerer 2006), with a matrix, covered by a loose network of ramified, thick-walled hyphae from which emanating hyphae and rhizomorphs originate; hyphae from the network mostly cylindrical, with frequent short protrusions and triangular inflation at ramifications, sometimes aculeate; ramifications Y-shaped or at 90°, sometimes polytomies with three branches; cells 2.2–3.5 µm diam.; cell walls light orange, smooth, irregularly thick (0.5–1.7 µm); septa simple, frequent, of the same thickness or thinner than the walls, sometimes incomplete; anastomoses lacking; hyphae deeper in the surface layer variably shaped; walls difficult to distinguish from the matrix at 1000X with transmission light microscope; walls and matrix yellowish. *Middle mantle layers* (Figs. 1f–g) a dense plectenchyma of ramified, cylindrical hyphae; cells 1.7–2.7 µm diam.; ramifications Y-shaped; cell walls smooth, yellowish, thin; septa simple, frequent, of the same thickness or thicker than the hyphal walls, sometimes incomplete; anastomoses scarce, short, open. *Inner mantle layers* (Fig. 1h) a dense plectenchyma of short, ramified hyphae, in a matrix; cells mostly cylindrical or of variable shape; walls thin, yellowish; septa simple, frequent, of the same thickness or thicker than the hyphal walls, sometimes incomplete; anastomoses scarce, short, open. *Very tip* with the same arrangement as in other parts of mantle.

Anatomical characters of emanating elements (Figs. 2a–d): *Rhizomorphs* (Figs. 2a–b) abundant, 5–25 µm diam., uniform-compact (uniform hyphae, densely packed and glued together, type B in Agerer 2006), often with ramified hyphae at distal end; hyphae 2.6–3.6 µm diam.; hyphal ramifications at rhizomorph distal end, Y-shaped, at 90°, or oriented backwards; walls yellowish, smooth, 0.7–1.5 µm thick, thin at hyphal distal end; septa simple, frequent, of the same thickness or thinner than the hyphal walls, sometimes incomplete; rhizomorph apex simple or with ramified, tortuous hyphae with protrusions, and with a protuberance at the tip; anastomoses open, short or with a bridge, only observed at rhizomorph distal end. *Emanating hyphae* (Figs. 2c–d) abundant, with or without short protrusions at proximal end, 2.8–3.2 µm thick; septa simple, frequent, evenly distributed, generally thinner than the hyphal walls, sometimes incomplete; hyphal ends simple; walls yellowish, 1–1.5 µm thick, sometimes thinner at distal end, surface smooth; ramifications lacking; anastomoses lacking; shorter emanating hyphae similar to bent awl-shaped *cystidia* (type A, Agerer 2006), 52–135 µm long, diam. between 1.7–2.5 and 2.6–3.7 µm, and wall thickness between 1–1.5 and 0.5–1 µm at proximal and distal end, respectively.

Anatomical characters, longitudinal section (Fig. 3): Mantle plectenchymatous, 15–25 µm wide, with two layers; outer layer 7–9 µm thick, with thick-walled

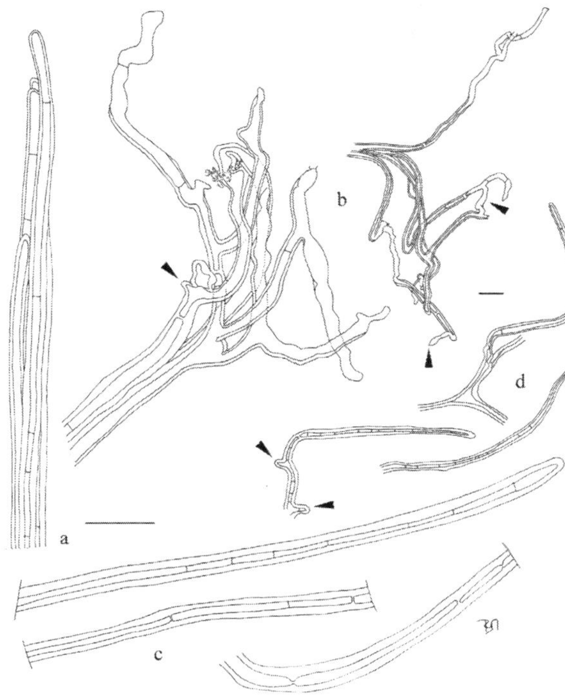

Fig. 2. Sebacinales BM03M3 + *Pakaraimaea dipterocarpacea* ssp. *nitida*. a–b. Rhizomorphs, thick-walled hyphae of uniform diameter running in parallel (a), apex simple (a) or with ramified, tortuous hyphae with protrusions and protuberance at the tip (b), septa frequent, of the same thickness or thinner than the walls (a, b), anastomosis open, short or with a hyphal bridge (a). c–d. Emanating hyphae, thick-walled, with or without short protrusion at proximal end (c), frequent simple septa, generally thinner than hyphal walls, shorter emanating hyphae similar to bent awl-shaped cystidia (d). (Bars = 10 µm).

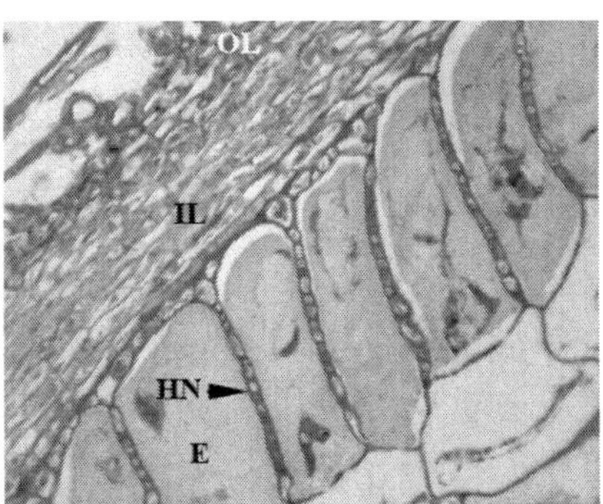

Fig. 3. Transmission light micrograph of Sebacinales BM03M3 + *Pakaraimaea dipterocarpacea* ssp. *nitida* in longitudinal section. OL: outer mantle layer, thick-walled cells imbedded in a matrix; IL: inner mantle layer, cells mostly cylindrical, matrix visible in innermost layer; HN: Hartig net paraepidermal, hyphal cells in one row; E: epidermal cell. (Bar = 10µm).

cells of different shape, imbedded in a matrix; inner layer 10–15 µm thick, with cells mostly cylindrical, tangentially 4–15 µm, radially 1.5–2 µm, matrix visible in innermost layer; remnants of calyptra observed. *Tannin cells* lacking. *Epidermal cells* rectangular, tangentially 7–20 µm, radially 24–33 µm. ECt (average tangential measure of epidermis cells) 12 µm, ECq (average ratios between tangential and radial measure of epidermis cells) 0.5. *Hartig net* paraepidermal, hyphal cells in 1 row; in plan view of palmetti type. *Haustoria* lacking.

Colour reactions with different reagents: Preparations of mantle: Melzer's reagent: cell walls orange brown.

Ultrastructure of ectomycorrhizae, longitudinal section (Figs. 4–8): Outer layer with thick-walled hyphae imbedded in an electron-dense matrix; inner layers with thinner-walled hyphae, matrix not very distinct; septal pores

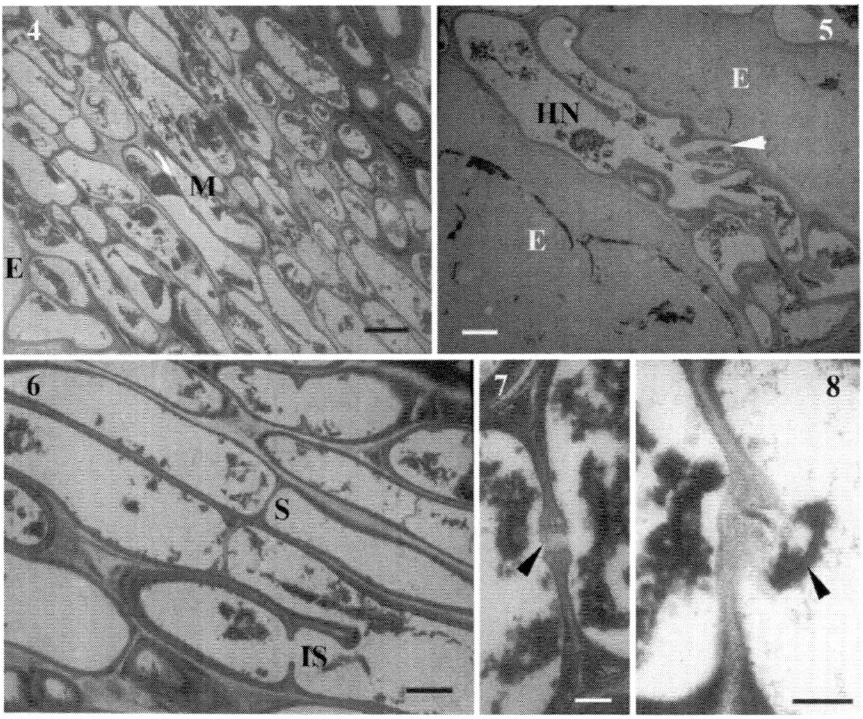

Figs. 4–8. Tansmission electron micrographs of Sebacinales BM03M3 + *Pakaraimaea dipterocarpacea* ssp. *nitida*. Figure 4: Section through the mantle. Thick-walled hyphae imbedded in a matrix in outer layer and thinner-walled hyphae in inner layer. E: portion of epidermal cell; M: mantle. Figure 5: Section through the Hartig net (HN), Palmetti type lobes. E: epidermal cells. Figure 6: Section through the inner mantle layers, complete (S) and incomplete septa (IS). Figure 7: detail of a dolipore. Figure 8: detail of imperforate parenthesome. Figures 4–5: Bar = 2 µm., Figure 6: Bar = 1 µm. Figures 7–8: Bar = 0.2 µm.

in the mantle and Hartig net, dolipore-parenthesome type, parenthesome imperforate with an electron-opaque middle lamella; Hartig net with thin-walled hyphae, in one row, palmetti type in plan-view.

Reference specimen: Venezuela, 4°20'N, 61°48'W, altitude 500 m, near Icabarú Village, in Gran Sabana, Estado Bolivar, tropical forest dominated by *Pakaraimaea dipterocarpacea* Maguire & Ashton ssp. *nitida* Maguire & Steyerm.; ectomycorrhizae collected by root tracing from a *P. dipterocarpacea* ssp. *nitida* adult tree (Moyersoen 2006), in top of mineral soil, below the litter, 25.11.2003, myc. coll. and isol. B. Moyersoen, reference specimen in glycerol/ethanol/water in Herb. B. Moyersoen BM03M3 (in LG).

Sebacinales PD10 + *Pakaraimaea dipterocarpacea*

Short description: The mycorrhizae are characterized by an irregularly monopodially-pinnate system with a smooth or cottony surface. Emanating hyphae are frequent and rhizomorphs are lacking. The mantle outer layer is plectenchymatous with squarrose branched hyphae; it is plectenchymatous with a matrix in middle layers. Septa are simple. Walls are thick or thin in outer mantle layer and they are colourless. Inflations are frequent in the outer mantle layer and they are triangular at ramifications. Thick-walled staghorn-like hyphae with polytomies with three to many finger-like outgrowths are frequent in the outermost layer. Emanating hyphae are thick-walled and smooth, with short protrusions. The mycorrhizae belong to the short distance exploration type.

Morphological characters (Fig. 9a): *Mycorrhizal systems* monopodial, irregularly pinnate. *Main axes* 0.2–0.3 mm thick. *Unramified ends* straight or bent, 0.5–2 mm long and 0.2–0.3 mm diam., colour not recorded. *Surface of unramified ends* smooth or cottony, with some woolly patches; emanating hyphae frequent, occurring throughout. Soil particles often sticking to the mantle surface and the emanating hyphae. *Exploration type*, short distance.

Anatomical characters of mantle in plan views (Figs. 9b–d, 10a–b): Outer mantle layers (Figs. 9b-d), a loose plectenchymatous network of squarrose branched hyphae (hyphae multiply ramified with short, frequently Y-shaped or almost rectangular branches, type E (Agerer 2006)) becoming more densely arranged in deeper layers; frequent staghorn-like hyphae, with many finger-like outgrowths and inflations in the outermost layer, hyphae mostly cylindrical in deeper layers; ramifications Y-shaped or at 90°, often with triangular inflation, polytomies with three to many finger-like outgrowths; cylindrical cells 2.2–6.4 µm diam.; cell walls colourless, smooth, either irregularly thick (0.5–2.5 µm), or thin; septa simple, not frequent, straight or curved, sometimes incomplete, thinner or of the same thickness as the hyphal walls; anastomoses not frequent, short, open; matrix lacking. *Middle*

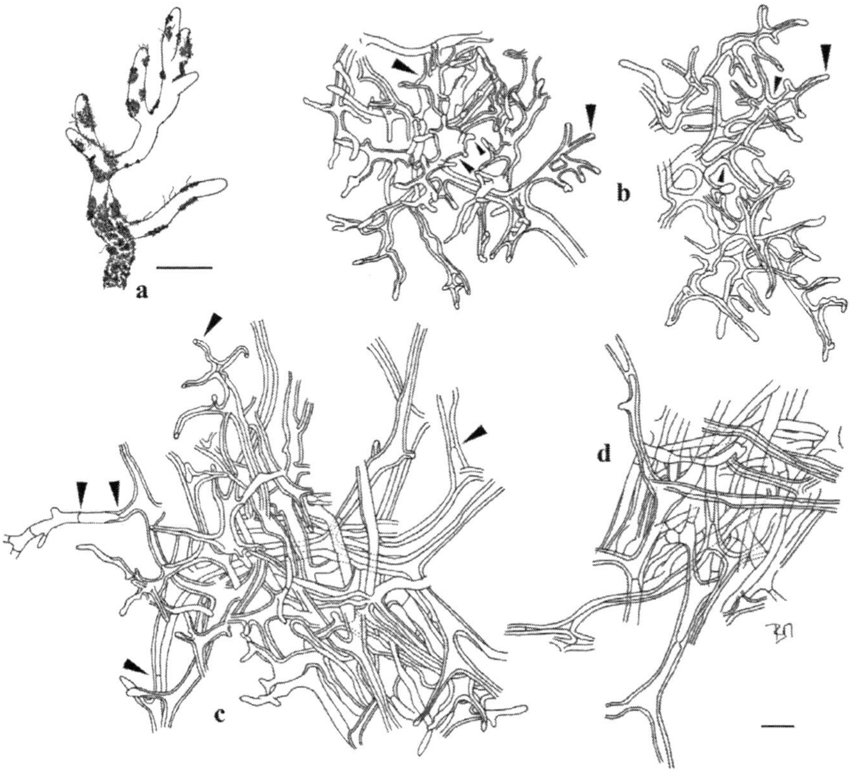

Fig. 9. Sebacinales PD10 + *Pakaraimaea dipterocarpacea* ssp. *nitida*. a. Habit of ectomycorrhiza, smooth or cottony surface with some woolly patches, emanating hyphae occurring throughout, soil particles sticking to the mantle surface. b–d. Plan views of outer layer; ramified, staghorn-like hyphae, with inflations either intercalary (b, c) or at ramifications (b), wall thickness irregular and variable (c, d), ramifications Y-shaped or at 90° (d), polytomies with three to many finger-like outgrowths (b, c), septa simple, not frequent, sometimes incomplete (c), thinner or of the same thickness as the hyphal walls (c). (Bar = 10 µm).

mantle layers (Fig. 10a), a dense plectenchymatous network of ramified, cylindrical hyphae, sometimes with constriction at septa; cells 1.3–2.6 µm diam.; ramifications Y-shaped or at 90°, often with triangular inflation; cell walls smooth, colourless, generally thin, some hyphae with thicker walls; septa simple; anastomoses short, open; matrix present. *Inner mantle layers* (Fig. 10b) a dense plectenchymatous network of ramified hyphae; hyphal shape cylindrical, sometimes with constriction at septa, hyphae sometimes with palmetti type lobes; cells 2.1–3.2 µm diam.; ramifications Y-shaped or at 90°, often with triangular inflation; walls smooth, thin, colourless; septa simple, frequent, sometimes incomplete; anastomoses not observed; matrix not observed. *Very tip* with the same arrangement as in other parts of the mantle.

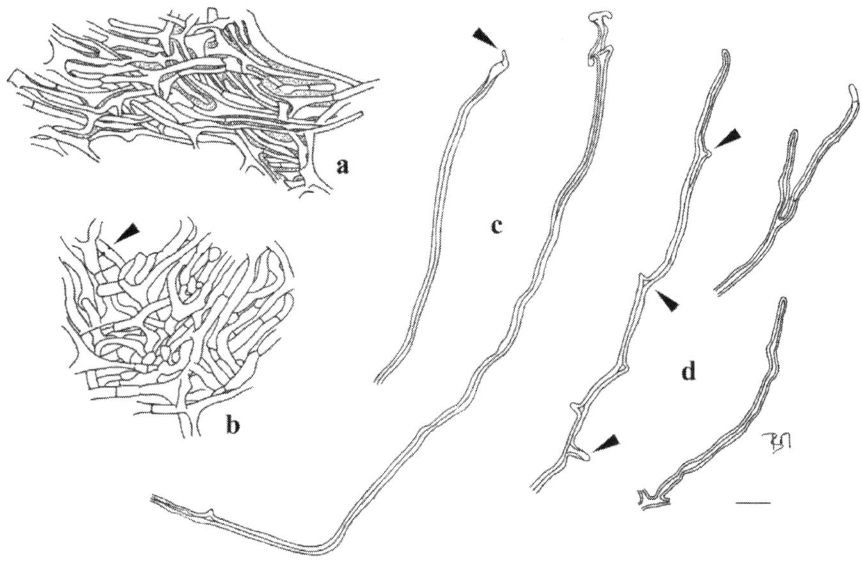

Fig. 10. Sebacinales PD10 + *Pakaraimaea dipterocarpacea* ssp. *nitida*. a. Plan view of middle mantle layer, dense plectenchyma of thinner-walled, ramified hyphae, with a matrix. b. Plan view of inner mantle layer, dense plectenchyma of ramified hyphae, septa frequent, sometimes incomplete. c–d. Emanating hyphae, thick-walled, distal end simple, sometimes thinner-walled and with a protuberance at the tip (c), septa infrequent, thinner than the hyphal walls, short protrusions occurring throughout (d), ramifications Y-shaped or at 90° with triangular inflation (d). (Bar = 10 µm).

Anatomical characters of emanating elements (Figs. 10c–d): *Rhizomorphs* lacking. *Emanating hyphae* abundant, 2.1–2.5 µm thick; septa simple, infrequent, thinner than the hyphal walls; hyphal ends simple or with a protuberance; walls 0.5–1 µm thick, even in thickness or thinner at distal end, surface smooth; ramifications Y-shaped or at 90°, often with triangular inflation; short protrusions occurring throughout; anastomoses lacking. *Cystidia* (i.e., hyphae with constant length) lacking, but staghorn-like hyphae in the outermost layer could be interpreted as cystidia.

Reference specimen for Pakaraimaea ectomycorrhiza: Venezuela, 4°20′ N, 61°48′ W, altitude 500 m, near Icabarú Village, in Gran Sabana, Estado Bolivar, tropical forest dominated by *Pakaraimaea dipterocarpacea* Maguire & Ashton ssp. *nitida* Maguire & Steyerm.; ectomycorrhizae collected by root tracing from a *P. dipterocarpacea* ssp. *nitida* adult tree (Moyersoen 2006), in top of mineral soil, below the litter, 20.07.2007, myc. coll. and isol. B. Moyersoen, reference specimen in cetyltrimethylammonium bromide (CTAB) in Herb. B. Moyersoen PD10 (in LG).

Sebacinales 6MM2 + *Pakaraimaea dipterocarpacea*

Short description: The mantle outer layer is a loose network of squarrose, branched, mostly cylindrical hyphae, becoming densely arranged in deeper layers and from which abundant emanating hyphae originate. A matrix is lacking both in outer and deeper layers. Septa are simple and they are often curved. Walls are thick in outer mantle layer, rhizomorphs and emanating hyphae and they are colourless. Inflations including triangular ones at ramifications and short protrusions are present both in outer mantle layer and emanating hyphae. Polytomies with three branches are present in the net. Rhizomorphs are frequent, not differentiated, and uniform compact. Emanating hyphae are smooth and some are similar to awl-shaped cystidia; clusters of hyphae are often observed emanating from the outer mantle layer. The mycorrhizae belong to a transition between short distance and smooth subtype of medium distance.

Morphological characters: not recorded. *Exploration type*, a transition between the short distance exploration type and the smooth subtype of medium distance exploration type (on basis of anatomy of emanating elements).

Anatomical characters of mantle in plan views (Figs. 11a–d): The mantle anatomy is difficult to observe for the great amount of thick-walled emanating hyphae covering the outer layer. *Outer mantle layers* (Figs. 11a–b), a loose plectenchymatous network of squarrose branched hyphae (mantle type E; Agerer 2006) becoming densely arranged in deeper layers, from which abundant emanating hyphae originate; hyphae mostly cylindrical, with frequent inflations and short protrusions; ramifications Y-shaped or at 90°, often with triangular inflation, polytomies with three branches, frequent clusters of hyphae emanating from the outermost layer; cells 2–3.6 µm diam.; cell walls colourless, smooth, irregularly thick (0.5–1.5 µm); septa simple, often thinner than the hyphal walls, often curved; anastomoses lacking; matrix not observed. *Middle mantle layers* (Fig. 11c) a dense plectenchyma of ramified, cylindrical hyphae; cells 2.2–3.4 µm diam.; ramifications Y-shaped; cell walls smooth, colourless, frequently thin, some hyphae with thicker walls (0.5–1 µm); septa simple, not frequent, of the same thickness as the hyphal walls; anastomoses not observed. *Inner mantle layers* (Fig. 11d) a transition between a plectenchyma and a pseudoparenchyma; cells cylindrical or of variable shape, sometimes with palmetti type lobes; between 21–24 cells in a square of 20 X 20 µm; walls thin, colourless; septa simple, frequent, sometimes incomplete; anastomoses open. *Very tip* with the same arrangement as in other parts of mantle.

Anatomical characters of emanating elements (Figs. 11e–i): *Rhizomorphs* (Figs. 11e–g) 10–50 µm diam., uniform-compact (type B, Agerer 2006); hyphae 2.4–3.2 µm diam.; hyphal ramifications infrequent, Y-shaped or

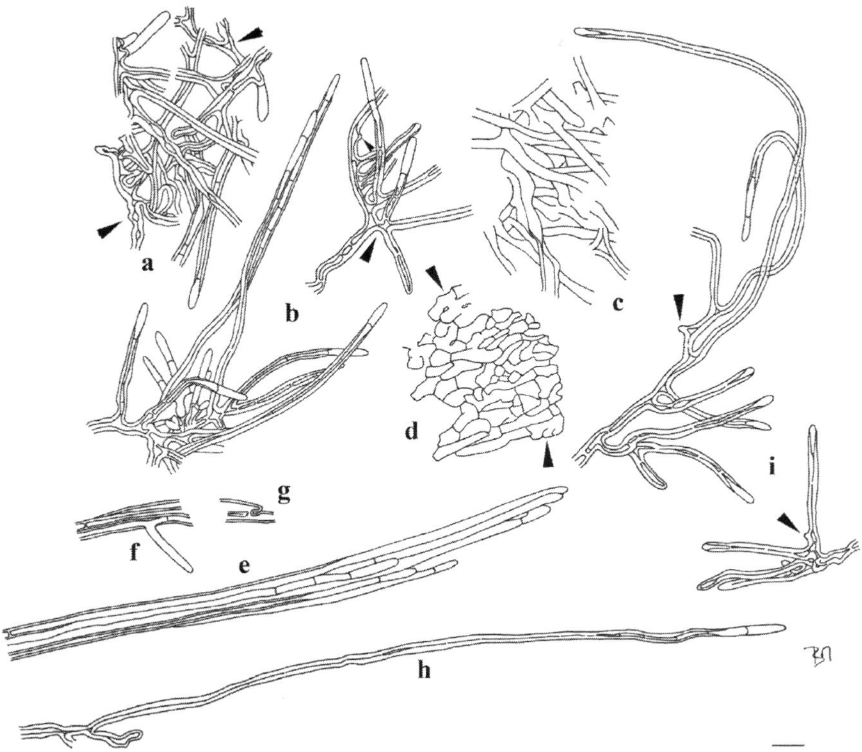

Fig. 11. Sebacinales 6MM2 + *Pakaraimaea dipterocarpacea*. a–b. Plan views of outer layer, ramified, thick-walled hyphae (a) from which many emanating hyphae originate (b), inflations intercalary or triangular at ramifications (a), short protrusions (b), ramifications Y-shaped or at 90° (a), polytomies with three branches (b), septa simple, frequent, of the same thickness or thinner than the hyphal walls (a–b). c. Plan view of middle mantle layer; dense plectenchyma, walls generally thinner than in outer layer. d. Plan view of inner mantle layer, a transition between a plectenchyma and a pseudoparenchyma; thin walls, palmetti type lobes, anastomosis short, open. e–g. Rhizomorph, thick-walled hyphae of uniform diameter running in parallel, apex simple, cell walls thinner at distal end, septa straight or curved, clustered at distal end, ramification Y-shaped (f), anastomosis open, short (g). h–i. Emanating hyphae, thick-walled, thinner-walled at distal end, with or without short protrusion at proximal end (i), simple septa frequent, of the same thickness or thinner than the hyphal walls, straight or curved, ramification Y-shaped, shorter emanating hyphae similar to awl-shaped cystidia (i). (Bar = 10 µm).

at 90°; walls colourless, smooth, 0.5–1 µm thick, thinner at hyphal distal end; septa simple, infrequent, of the same thickness as the hyphal walls, straight or curved, often one to three septa at hyphal distal end; hyphal end at rhizomorph apex simple; anastomosis short, open. *Emanating hyphae* (Figs. 11h–i) abundant, 2.6–3.1 µm thick; septa simple, frequent, evenly distributed, of the same thickness or thinner than the hyphal walls; hyphal ends simple; walls colourless, 0.7–1 µm thick, thinner at distal end, surface smooth; ramifications Y-shaped or at 90°, often with triangular inflation;

anastomoses lacking; intrahyphal hyphae present. *Cystidia* lacking, but emanating hyphae originating from the superficial net similar to awl-shaped cystidia with basal dichotomous or trichotomous ramifications (Agerer 2006).

Reference specimen for Pakaraimaea ectomycorrhiza: Venezuela, 4°20′ N, 61°48′ W, altitude 500 m, near Icabarú Village, in Gran Sabana, Estado Bolivar, tropical forest dominated by *Pakaraimaea dipterocarpacea* Maguire & Ashton ssp. *nitida* Maguire & Steyerm. ectomycorrhizae collected from a soil core collected near a *P. dipterocarpacea* ssp. *nitida* adult tree (Moyersoen 2012a), in first 5 cm organic soil, 27.07.2008, myc. coll. and isol. B. Moyersoen, DNA extract in Herb. B. Moyersoen 6MM2 (in LG).

3. Molecular Identification with Sequences from GenBank and UNITE

For the molecular identification of BM03M3, PD10 and 6MM2, DNA protocols used have been described by Moyersoen (2006, 2012a). Fungal sequences from both ITS and LSU were obtained using the primers ITS1f, ITS4b, LROR, LR21 and LR6, depending upon the success of PCR amplifications. GenBank accession number for ITS are JQ063056, JQ063057, and JQ063058 for specimens BM03M3, PD10, and 6MM2, respectively. The host of specimen 6MM2 was identified by DNA sequencing using the primers *rbc*LN and *rbc*LR as described by Moyersoen (2006, 2012a) and host identity of specimen BM03M3 was confirmed by *rbc*L DNA RFLP using *Hinf*1 as restriction enzyme (Moyersoen 2006). ITS sequences were assigned to molecular species on the basis of arbitrary 3% similarity cut-off value (Nilsson et al. 2008). Sequence comparisons were performed using the BlastN (Altschul et al. 1997) in GenBank and UNITE (Kõljalg et al. 2005, http://unite.ut.ee).

None of the *Pakaraimaea* Sebacinales ITS and LSU sequences matched identified Sebacinales species (data not shown, analysis 22 January 2013). A significant match with non-described species was found between specimen BM03M3 ITS sequence and uncultured Sebacinales voucher EcM17 (JN168754) (497/514, 97%) and specimen 6MM2 ITS with uncultured Sebacinales voucher EcM84 (JN168756) (251/258, 97%). Both matching sequences were from a *Dicymbe* dominated forest in Guyana (Moyersoen 2012a). ITS best match for specimen PD10 was with Sebacinales isolate L7604_Seb MAD02 (FR731328) (518/568, 91%).

4. Analysis of Specimens based on Morpho-anatomical and Ultrastrutural Features Together with DNA Sequences

Ultrastructural characteristics (investigated in one anatomotype) and morpho-anatomical features, together with DNA sequences indicated that BM03M3, PD10 and 6MM2 Sebacinales species were involved in *P. dipterocarpacea* EcM.

Ultrastructurally, the dolipore parenthesome architecture investigated in specimen BM03M3 was consistent with that of Sebacinales (Urban et al. 2003). The ultrastructure of septa could not be studied in specimens PD10 and 6MM2, but the presence of only simple septa was in agreement with the sequenced Sebacinales EcM.

Morphological and anatomical features observed on temperate Sebacinales EcM are the following: a plectenchymatous mantle with squarrose branched hyphae (type E/C, Agerer 2006) or a transition between a plectenchyma of ramified hyphae and a pseudoparenchyma, a matrix in the outer mantle layer, thick-walled hyphae (at least in the outer mantle layer and/or the emanating elements), ramifications with more than two branches (referred to as polytomies), inflations including triangular inflations at ramifications, the absence of hydrophobicity, cystidia (when present) awl-shaped, smooth and colourless cell walls. Rhizomorphs were lacking in most temperate Sebacinales EcM anatomotypes, and the EcM were classified as short distance exploration type by Agerer (2006). Most recently, infrequent, undifferentiated rhizomorphs with loosely arranged hyphae in a matrix have been reported in *Pinirhiza nondextrinoidea* (Wei and Agerer 2011). A dextrinoid colour reaction was observed in *Pinirhiza multifurcata* (Wei and Agerer 2011) and *Quercirhiza dendrohyphidiomorpha* + *Quercus suber* L. (Azul et al. 2006).

Specimens BM03M3, PD10 and 6MM2 shared morphological and anatomical features with temperate Sebacinales. The mantle was a plectenchyma of squarrose branched hyphae in PD10 and 6MM2 or a transitional type to a pseudoparenchyma in BM03M3. Hyphae in the outer mantle layers and emanating elements were thick-walled. Cell walls were smooth and hyphal inflations were conspicuous in the outer mantle layer and emanating elements. A matrix was present in BM03M3 and PD10. Polytomies were frequent either in the superficial hyphal net (BM03M3 and PD10) or in the emanating hyphae (6MM2). Both BM03M3 and PD10 were hydrophilic (this feature was not recorded in 6MM2). Colour reaction to Meltzer was only recorded in BM03M3 and a slight dextrinoid reaction was observed.

Specimens BM03M3, PD10 and 6MM2 could be distinguished by several morphological and anatomical features. BM03M3 was woolly and it was cottony in PD10. Cell walls were yellowish in BM03M3 whereas they were colourless in the remaining two anatomotypes. Rhizomorphs were observed only in BM03M3 and 6MM2. The transition between a plectenchyma and a pseudoparenchyma in BM03M3 was different from the plectenchymatous mantle in PD10 and 6MM2. The cell diameter in outer layer was more variable and greater on average in PD10 than in 6MM2 and BM03M3. A matrix was present in BM03M3 outer mantle layer, in PD10 in deeper mantle layers and it was absent in 6MM2. Hyphae from outermost layer were variably shaped in BM03M3 and PD10 and they were mostly cylindrical in 6MM2. Staghorn-like hyphae were observed only in PD10. Ramifications included polytomies with up to three branches in BM03M3 and 6MM2; they could ramify in three to many branches in PD10. Clusters of emanating hyphae were only observed in 6MM2's outer mantle layer. Simple septa were frequently curved in 6MM2, either straight or curved in PD10 and straight in BM03M3. Hyphal ends in emanating hyphae were thin-walled in 6MM2 and either thick or thin-walled in both, BM03M3 and PD10. Emanating elements similar to awl-shaped cystidia were lacking in PD10; they were scattered in BM03M3 and often in clusters in 6MM2.

BM03M3 anatomotype was particularly close to the unidentified sample Y62M1 (Urban et al. 2003, Agerer and Rambold 2004–2011). Both are yellowish and hydrophilic. Shared anatomical features between the two anatomotypes included the transition between a dense plectenchyma of thick-walled ramified hyphae and a pseudoparenchyma in the outer layer (type E/C/H, Agerer 2006). This mantle layer organization was also observed in Sebacinoid ectomycorrhiza M2 (Selosse et al. 2002b). Remaining shared features between BM03M3 and sample Y62M1 (Urban et al. 2003) included: the superficial, loose hyphal net with variably shaped, thick-walled hyphae with protrusions and inflations, the emanating hyphae originating from the net and the plectenchymatous deeper layers. The main differences between BM03M3 and sample Y62M1 included: the yellowish cell walls in BM03M3, the rhizomorphs lacking in Y62M1, the greater size (3.5–7 versus 2.2–3.5 µm) and the different shape of hyphae in Y62M1 superficial net, the location of polytomies, either in the net (BM03M3) or the emanating hyphae (Y62M1) and the variable diameter of emanating hyphae only in Y62M1 (1–4 versus 2.8–3.2 µm). Thick-walled, straight or wavy emanating hyphae with polytomies in Y62M1 (Urban et al. 2003) were classified by Agerer and Rambold (2004–2011) as awl-shaped (type A) or ramified (type E) cystidia. Shorter emanating hyphae in BM03M3 were similar to bent awl-shaped cystidia; the frequent simple septa and the lack of anatomical differences with longer emanating hyphae suggest that they are a transition between emanating hyphae and cystidia.

PD10 anatomotype shared the plectenchymatous mantle with squarrose branched hyphae and the lack of rhizomorphs observed in most Sebacinales EcM. The matrix previously observed in Sebacinales anatomotypes was superficial; in PD10, it was covered by a matrix-free outer layer. Superficial hyphal net with variably shaped hyphae as in PD10 was also reported on *Sebacina incrustans* + *Picea abies* (Urban et al. 2003), sample Y62M1 (Urban et al. 2003), *Quercirhiza dendrohyphidiomorpha* (Azul et al. 2006) and *Pinirhiza multifurcata* (Wei and Agerer 2011). No structures similar to PD10 staghorn-like hyphae (thick-walled, often swollen hyphae with numerous finger-like outgrowths and ramifications including polytomies with numerous branches) have been previously described on Sebacinales EcM. Schematic representations of cystidia with finger-like outgrowths (type E and I) were reported by Agerer (1991). The term hyphae is preferred for PD10 outermost mantle hyphal cells because these structures were bearing a variable number of outgrowths, with variable length and sometimes septa. PD10 variably shaped hyphae are called staghorn-like hyphae for the similarity with staghorn hyphae on *Pseudotsuga menziesii* (Mirb.) Franco + *Byssoporia* (*Poria*) *terrestris* var. *lilacinorosea* M.J. Larsen & Zak EcM anatomotype (Zak and Larsen 1978, Agerer and Rambold 2004–2011). Differences between *B. terrestris* and PD10 staghorn-like hyphae included the thin walls, and the lack of inflations and conspicuous polytomies in the first anatomotype.

6MM2 anatomotype shared features of *Sebacina incrustans* + *Picea abies* (sample Y129M5) (Urban et al. 2003, Agerer and Rambold 2004–2011) and Sebacinaceae sp. + *Betula papyrifera* Marshall (Twieg and Durall 2009). A matrix was lacking and the thick-walled plectenchymatous outer mantle layer was covered by numerous thick-walled emanating elements in these three anatomotypes. Rhizomorphs were lacking in *Sebacina incrustans* + *Picea abies* (L.) H. Karst. (sample Y129M5) (Urban et al. 2003) and Sebacinaceae sp. + *Betula papyrifera* (Twieg and Durall 2009). Emanating elements in *Sebacina incrustans* + *Picea abies* (Urban et al. 2003, Agerer and Rambold 2004–2011) and Sebacinaceae sp. + *Betula papyrifera* (Twieg and Durall 2009) were interpreted as awl-shaped cystidia and emanating hyphae. 6MM2 shorter emanating hyphae were similar to awl-shaped cystidia (Agerer 2006). The thinner wall at cell distal end and the frequent septa suggest they are a transitional growing stage of emanating hyphae. Despite similarities with *Sebacina incrustans* anatomotype, the presence of rhizomorphs and the absence of cystidia indicate that 6MM2 belongs to a different species.

5. Conclusions

This study confirmed that morpho-anatomical features are important information, together with ultrastructure and DNA sequences to confirm the EcM status of Sebacinales species. This first study of tropical Sebacinales EcM anatomotypes corroborated the earlier conclusion (Agerer 2006) that Sebacinales anatomotypes share general morpho-anatomical features.

Specimens BM03M3, PD10 and 6MM2 could be clearly distinguished from each other. These results are in agreement with the view that EcM morpho-anatomical features are useful for an integrative taxonomy of this diverse fungal group for which the study of fruitbodies is difficult (Wei and Agerer 2011). Further, combined morpho-anatomical and molecular-phylogenetic studies are necessary to define affiliations of the three *P. dipterocarpacea* Sebacinales EcM anatomotypes to species already known to science, or to hint at still undetected new species.

This study demonstrated the need to expand Sebacinales anatomotype morpho-anatomical studies. Comparisons with previously described anatomotypes demonstrated that BM03M3, PD10 and 6MM2 are new Sebacinales EcM anatomotypes to science. Yellowish cell walls in BM03M3 were observed for the first time in Sebacinales EcM. Infrequent rhizomorphs were already described from the Chinese anatomotype *Pinirhiza nondextrinoidea* (Wei and Agerer 2011), but their characteristics (rhizomorph type A/B, hyphae imbedded in a matrix) were different from the type B rhizomorphs in BM03M3 and 6MM2. Rhizomorph types can differ in Sebacinales and the capacity to form rhizomorphs in this fungal group is more widespread amongst species than previously assumed (comp. Agerer 2006). These results are not only important for the taxonomy of Sebacinales EcM but also for the understanding of Sebacinales ecology. BM03M3 and 6MM2 EcM clearly belong to the medium distance exploration type (Agerer 2001). PD10 staghorn-like hyphae were observed for the first time in Sebacinales. This new feature in Sebacinales anatomotypes demonstrates the importance of tropical studies for a more comprehensive taxonomy of EcM.

Ackowledgements

The Instituto Venezolano de Investigaciones Científicas and University of Tübingen provided support including lab facilities. I am particularly grateful to R Bauer for his help in ultrastructural analyses, and M. Weiß and I. Kottke for facilitating molecular and light microscope work at an early stage of this project. Thanks to R. Agerer, anonymous reviewers for helpful comments on the manuscript and Krista L. McGuire and Caitlyn Gillikin for improving the language.

References

Agerer, R. 1991. Characterization of ectomycorrhiza. Methods Microbiol. 23: 25–73.
Agerer, R. 2001. Exploration types of ectomycorrhizae. A proposal to classify ectomycorrhizal mycelial systems according to their patterns of differentiation and putative ecological importance. Mycorrhiza 11: 107–114.
Agerer, R. 2006. Fungal relationships and structural identity of their ectomycorrhizae. Mycol. Progress 5: 67–107.
Agerer, R. and G. Rambold. 2004–2011. DEEMY—an information system for characterization and determination of ectomycorrizae. www.deemy.de München, Germany.
Altschul, S.F., LM. Thomas, A.S. Alejandro, Z. Jinghui, M. Webb and J.L. David. 1997. Gapped BLAST and PSI-BLAST: a new generation of protein database search programs. Nucleic Acids Research 25: 3389–3402.
Azul, A.M., R. Agerer and H. Freitas. 2006. "Quercirhiza dendrohyphidiomorpha" + Quercus suber L. Descr. Ectomyc. 9/10: 87–91.
Dayrat, B. 2005. Towards integrative taxonomy. Biol. J. Linn. Soc. 85: 407–415.
Henkel, T.W., P. Roberts and M.C. Aime. 2004. Sebacinoid species from the Pakaraima mountains of Guyana. Mycotaxon 89(2): 433–439.
Hynson, N.A., M. Weiß, K. Preiss, G. Gebauer and K.K. Treseder. 2013. Fungal host specificity is not a bottleneck for the germination of Pyroleae species (Ericaceae) in a Bavarian forest. Mol. Ecol. doi 10.1111/mec.12180.
Jairus, T., R. Mpumba, S. Chinoya and L. Tedersoo. 2011. Invasion potential and host shifts of Australian and African ectomycorrhizal fungi in mixed eucalypt plantations. New Phytol. 192: 179–187.
Kennedy, A.H., D.L. Taylor and L.E. Watson. 2011. Mycorrhizal specificity in the full mycoheterotrophic Hexalectris Raf. (Orchidaceae: Epidendroideae). Mol. Ecol. 20: 1303–1316.
Kõljalg, U., K.-H. Larsson, K. Abarenkov, R.H. Nilsson, I.J. Alexander, U. Eberhardt, S. Erland, K. Høiland, R. Kjøller, E. Larsson, T. Pennanen, R. Sen, A.F.S. Taylor, L. Tedersoo, T. Vrålstad and B.M. Ursing. 2005. UNITE: a database providing web-based methods for the molecular identification of ectomycorrhizal fungi. New Phytol. 166(3): 1063–1068.
Maguire, B., P.S. Ashton, C. de Zeeuw, D.E. Giannasi and K.J. Niklas. 1977. Pakaraimoideae, Dipterocarpaceae of the Western hemisphere. Taxon 26(4): 341–385.
Maguire, B. and P.S. Ashton. 1980. *Pakaraimaea dipterocarpacaea* II. Taxon 29(2/3): 225–231.
Maguire, B. and J.A. Steyermark. 1981. *Pakaraimaea dipterocarpacea*. III. Memoires of the New York Botanical Garden 32: 306–309.
Morris, M.H., M.A. Pérez-Pérez, M.E. Smith and C.S. Bledsoe. 2008. Multiple species of ectomycorrhizal fungi are frequently detected on individual oak root tips in a tropical cloud forest. Mycorrhiza 18: 375–383.
Morris, M.H., M.A. Pérez-Pérez, M.E. Smith and C.S. Bledsoe. 2009. Influence of host species on ectomycorhizal communities associated with two co-occurring oaks (*Quercus* spp.) in a tropical cloud forest. FEMS doi:10.1111/j.1574-6941.2009.00704.x.
Moyersoen, B. 2006. *Pakaraimea dipterocarpacea* is ectomycorrhizal, indicating an ancient Gondwanaland origin for the ectomycorrhizal habit in Dipterocarpaceae. New Phytol. 172: 753–762.
Moyersoen, B. 2012a. Dispersion, an important radiation mechanism for ectomycorrhizal fungi in neotropical lowland forests? *In*: P. Sudarshana, M. Nageswara–Rao and J.R. Soneji (eds.). Tropical Forests, pp. 93–116.
Moyersoen, B. 2012b. *Clavulina* sp. + *Pakaraimaea dipterocarpacea* Maguire & Ashton ssp. nitida Maguire & Steyerm. Descr. Ectomyc. 13: 16–23.
Moyersoen, B. 2012c. *Coltriciella* sp. + *Pakaraimaea dipterocarpacea* Maguire & Ashton ssp. nitida Maguire & Steyerm. Descr. Ectomyc. 13: 24–30.

Moyersoen, B. 2012d. *Inocybe* sp. + *Pakaraimaea dipterocarpacea* Maguire & Ashton ssp. nitida Maguire & Steyerm. Descr. Ectomyc. 13: 42–47.

Nilsson, R.H., E. Kristiansson, M. Ryberg, N. Hallenberg and K.-H. Larsson. 2008. Intraspecific ITS variability in the kingdom *fungi* as expressed in the international sequence databases and its implications for the molecular species identification. Evol. Bioinf. 4: 193–201.

Peay, K.G., P.G. Kennedy, S.J. Davies, S. Tan and T.D. Bruns. 2009. Potential link between plant and fungal distributions in a dipterocarp rainforest: community and phylogenetic structure of tropical ectomycorrhizal fungi across a plant and soil ecotone. New Phytol. 185: 529–542.

Pirozynski, K.A. 1983. Pacific mycogeography: an appraisal. Aust. J. Bot., Suppl. Ser. 10: 137–159.

Rivière, T., A.G. Diedhiou, M. Diabate, G. Senthilarasu, K. Natarajan, A. Verbeken, B. Buyck, B. Dreyfus, G. Bena and A.M. Bâ. 2007. Genetic diversity of ectomycorrhizal Basidiomycetes from African and Indian tropical rain forests. Mycorrhiza 17: 415–428.

Roy, M., S. Watthana, A. Stier, F. Richard, S. Vessabutr and M.-A. Selosse. 2009. Two mycoheterotrophic orchids from Thailand tropical dipterocapacean forests associate with a broad diversity of ectomycorrhizal fungi. BMC Biology 7: 51 doi:10.1186/1741-7007-7-51.

Selosse, M.-A., M. Weiß, J.-L. Jany and A. Tillier. 2002a. Communities and populations of Sebacinoid basidiomycetes associated with the achlorophyllous orchid *Neottia nidus-avis* (L.) L.C.M. Rich. and neighbouring tree ectomycorrhizae. Mol. Ecol. 11: 1831–1844.

Selosse, M.-A., R. Bauer and B. Moyersoen. 2002b. Basal hymenomycetes belonging to the Sebacinaceae are ectomycorrhizal on temperate deciduous trees. New Phytol. 155: 183–195.

Setaro, S.D., S. Garnica, P.I. Herrera, J.P. Suárez and M. Göker. 2012. A cluster optimization strategy to estímate species richness of *Sebacinales* in the tropical Andes based on molecular sequences from distinct DNA regions. Biodivers. Conserv. 21: 2269–2285.

Smith, M.E., T.W. Henkel, M.C. Aime, A.K. Fremier and R. Vilgalys. 2011. Ectomycorrhizal fungal diversity and community structure on three co-occurring leguminous canopy tree species in a Neotropical rainforest. New Phytol. doi: 10.1111/j.1469-8137.2011.03844.x.

Smith, M.E., T.W. Henkel, J.K. Uehling, A.K. Fremier, H.D. Clarke and R. Vilgalys. 2013. The ectomycorrhizal fungal community in a Neotropical forest dominated by the endemic dipterocarp *Pakaraimaea dipterocarpacea*. PLoS one 8(1): e55160. doi:10.1371/journal.pone.0055160.

Spurr, A.R. 1969. A low-viscosity epoxy resin embedding medium for electron microscopy. J. Ultrastruct. Res. 26: 31–43.

Tedersoo, L., R.H. Nilsson, K. Abarenkov, T. Jairus, A. Sadam, I. Saar, M. Bahram, E. Bechem, G. Chuyong and U. Kõljalg. 2010a. 454 Pyrosequencing and Sanger sequencing of tropical mycorrhizal fungi provide similar results but reveal substantial methodological biases. New Phytol. doi:10.1111/j.1469-8137.2010.03373.x.

Tedersoo, L., A. Sadam, M. Zambrano, R. Valencia and M. Bahram. 2010b. Low diversity and high host preference of ectomycorrhizal fungi in Western Amazonia, a neotropical biodiversity hotspot. The ISME J. 4: 465–471.

Tedersoo, L., M. Bahram, T. Jairus, E. Bechem, S. Chinoya, R. Mpumba, M. Leal, E. Randrianjohany, S. Razafimandimbison, A. Sadam, T. Naadel and U. Kõljalg. 2011. Spatial structure and the effects of host and soil environments on communities of ectomycorrhizal fungi in wooded savannas and rain forests of Continental Africa and Madagascar. Mol. Ecol. 20: 3071–3080.

Twieg, B. and D. Durall. 2009. Sebacinaceae sp. + Betula papyrifera Marsh. *In*: D.M. Goodman, D.M. Durall, J.A. Trofymow and S.M. Berch (eds.). Concise descriptions of North American Ectomycorrhizae. Mycologue Publications, and Canada-B. C. Forest Resource Development Agreement, Canadian Forest Service, Victoria, B. C., pp. CDE28.

Urban, A., M. Weiß and R. Bauer. 2003. Ectomycorrhizas involving sebacinoid mycobionts. Mycol. Res. 107(1): 3–14.

Wei, J. and R. Agerer. 2011. Two Sebacinoid ectomycorrhizae on Chinese pine. Mycorrhiza 21: 105–115.

Weiß, M., M.A. Selosse, K.L. Rexer, A. Urban and F. Oberwinkler. 2004. *Sebacinales*: a hitherto overlooked cosm of heterobasidiomycetes with a broad mycorrhizal potential. Mycol. Res. 108(9): 1003–1010.

Weiß, M., Z. Sýkorová, S. Garnica, K. Riess, F. Martos, C. Krause, F. Oberwinkler, R. Bauer and D. Redecker. 2011. Sebacinales everywhere: previously overlooked ubiquitous fungal endophyes. PLos ONE 6(2): e16793. doi: 10.1371/journal.pone.0016793.

Wells, K. and F. Oberwinkler. 1982. *Tremelloscypha gelatinosa*, a species of a new family Sebacinaceae. Mycologia 74(2): 325–331.

Zak, B. and M.J. Larsen. 1978. Characterization and classification of mycorrhizae of Douglas-fir. III. *Pseudotsuga menziesii* + *Byssoporia* (*Poria*) *terrestris* vars. *lilacinorosea, parksii,* and *sublutea*. Can. J. Bot. 56: 1416–1424.

CHAPTER 6

Abundance, Distribution, and Function of *Pisolithus albus* and other Ectomycorrhizal Fungi of Ultramafic Soils in New Caledonia

Philippe Jourand,[1,*] *Fabian Carriconde,*[2] *Marc Ducousso,*[3] *Clarisse Majorel,*[1] *Laure Hannibal,*[1] *Yves Prin*[3] and *Michel Lebrun*[4]

1. Introduction

Ultramafic soils, also known as "serpentine soils" in literature, are a weathered product from ultramafic bedrock that covers less than 1% of the earth's surface (Coleman and Jove 1992). These soils are characterized by high concentrations of iron oxides (up to 85% w/w), unbalanced calcium-

[1]IRD, UR040 LSTM, Centre IRD, BPA5, Promenade Roger Laroque, 98848 Nouméa Cedex, Nouvelle-Calédonie.
[2]Institut Agronomique néo-Calédonien (IAC), Axe 2 'Diversités biologique et fonctionnelle des écosystèmes terrestres', Nouméa IRD research centre, BP18239, 98857 Nouméa, Nouvelle-Calédonie.
[3]CIRAD, UMR LSTM, TA A-82/J Campus International de Baillarguet, 34398 Montpellier Cedex 5 France.
[4]Université Montpellier 2, UMR28 LSTM, TA A-82/J Campus International deBaillarguet, 34398 Montpellier Cedex 5, France.
*Corresponding author: philippe.jourand@ird.fr

to-magnesium ratio (up to 1/30 that consequently may influence both Mg and Ca plant nutrition), and the presence of various heavy metals at high concentrations, such as chromium, cobalt, manganese and nickel, all of which are mostly toxic for many plants (Brooks 1987). They are also extremely deficient in elements that are essential for plant nutrition, including nitrogen, phosphorus and potassium (Brooks 1987, Chiarucci and Baker 2007). Previous studies have shown that ultramafic soils are characterized by a high biological diversity of plants as described in Proctor (2003) and micro organisms that use various mechanisms to cope with the extreme edaphic conditions, in particular adaptation to toxic heavy metals (Brady et al. 2005, Kazakou et al. 2008, Rajkumar et al. 2009). Recently, major data about the ecological traits of ultramafic soils have been reviewed to propose these soils as a model system in ecology and conservation,mostly because of their high plant diversity (Harrisson and Rajakaruna 2011).

In ultramafic ecosystems, it is well known that most plants tolerant to these extreme soils are involved in mycorrhizal associations, which may greatly enhance plant nutrition (such as P assimilation) and reduce metal toxicity on plants (Alexander et al. 2007, Smith and Read 2008). Studies carried out on ectomycorrhizal (ECM) fungal communities in ultramafic soils showed a high diversity of fungal species developing ECM symbioses with plants growing on these substrates (Moser et al. 2005, Urban et al. 2008). In addition, it was recently demonstrated that ultramafic soils do not limit, and can even promote, the ectomycorrhizal fungal diversity (Moser et al. 2009, Branco and Ree 2010, Branco 2010). However, the comparison of ECM fungal diversity between serpentine and non-serpentine soils showed differences within the fungal population structure (Brealey et al. 2006), sometimes with the presence of unique species (Moser et al. 2005). Moreover, studies about physiological behaviour such as metal tolerance within a same fungal species present on both serpentine and non-serpentine soils have suggested adaptive evolution, raising questions about the adaptation and evolution of fungal species on these soils (Gonçalves et al. 2007, 2009).

Here, we have presented a review carried out on ECM fungi collected from ultramafic soils in New Caledonia, which is a tropical archipelago located in the South Pacific Ocean (Fig. 1). In New Caledonia, these soils cover one-third of the main island due to geological evolution (Fig. 1). As a result of the presence of such ultramafic outcrops, numerous endemic ecosystems have developed (Jaffré 1992), making the main island a biodiversity hot spot (Myers et al. 2000). In the first section, we have summarized results from existing studies and new results of ECM fungal diversity found on these extreme soils. In the second section, we have gathered data about the ECM *Pisolithus albus* (Cooke and Massee) isolated from ultramafic soils in New Caledonia: diversity, metal-tolerance and symbiotic interactions with its host plant are presented.

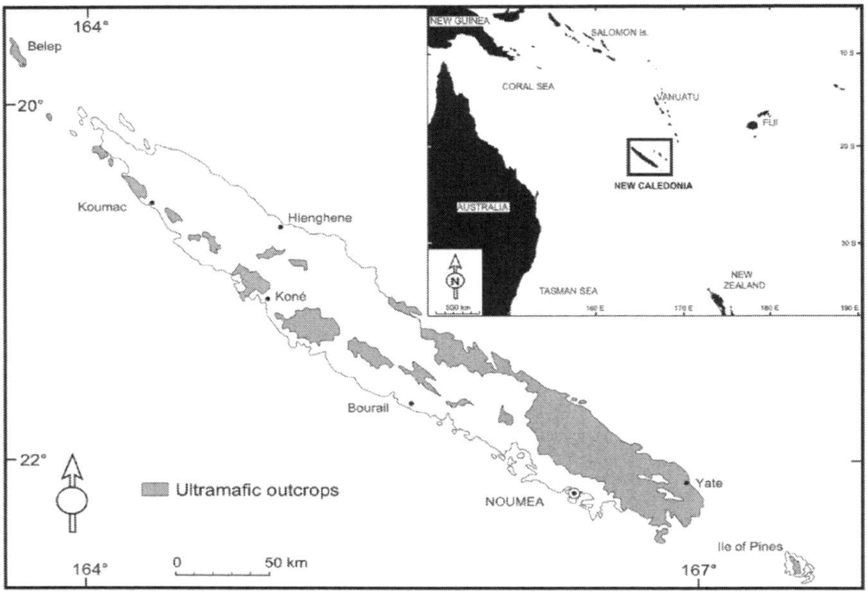

Fig. 1. Geographical map of the New Caledonian archipelago in the South Pacific Ocean with location of ultramafic massifs (in grey). Data from Perrier et al. (2006a).

2. Plant ECM Status and Fungal Diversity in Ultramafic Soils of New Caledonia

Considering the vascular plants:fungi ratio of 1:6 as reported by Hawksworth (1991, 2001) and the number of vascular plants of 3,371 species identified in New Caledonia (T. Jaffré, personal communication), we could hypothesize that at least 20,000 fungal species inhabit the archipelago. Referring to the available literature and herbarium data, the mycologists Horak and Mouchacca listed about 420 Ascomycota and Basidiomycota taxa in New Caledonia (Horak and Mouchacca 1998, Mouchacca 1998, Mouchacca and Horak 1998), which would indicate that approximately 2% of the species have been inventoried.

The studies undertaken by Perrier, though preliminary, are to date the only ones that have characterized the ECM status of some New Caledonian plant species and the related ECM fungal diversity (Perrier 2005, Perrier et al. 2006a,b). In New Caledonia, two main plant formations are basically distinguished on ultramafic rocks: sclerophyllous scrubland formations, called "maquis" or "maquis minier", and rain forest formations (Jaffré and L'Huillier 2010). According to the type of soil, the altitude, and the floristic composition, many groups are further identified. The plant formations studied by Perrier were located on the ultramafic Koniambo Massif and correspond to four distinct vegetation groups (Fig. 2): a maquis with emerging *Araucaria* trees, a lingo-herbaceous maquis, a *Tristaniopsis* spp. maquis and a rain forest dominated by *Nothofagus balansae* with patches of *N. codonandra*. Investigation of the root systems of 19 species revealed that two *Tristaniopsis* species, *T. calobuxus* and *T. guillainii*, are involved in ECM symbioses. These species belong to the Myrtaceae (Leptospermoideae group), a well-known plant family frequently found to be associated with ECM fungi (Smith and Read 2008, Wang and Qiu 2006). *Nothofagus balansae* and *N. codonandra* roots were also characterized by the presence of a fungal mantle and a Hartig net. Another New Caledonian species, *N. aequilateralis*, has also been previously shown to be able to develop ECM associations (McCoy 1991). *Nothofagus* are indirectly, by the presence of putative ECM fungal fruit bodies, and/or directly, by investigation of the root system, defined in other regions of the world (i.e., Australia, New Zealand, Papua New Guinea and South America) as ECM trees (Horak and Wood 1990, Garnica et al. 2003, Tedersoo et al. 2008, Dickie et al. 2010), and subsequently recognized as an important ECM genus in the Southern Hemisphere (Smith and Read 2008).

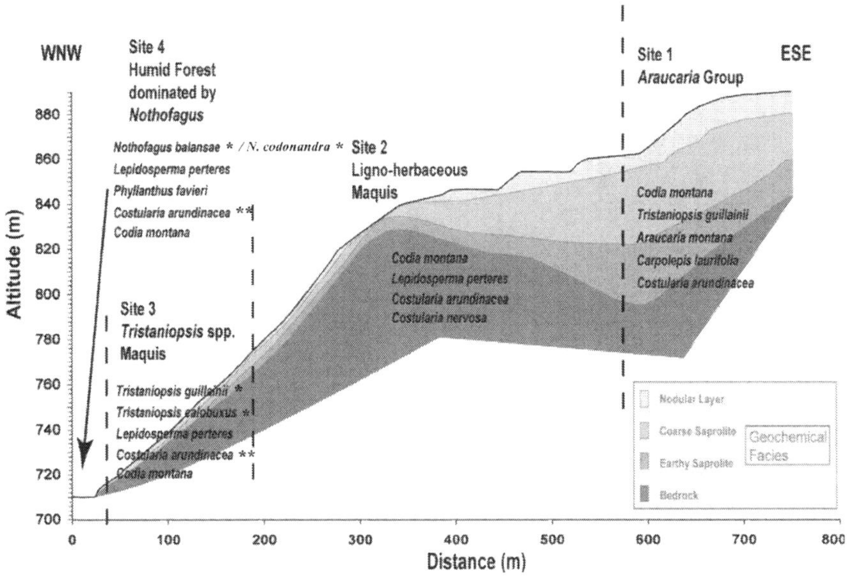

Fig. 2. Repartition of the four distinct plant formations (sites 1 to 4) on the topographic sequence studied by Perrier et al. (2006) at the Koniambo Massif. Most abundant and potential species for restoration purpose within each vegetation type are shown. Plants with ECM structures or ECM-like-structures on their root systems are indicated by an asterisk or two asterisks, respectively. The distance in meters (m) from the valley to the plateau and the altitude are given in abscissa and ordinate, respectively. Modified data from Perrier et al. (2006a).

Surprisingly, Perrier et al. (2006a) observed ECM-like structures on Cyperaceae roots as *Costularia arundinacea*, with the presence of a fungal mantle but the absence of a Hartig net. Such observation has already been done on two other Cyperaceae species belonging to the genus *Carex* (Harrington and Mitchell 2002). However, the colonization of *C. arundinacea* roots was only observed on sites 3 and 4, dominated by *T. guillainii* and *N. balansae* respectively (Fig. 2). The vicinity of both ECM plants may thus explain the colonization of *C. arundinacea* root systems. Further investigations led to the identification of other plant species as ECM (Table 1) (Amir and Ducousso 2010, F. Carriconde personal communication). Regarding the Myrtaceae, four additional *Tristaniopsis* species, two *Melaleuca* species and the monospecific and endemic genus *Arillastrum* are involved in such associations (Table 1). The Fabaceae *Acacia spirorbis* has also been identified as ECM (Ducousso et al. 2012). Finally, in New Caledonia, the ECM status has been characterized for only 13 plant species among Fabaceae, Myrtaceae and Nothofagaceae (Table 1). Giving the large representation of the Myrtaceae family, especially the Leptospermoidae group, and the Fabaceae family in New Caledonia (Morat et al. 2012), we

Table 1. Plant families and species in New Caledonia characterized as ECM. The biogeographical native status is given according to Jaffré et al. (2001) (N: native, i.e., species for which their natural distribution area extend beyond the boundaries of New Caledonia; E: endemic species; EE: endemic genus). The types of soils on which species are encountered are also indicated (C: calcareous; UM: ultramafic soils; VS: volcano-sedimentary). For species known to be present on more than one type of soil, the predominant types are highlighted in bold. Modified data from Amir and Ducousso (2010).

Family	Species	Biogeographical status	Type soil
Fabaceae	*Acacia spirorbis*	N	C, **VS**, **UM**
Myrtaceae	*Arillastrum gummiferum*	EE	UM
	Melaleuca pancheri	E	UM
	Melaleuca quinquenervia	N	C, **VS**, UM
	Tristaniopsis calobuxus	E	**UM**, VS
	Tristaniopsis glauca	E	UM
	Tristaniopsis guillainii	E	UM
	Tristaniopsis macphersonii	E	UM
	Tristaniopsis ninndoensis	E	VS
	Tristaniopsis vieillardii	E	UM
Nothofagaceae	*Nothofagus aequilateralis*	E	UM
	Nothofagus balansae	E	UM
	Nothofagus codonandra	E	UM

could expect a large number of species to be involved in such mutualistic interaction. This clear lack of knowledge is furthermore well-illustrated by the fact that the ECM status remains unknown for two *Nothofagus* species in New Caledonia: *N. baumanniae* and *N. discoidea*.

The aboveground and belowground fungal diversity has been, to some extent, investigated on the topographic sequence at the Koniambo Massif (Fig. 2) by collecting sporocarps, ectomycorrhizal root tips and hyphal mats in the soil (Table 2, Perrier 2005). Molecular identification has been carried out by sequencing the nuclear ribosomal DNA (rDNA) internal transcribed spacer (ITS), a widely used marker in mycology and recently defined as the reference region for fungal DNA barcoding (Schoch et al. 2012). Twenty-nine sporocarps, 11 ECM root tips and 7 hyphae collected from soil cores were successfully sequenced (Table 2, Perrier 2005). Comparison of the generated ITS sequences to the international available database GenBank using the BLAST algorithm (Altschul et al. 1990) showed the presence of several genera (Table 2). Interestingly, out of the total of 47 samples, 45 presented a percentage of similarity less than 97% (Table 2), a value commonly used to differentiate ECM species (e.g., Tedersoo et al. 2003, Izzo et al. 2005, Smith et al. 2007). Two sporocarps, K66C and KC03C, had a percentage of similarity >97% with samples from Australia and New Zealand, respectively. Overall, these results suggest there is a diverse and unique ECM fungal assemblage at these study sites and possibly across New Caledonia at a regional scale.

Indeed, the description in the last few years of new putative ECM species, such as the impressive *Podoserpula miranda* (Fig. 3), thought to be associated with *Arillastrum gummiferum* in the South of New Caledonia (Ducousso et al. 2009), or the chanterelle, *Cantharellus garnieri* (Fig. 3) collected under distinct potential host trees in different localities and type of soils (Ducousso et al. 2004), strengthened the idea of the high fungal diversity in the archipelago. Regarding the abundance of the different fungal genera at Koniambo's sites, samples belonging to the *Cortinarius* genus were largely represented. Indeed, out of the 29 sporocarps, 11 ECM root tips and 7 hyphal mats collected, 11 (~38%), 6 (~55%), and 5 (~71%) were assigned to this genus. The large belowground representation of *Cortinarius* has already been highlighted in *Nothofagus* forests in Australia and New Zealand (Tedersoo et al. 2008, Dickie et al. 2010). Co-evolution between Cortinariaceae and *Nothofagus* in Australia has been suggested (Bougher et al. 1994), and could thus be one of the main driving forces that may have led to the diversification of this fungal group in the Pacific region.However, the limited sampling size of Perrier's study (in total only 47 samples), and particularly the very restricted number of studies undergone to date on fungal diversity, do not allow us to draw any conclusions on the diversity level and the structure of this diversity on the archipelago.

Table 2. Sporocarps, ECM root tips and hyphae samples collected in the *Tristaniopsis* spp. maquis (site 3) and the rain forest dominated by *Nothofagus balansae* (site 4), located on the topographic sequence at the Koniambo Massif and genotyped by sequencing of the ITS region. The host plant (putative), the morphospecies when available, the ITS sequence length, the closest BLAST match and the related information are presented. ITS sequence data generated by Perrier (2005) were recently analyzed.

Sample reference	Sample type	Plant formation	Host plant (putative) †	Morphospecies	GenBank accesion number	Sequence lenght (bp)	Closest species BLAST match ‡	Bases matched	% Similarity	Best match GenBank accesion number
K66C	Sporophore	3	*Tristaniopsis guillainii*	*Pisolithus sp*	FJ656011	527	*Pisolithus sp*	520/526	99%	AF270787
K02C	Sporophore	4	*Nothofagus balansae*	*Boletus sp*	FJ656001	609	*Boletus sp*	384/449	86%	EU569234
K05C	Sporophore	4	*Nothofagus balansae*	nd	FJ656002	551	*Cortinarius subgemmeus**	498/561	89%	JX000354
K06C	Sporophore	4	*Nothofagus balansae*	-	FJ656003	605	*Phellodon sp*	544/597	91%	GU222318
K09C	Sporophore	4	*Nothofagus balansae*	nd	FJ656004	547	*Austrogautieria macrospora**	438/504	87%	GQ981492
K10C	Sporophore	4	*Nothofagus balansae*	nd	FJ656005	670	*Tricholoma imbricatum*	628/668	94%	AY573537
K10C	Sporophore	4	*Nothofagus balansae*	nd	FJ656005	670	*Tricholoma imbricatum*	626/668	94%	AY573537
K12C	Sporophore	4	*Nothofagus balansae*	nd	FJ656006	572	*Cortinarius austrovenetus*	534/573	93%	GQ890318
K14C	Sporophore	4	*Nothofagus balansae*	*Inocybe sp*	FJ656007	506	*Dermocybe largofulgens**	483/504	96%	GU233324
K16C	Sporophore	4	*Nothofagus balansae*	nd	FJ656008	706	*Lactarius scrobiculatus*	660/719	92%	EU597079
K18C	Sporophore	4	*Nothofagus balansae*	Lactaroides	FJ656009	598	*Russula zonaria **	526/569	92%	DQ421990
KC02C	Sporophore	4	*Nothofagus balansae*	nd	FJ656012	436	*Cortinarius lividus*	390/433	90%	AF539734
KC05C	Sporophore	4	*Nothofagus balansae*	nd	FJ656014	580	*Inocybe aeruginascens**	491/569	86%	GU949591
KC08C	Sporophore	4	*Nothofagus balansae*	nd	FJ656016	670	*Lactarius olympianus*	624/684	91%	EF685079
KC11C	Sporophore	4	*Nothofagus balansae*	nd	FJ656018	609	*Russula sp*	577/608	95%	GU222292
KC12C	Sporophore	4	*Nothofagus balansae*	nd	FJ656019	518	*Cortinarius flammuloides*	460/540	85%	AF539716
KC16C	Sporophore	4	*Nothofagus balansae*	nd	FJ656020	574	*Cortinarius multiformis*	518/589	88%	AF389135
KC17C	Sporophore	4	*Nothofagus balansae*	nd	FJ656021	574	*Leratiomyces ceres*	394/411	96%	HQ604750
KC19C	Sporophore	4	*Nothofagus balansae*	nd	FJ656022	583	*Cortinarius singularis**	508/584	87%	JQ287672
KC22C	Sporophore	4	*Nothofagus balansae*	nd	FJ656023	528	*Austrogautieria macrospora**	442/503	88%	GQ981492
KD37C	Sporophore	4	*Nothofagus codonandra*	nd	FJ656038	582	*Cortinarius eutactus**	559/582	96%	JX000366
KC23C	Sporophore	4	*Tristaniopsis guillainii*	nd	FJ656024	414	*Cortinarius elaiops**	241/262	92%	JX000369
K01C	Sporophore	4	*Nothofagus balansae*	nd	FJ656000	701	*Tricholoma ustale*	632/713	89%	AF458435
K22C	Sporophore	4	*Nothofagus balansae*	nd	FJ656010	582	*Inocybe aeruginascens**	491/569	86%	GU949591
KC03C	Sporophore	4	*Nothofagus balansae*	nd	FJ656013	612	*Russula sp*	593/613	97%	GU222292
KC06C	Sporophore	4	*Nothofagus balansae*	nd	FJ656015	436	*Phaeocollybia redheadii*	394/411	96%	JN102541
KC10C	Sporophore	4	*Nothofagus balansae*	nd	FJ656017	447	*Cortinarius aff. austrosanguineus*	427/454	94%	GQ890317
KD36C	Sporophore	4	*Nothofagus codonandra*	nd	FJ656037	585	*Cortinarius elaiops**	554/586	95%	JX000369
KD42C	Sporophore	4	*Nothofagus codonandra*	nd	FJ656039	485	*Tricholoma ustale*	402/455	88%	AF458435
KE01-2M	ECM	3	*Tristaniopsis guillainii*	-	FJ656040	457	*Piloderma sp*	400/450	89%	JQ711951
KE02M	ECM	3	*Tristaniopsis guillainii*	-	FJ656041	511	*Cortinarius vernicifer**	387/449	86%	JX000370
KE04M	ECM	3	*Tristaniopsis guillainii*	-	FJ656042	438	*Piloderma sp*	383/430	89%	JQ711951
KD10M	ECM	4	*Nothofagus balansae*	-	FJ656025	552	*Oidiodendron chlamydosporicum*	477/519	92%	AF062789
KE06M	ECM	4	*Nothofagus balansae*	-	FJ656043	584	*Cortinarius amoenus*	544/590	92%	AF389160
KE12-1M	ECM	4	*Nothofagus balansae*	-	FJ656045	620	*Tricholoma ustale*	557/640	87%	AF458435
KD18M	ECM	4	*Nothofagus codonandra*	nd	FJ656026	474	*Cortinarius calyptratus**	425/476	89%	EU525980

KD29-2M	ECM	4	*Nothofagus codonandra*	-	FJ656033	608	*Cortinarius elaiops**	558/606	92%	JX000369
KD31"-2M	ECM	4	*Nothofagus codonandra*	-	FJ656034	570	*Cortinarius elaiops**	518/600	86%	JX000369
KD31'M	ECM	4	*Nothofagus codonandra*	-	FJ656035	682	*Tomentellopsis submollis*	642/684	94%	JQ711898
KD36-2M	ECM	4	*Nothofagus codonandra*	-	FJ656036	584	*Cortinarius singularis**	509/584	87%	JQ287672
KD19 1S	Hyphae	3	*Tristaniopsis guillainii*	-	FJ656027	698	*Lycoperdon sp*	682/726	94%	JX029934
KD19 2S	Hyphae	3	*Tristaniopsis guillainii*	nd	FJ656028	592	*Cortinarius eutactus**	560/597	94%	HQ533023
KD19 9S	Hyphae	3	*Tristaniopsis guillainii*	-	FJ656029	410	*Cortinarius sp*	350/402	87%	JQ287690
KD20 5S	Hyphae	3	*Tristaniopsis guillainii*	-	FJ656031	585	*Cortinarius sp*	543/594	91%	JN942302
KD20 6S	Hyphae	3	*Tristaniopsis guillainii*	nd	FJ656032	615	*Cortinarius sp*	552/621	89%	JN942302
KE12 2S	Hyphae	4	*Nothofagus balansae*	-	FJ656046	608	*Tricholoma ustale*	554/629	88%	AF458435
KE18 2S	Hyphae	4	*Nothofagus codonandra*	-	FJ656047	540	*Cortinarius subgemmeus**	477/560	85%	JX000354

† ECM root tips were sampled by tracing the roots from the tree trunks.
‡ Voucher specimens are indicated by an asterisk.

Fig. 3. (A) *Podoserpula miranda* and (B) *Cantharellus garnieri*. Photos by courtesy from Ducousso Marc, CIRAD.

Color image of this figure appears in the color plate section at the end of the book.

Although preliminary analysis of the ECM diversity has been achieved, a thorough description of ECM and fungal communities and the interaction with host-plants in New Caledonia should be carried out. In order to really investigate such fungal diversity and better understand the mechanisms involved, molecular ecology studies on ECM communities by sequencing sporocarps and ectomycorrhizas using the classical Sanger approach, complemented by the use of next generation sequencing on soil cores, should be undertaken.

3. *Pisolithus albus* from Ultramafic Soils in New Caledonia: Diversity and Physiological Response to Nickel

Pisolithus albus (Cooke and Massee) is a fungal species belonging to *Pisolithus* Alb. and Schwein known to be one of the major ectomycorrhizal *Boletale* distributed on a worldwide scale that forms ectomycorrhizal symbioses with a broad range of angiosperm and gymnosperm tree species (Marx 1977, Martin et al. 2002). *Pisolithus* is also regarded as an early colonizer that persists on sites subject to edaphic stresses (Anderson et al. 1998). In New Caledonia, *P. Albus* fruit bodies are very abundant. The species also develops ectomycorrhizal associations with many endemic plants belonging to various genera of the *Myrtaceae* such as *Melaleuca*, *Tristaniopsis* and *Sannantha*, and one *Mimosaceae*, i.e., *Acacia spirorbis* (Perrier 2005). In New Caledonia most of the plants able to form ECM with *P. albus* dominate specific zones in their respective ecosystem: for example, *Tristaniopsis* genus colonizes specific zones of the ultramafic ecosystem at an altitude from 400 to 900 meters (L'Huillier et al. 2010). Altogether, the abundance of *P. albus* and its ability to develop ECM symbioses with endemic plants that colonize specific ecosystems in New Caledonia has led to the study of the genetic diversity of *P. albus* in New Caledonia.

3.1 Diversity of Pisolithus albus and their symbioses in New Caledonia

Isolates of ectomycorrhizal *P. albus* were sampled from both ultramafic and non-ultramafic soils in New Caledonia in order to investigate the relationships between (i) genetic diversity and (ii) the edaphic constraintssuch as the deficiency of major nutrient elements (N, K and P), the unbalanced Ca/Mg ratio and the presence of heavy metals at high concentrations (Jourand et al. 2010a). Fruiting body description, spore morphology (Fig. 4) and phylogenetic analysis based on internal transcribed spacer (ITS) rDNA (as previously reported by Martin et al. 2002) sequences confirmed that all isolates belong to *P. albus* and are closely related to other Australasian specimens (Fig. 5). In addition, the ecology of *P. albus* isolated from New Caledonia confirmed the dominant association with endemic plants belonging to genera of the Myrtaceae family (e.g., *Melaleuca*, *Sannantha*, *Tristaniopsis*) or the Fabaceae family (e.g., *Acacia*).

Altogether, the ecological and molecular data of *P. albus* isolated in New Caledonia were in agreement with the phylogeography of the ectomycorrhizal *Pisolithus* genus inferred from rDNA-ITS sequences, suggesting that (1) evolutionary lineages within *Pisolithus* are related to the biogeographical origin of their plant hosts (Martin et al. 2002) and (2) a long-distance dispersal event of ectomycorrhizal fungi from Australia

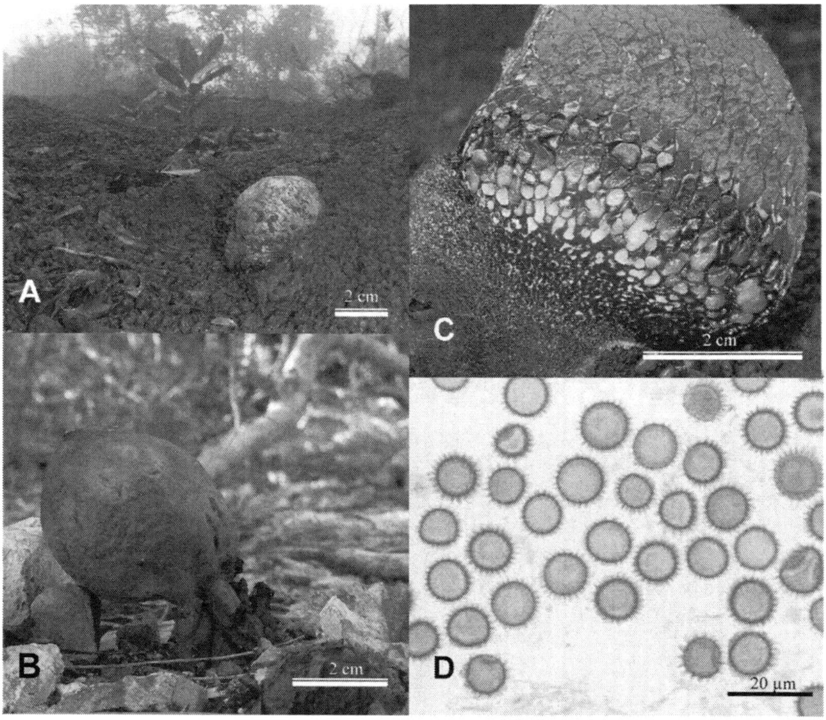

Fig. 4. *Pisolithus albus* from New Caledonia. A: *Pisolithus albus* MD07-117 from the Koniambo massif; B: *Pisolithus albus* MD07-228 from the Ouen-Toro, Noumea; C: cross section of *Pisolithus albus* MD07-166 from Pindjen water-fall and D: globose spores (8.77 to 9.62 µm) of *Pisolithus albus* MD06-379 from Poum, erected spines (1.2 µm) are clearly visible. From Jourand et al. (2010a).

Color image of this figure appears in the color plate section at the end of the book.

might explain the introduction of *Pisolithus* species in the South Pacific zone (Moyersoen et al. 2003). Interestingly, the use of other molecular tools such as ITS-restriction fragment length polymorphism (Fig. 6A) and amplified fragment length polymorphism markers (AFLP) (Fig. 6B), showed the existence of one genotype within *P. albus* grouping isolates from ultramafic soils (Jourand et al. 2010a). Such results raised the question of the presence a fungal ecotype on ultramafic soils, as described for plants found on these soils (Harrison and Rajakaruna 2011). They also contribute to the hypothesis of a link between the phylogenetic population structure and the ecological adaptation due to the particular mineral constraints,

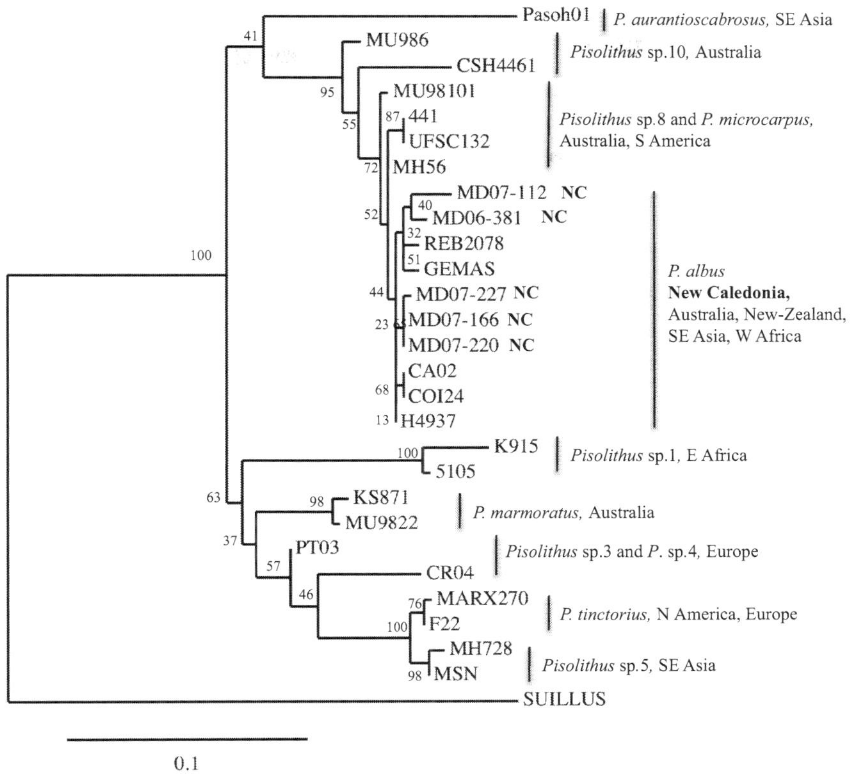

Fig. 5. Phylogenetic synthetic relationships among representative *Pisolithus* sp. from New Caledonia collection sites and worldwide reference isolates. The phylogeny is based on the analysis of the rDNA ITS1, 5.8S and ITS2 sequences. Tree shown is a 50% Majority rule consensus of the most parsimonious trees (Tree Length = 3733) obtained with PAUP4 (see Materials & Methods). Values indicated at tree nodes are percentage values of 1000 bootstrap replicates under MP criterion using fast stepwise addition (only values >50% are shown). The tree was rooted with *Suillus luteus* ITS sequences. Significant bootstrap frequencies are indicated. Abbreviations: S America: South America; SE Asia: South East Asia; W Africa: West Africa, E Africa: East Africa.

in particular ultramafism, as observed in ectomycorrhizal communities from other ultramafic soils (Urban et al. 2008). To further investigate this hypothesis, considering that nickel is one of the most toxic and bioavailable metal found at high concentrations in these soils (Echevarria et al. 2006), *P. albus* molecular and physiological responses to nickel were assessed in a further study.

A) *P. albus* ITS RFLP profiles

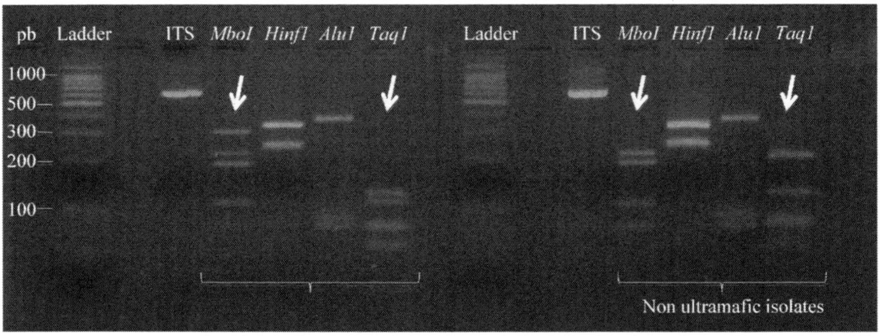

B) *P. albus* AFLP profile relationship

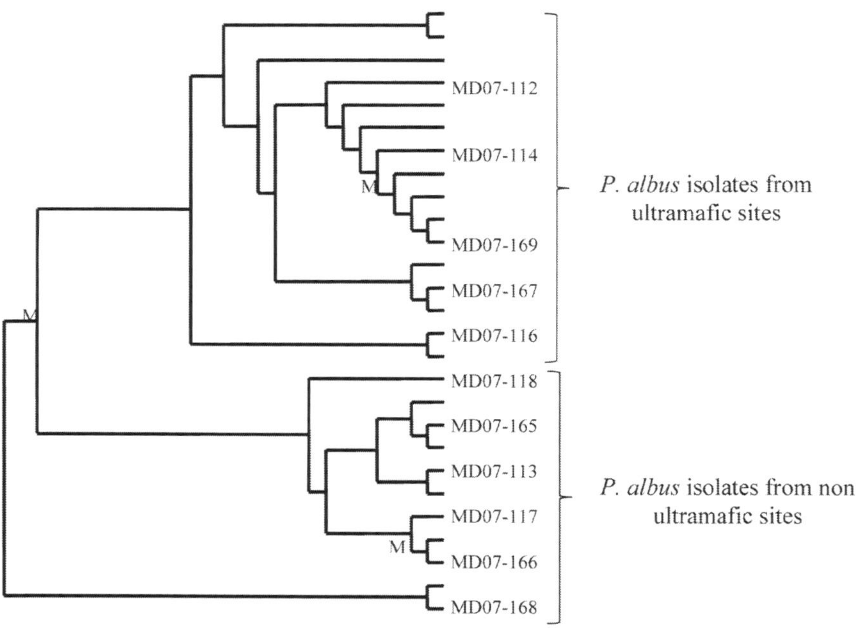

Fig. 6. A) Representative patterns of ITS restriction fragment length polymorphism (RFLP) profiles of *Pisolithus albus* isolates from both ultramafic and volcano sedimentary soils compared to both undigested amplified ITS and 100 pb DNA Ladder (Promega). Arrows highlight major differences between profiles. B) Genetic relationship within *P. albus* isolates from New Caledonia according to AFLP analysis. Bootstrap consensus UPGMA tree obtained for 882 AFLP scored fragments obtained with the 9 selective primers pairs on the 27 *P. albus* isolates (100 replicates). Data from Jourand et al. (2010a).

3.2 Tolerance and adaptation to nickel of *Pisolithus albus* from New Caledonia

In ultramafic soils, nickel (Ni) is one the most bioavailable and phytotoxic element: nickel content may reach up to 10 g/kg in ultramafic soils when compared with the average 50 mg/kg in cultivated soils (Wenzel and Jockwer 1999, Echevarria et al. 2006). This mineral element is a crucial selecting factor for plant survival on ultramafic soils: to grow on such high concentrations of nickel as found in serpentine environments (often coinciding with high concentrations of other heavy metals), plants had to develop major adaptations that include exclusion of the absorption of the toxic metal by the roots and/or metal hyperaccumulation with internal complexation and compartmentation (Kazakou et al. 2008). In addition, ECM symbioses might contribute to limit the metal accessibility and uptake by the plant (Colpaert et al. 2011).

3.2.1 *Pisolithus albus* nickel content and *in vitro* tolerance

In the previous study, the nickel concentration in fruiting body tissues of *Pisolithus albus* isolates from New Caledonia was assessed, as well as the *in vitro* nickel tolerance of cultivated mycelia from isolates collected from soil type (ultramafic vs non-ultramafic) where *P. albus* were collected (Jourand et al. 2010a). In fruiting bodies of *P. albus* from ultramafic soils, the nickel concentration reached an average of 5.7µg/g of dried tissue. In contrast, tissue of carpophores of isolates collected from non-ultramafic soils contained 2.5 times less nickel. In addition, *P. albus* mycelia from ultramafic soils included isolates with high variations of *in vitro* nickel-tolerance, with both nickel-tolerant isolates (with an average that half the maximal effective concentration of Ni that reduced fungal growth by 50% was 575 mM) and nickel-sensitive isolates (average Ni EC_{50} 37 mM). In contrast, all isolates from non-ultramafic soils were found to be nickel-sensitive (average Ni EC_{50} at 32 mM).

Within *Pisolithus* spp., previous studies have showed that some isolates were able to tolerate high concentrations of nickel. For example isolates of *Pisolithus tinctorius* were found to tolerate nickel with a Ni EC_{50} ranging from 126 to 170 mM (Tam 1995). Aggangan et al. (1998) also described one isolate of *P. tinctorius* from ultramafic soils in New Caledonia able to grow on nickel from 20 to 200 µM. More recently, Blaudez et al. (2000) and Ray et al. (2005) reported isolates of *P. tinctorius* that are able to grow on medium with nickel concentrations ranging from 17 to 350 µM. The mycelia from *P. albus* isolates from New Caledonian ultramafic soils displayed both *in vitro* nickel-sensitive and nickel-tolerant phenotypes. In addition, the nickel-tolerant isolates presented a noteworthy tolerance to Ni with an average Ni EC_{50} two to three times higher than the Ni EC_{50} already reported for other *Pisolithus* spp. mentioned above. To explain the high variability in nickel-tolerance observations, it was first hypothesized that such variations could be correlated to high real fluctuations of bioavailable nickel content in ultramafic soils, which is assessed as the DTPA-Ni fraction according to Echevarria et al. (2006). Perrier et al. (2006a) reported that the nickel-DTPA concentrations in ultramafic soils varied in a range from 17 to 980 µmol/kg. Assuming that the average nickel-DTPA concentration does not reflect real fluctuations of bioavailable nickel in ultramafic soils, and considering the range of nickel-DTPA concentrations in ultramafic soils reported by Perrier et al. (2006a,b), it is not surprising to find isolates of *P. albus* with high variations in nickel tolerance from the same ultramafic site. Similar variations in metal-tolerant fungal populations in correlation to metal-soil content have already been reported. For instance, in Suilloid fungi, populations displayed zinc tolerance relative to zinc concentrations in polluted soils, suggesting an evolutionary adaptation of fungi to the soil environment (Colpaert et al. 2004). More recently, evidence of adaptation to nickel was provided in isolates of *Cenococcumgeophilum* from ultramafic soils in Portugal and the USA (Gonçalves et al. 2009). No clear relationship between the phenotypic physiological response to nickel and the population genetic differentiation observed within *P. albus* from soils could be established as the nickel-tolerant isolates from ultramafic soils did not cluster in a homogeneous group. It was thus tempting to speculate that the capacity of some *P. albus* to tolerate high nickel concentrations reflects the expression of an adaptive response to high concentrations of bioavailable nickel in soils as suggested for other fungi in response to high heavy metal levels (Hartley et al. 1997, Colpaert et al. 2004, Gonçalves et al. 2009). However, if New Caledonian population of *P. albus* seems to be structured into one ecotype, nickel tolerance alone might not be a sufficient feature to explain such results. Thus, the ultramafic constraint should be considered as a whole, even if each factor (N, P, K contents, Ca/Mg imbalance, heavy metal presence) is studied separately, as suggested by Kazakou et al. (2008).

3.2.2 *Pisolithus albus* transcriptomic response to nickel

In another study on nickel-tolerant *Pisolithus albus* isolated from ultramafic soils in New Caledonia, the comparison of the transcriptomes of a nickel-tolerant isolate in the presence and absence of nickel was monitored by using pyrosequencing and quantitative polymerase chain reaction (qPCR) approaches in order to identify genes involved in the specific molecular response to nickel and to quantify their expression (Majorel et al. 2012). As a result of the experiment, two non-normalized cDNA libraries were obtained from one nickel-tolerant *P. albus* isolate grown in the presence and absence of nickel. A total of 19,518 genes could be obtained through the de novo assembly of the sequence reads from the two non-normalized cDNA libraries. The expression of 30% of these genes was regulated by nickel. Further analysis identified 4,211 genes (21%) that were up-regulated by nickel and 1763 genes (9%) that were down-regulated by nickel. The global statistical distribution of these 19,518 genes is presented on a scatter plot in Fig. 7A. The genes, for which expression was induced most markedly by nickel, encoded products that were putatively involved in a variety of biological functions, such as the modification of cellular components (53%) and the regulation of biological processes (27%) and molecular functions (20%) (Fig. 7B). Compared to most previous studies conducted on ECM samples isolated from soils polluted with heavy metals as a result of human activities (Jacob et al. 2004, Muller et al. 2007, Ruytinx et al. 2011), this study was the first repository of its kind. These results clearly suggested a positive transcriptomic response of the fungus to nickel-rich environments, which may contribute to the tolerance of the fungus to the extreme conditions as found in New Caledonia. The analysis of the results based on gene ontology (GO) analysis and functional genetic tools also suggests the role of these genes as putative adaptive mechanisms of nickel tolerance in *P. albus*. The majority of genes up-regulated by nickel belonged to the GO category 'cellular component'. Information on the annotations of these genes is valuable for the further investigation of gene functions, cellular structures and biological processes that might be involved in the tolerance of fungi to nickel via extracellular and intracellular mechanisms, as suggested by Bellion et al. (2006).

In the second step of the experiment, ten genes that were analyzed as the most nickel-induced in pyrosequencing were characterized by qPCR analysis in both nickel-tolerant and nickel-sensitive *P. albus* isolates from ultramafic soils. Among them, six genes were expressed exclusively in nickel-tolerant isolates as well as in ECM samples *in situ*. In addition, in the nickel-tolerant isolates, the presence of nickel increased their level of expression by between one- and nine-fold (Fig. 8). Their functional classification showed that these genes encoded for putative proteins involved either in chitin cell wall rearrangements as GPI-anchor-like protein and class III chitinase, or biological regulations as vacuolar protein sorting and APC amino acid permease, suggesting a possible role of fungi in metal immobilization and consequently in reducing metal toxicity when in symbiosis with plants. In previous studies involving fungi, many genes involved in the response to stress induced by heavy metals were found to encode proteins that function as metal transporters or metal-binding proteins (Jacob et al. 2004, Bellion et al. 2006, Bolchi et al. 2011). However, in Majorel et al. (2012), most of the genes overexpressed in the presence of nickel did not encode proteins that are generally involved in metal-stress responses. This suggested that the

A) Gene expression level repartition in Ni-tolerant *Pisolithus albus*

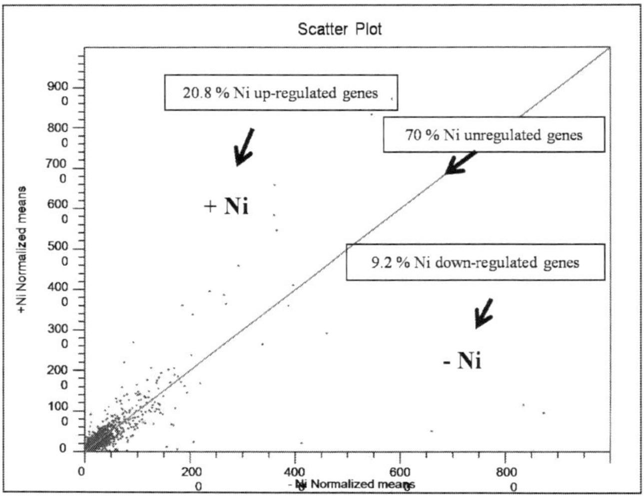

B) Gene functional GO terms assignment and distribution in Ni-tolerant *P. albus*

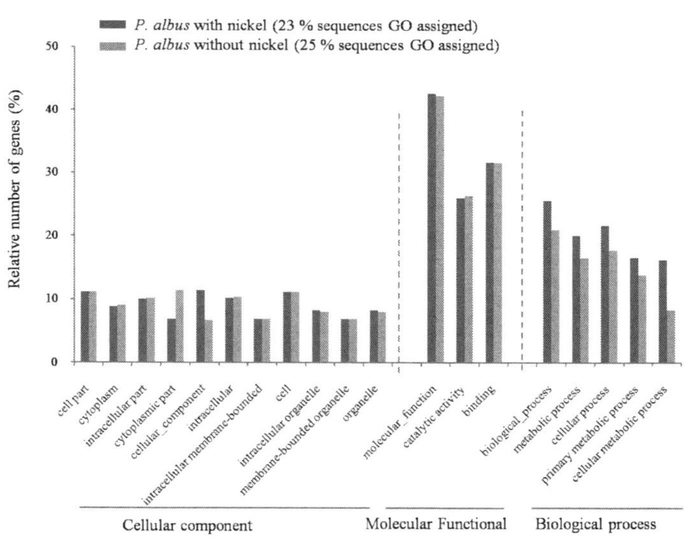

Fig. 7. A) Scatter plot presenting gene expression levels in *Pisolithus albus* Ni-tolerant ecotypefree-living mycelium grown without or with Ni at 250 µM. The expression levels of genes were normalized using a scale of 0 to 10,000. Each circle in the plot represents expression of one gene. B) Functional GO terms assignment and distribution of total sequences of two transcriptomes of Ni-tolerant *P. albus* with (+250 µM) and without nickel, among Gene Ontology (GO) biological process, molecular function and cellular component. From Majorel et al. (2012).

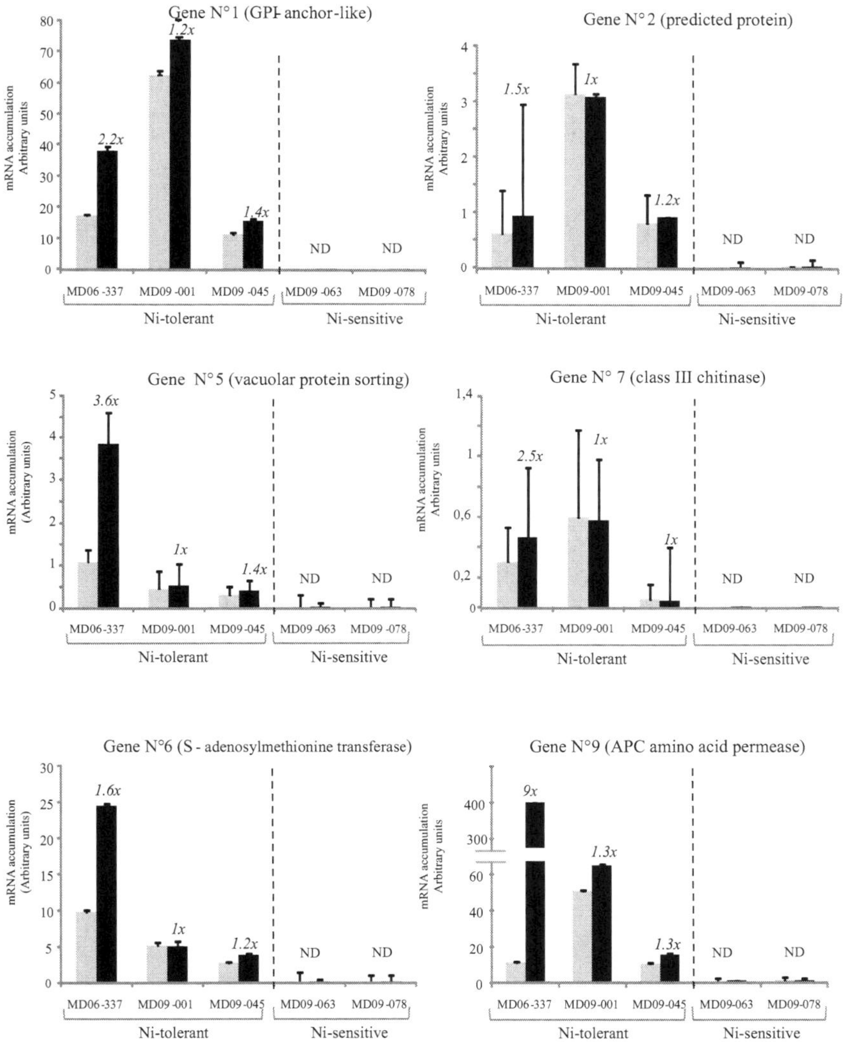

Fig. 8. Comparison of mRNA accumulation profiles for six selected Ni up-regulated genes in five *P. albus* isolates from ultramafic soil in presence of nickel 50 μM (black columns) and in absence of nickel (grey columns). Three nickel-tolerant isolates (MD06-337, MD09-045, and MD07-001) and two nickel-sensitive isolates (MD09-078 and MD09-063) were compared. Transcript accumulation was quantified by qPCR using $2^{-\Delta\Delta CT}$ method with normalization to two reference genes, GAPDH and EF4α, and is expressed as arbitrary units. The data indicate mean values ± S.D. values, calculated from three technical replicates with triplicate biological samples. The fold induction by nickel is presented above the black columns in italics. ND: mRNA non-detected (Ct values >37). From Majorel et al. (2012).

mechanisms that underlie the nickel tolerance in *P. albus* from ultramafic soils might differ from those of other fungi. In particular, that might reflect a long-term adaption to nickel in natural environment, in contrast to short-term adaptation on metal contaminated soils. Among the genes in which expression was remarkably induced in presence of nickel, and exclusively expressed in nickel-tolerant, it was interesting to identify genes that encode chitinase-like and glycosylphosphatidylinositol (GPI) cell-wall structural proteins that are involved in extracellular processes and encode putative cell-wall proteins. Recently, it was suggested that modifications of structural elements of the cell wall, such as the rearrangement of chitin and biosynthesis of glucan- or galactosamine-containing polymers, might play a key role in modulating the integrity of the cell wall and its capacity to immobilize heavy metals. In this way, such modifications could confer tolerance to metals and affect the ability of fungi to survive in stressful environments (Meharg 2003, Bellion et al. 2006, Fuchs and Mylonakis 2009).

Altogether, these results evidenced a strong and specific transcriptomic response to nickel of ultramafic-adapted *P. albus* both *in vitro* and *in situ*. This led the authors to hypothesize that the presence of both nickel-tolerant and nickel-sensitive fungal phenotypes in ultramafic soils might reflect environment-dependent phenotypic responses to variations in the effective concentrations of nickel in heterogeneous ultramafic habitats (Majorel et al. 2012).

3.3 Role of ECM symbiose between nickel-tolerant Pisolithus albus and its host plant Eucalyptus globulus exposed at toxic nickel concentrations

As ECM symbioses are known to play a major role in the fitness of plants in the presence of heavy metals (Jentschke and Godbold 2000), experiments were carried out to analyse the symbiotic interactions between *P. albus* and one of its host plants in the presence of nickel. Ectomycorrhizal *Pisolithus albus* isolated in nickel-rich ultramafic soils from New Caledonia and showing *in vitro* adaptive nickel tolerance were inoculated to *Eucalyptus globulus* Labill used as a *Myrtaceae* plant-host model to study ectomycorrhizal symbiosis. Plants were then exposed to a nickel dose-response experiment with increased nickel treatments up to 60 mg/kg soil as maximum extractable nickel content found in ultramafic soils (Perrier et al. 2006a). Results showed that plants inoculated with ultramafic ECM *P. albus* were able to tolerate high and toxic concentrations of Ni (up to 60 mg/kg) while uninoculated controls were not (Fig. 9). At the highest nickel concentration tested, root growths were more than 20-fold higher and shoot growths more than 30-fold higher in ECM plants compared with

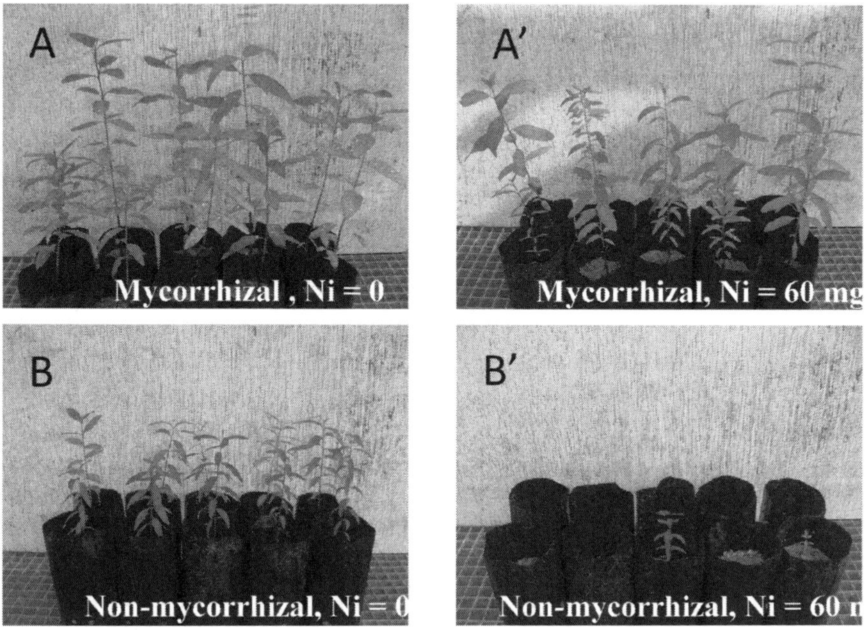

Fig. 9. *Eucalyptus globulus* seedlings after 12-weeks growth. A and A′ mycorrhizal; B and B′: non-mycorrhizal (controls). A and B: no nickel added; A′ and B′ seedlings treated with Ni. From Jourand et al. (2010b).

Color image of this figure appears in the color plate section at the end of the book.

control plants. Ergosterol was also measured in roots as it is a major sterol in fungi and is a good indicator of the level of mycorrhizal colonization of roots (Martin et al. 1990). Without nickel, roots had a mean level of 19.7% ectomycorrhization. At low nickel concentrations (0.6 and 6 mg/kg), the level of root ECM colonization varied from 15.6 to 27.8%. At high and toxic nickel concentrations (30 and 60 mg/kg), the level of root colonization was significantly reduced to around 9%, but confirmed the presence of viable ECM. At the highest nickel concentration tested, the improved growth in ECM plants was also associated witha 2.4-fold reduction in root nickel concentration but a massive 60-fold reduction in transfer of nickel from root to shoots, while for all other major plant nutrient elements analyzed, i.e., N, P, K, Ca and Mg, no significant differences in concentration were noted in either shoots or roots in response to nickel treatments or fungal treatment. To determine whether nickel tolerance was related to the release of metal binding compounds, exudates from roots were analyzed. The two principal chemical components of the exudate solution were non-protein thiols and oxalate. Control plants excreted significantly more thiols and oxalate than plants developing ECM symbiose with *P. albus*, with the increase being more evident at higher nickel concentrations: control plants released 5-fold more thiols at 30 and 60 mg/kg of nickel, and 12- and 8-fold more oxalate at 30 and 60 mg/kg nickel, respectively. All these results confirmed that the nickel tolerance of the ECM has a substantial beneficial effect on the plant host. Ultramafic ECM isolates produced significant increases in growth in both the absence and low concentrations of nickel (from 0.6 to 6 mg/kg). Plant root surface was greatly increased, and the high level of mycorrhizal colonization is consistent with previous data on the interaction between *Pisolithus* and *Eucalyptus* (Martin et al. 1990, Brundrett et al. 1996). At low nickel concentrations, the increase in both shoot and root biomass observed in ECM plants compared with non-inoculated plants is probably a consequence of better mineral nutrition (Marschner 1995, Finlay 2004). However, at toxic levels of nickel the contribution of the ECM symbiosis with ultramafic *P. albus* to host nickel tolerance was more substantial. Interestingly, *P. albus* isolates could withstand *in vitro* high nickel concentrations but accumulated very little nickel in its tissue (Jourand et al. 2010b). The lower nickel uptake by mycorrhizal plants could not be explained by increased release of metal-complexing chelates since these were 5- to 12-fold lower in mycorrhizal plants at high nickel concentrations. It was proposed that the fungal sheath covering the plant roots acts as an effective barrier to limit transfer of nickel from soil into the root tissue.

4. Conclusions

Overall, the observations about ECM diversity found on ultramafic soils in New Caledonia raise very compelling questions about the evolutionary processes involved in fungal diversification in New Caledonia and at a regional scale. The focus on ECM *Pisolithus albus* isolated from soils in New Caledonia highlighted the identification of an ultramafic nickel-tolerant ecotype as reported in Jourand et al. (2010a), showing specific and adaptive molecular response to this metal (Majorel et al. 2012), and having a key role in plant host adaptation to toxic nickel concentrations as found in these soils (Jourand et al. 2010b). Together, these results constitute an important step in evaluating the potential of ECM symbioses for plant adaptation to ultramafic soils containing high concentrations of heavy metals, which is a prerequisite for their use in strategies for ecological restoration of mine sites as suggested by Reddell et al. (1999), Perrier et al. (2006a) and, more recently, Khosla and Reddy (2008). Further characterization of ECM fungal communities in New Caledonia would increase knowledge about fungal diversity and identify fungal species that might be relevant for plant inoculation purpose and their direct implications in restoration strategies.

Acknowledgements

Most of these studies were supported by (i) the GIP CNRT "Nickel and its Environment" [grant number GIPCNRT98] and (ii) the ANR ECCO2005 and BIODIV2007 research programs entitled "Niko" and "Ultrabio" respectively. The authors wish to thank Pr R. Reid (University of Adelaide, South Australia), Dr T. Jaffré (IRD, Nouméa, New Caledonia), S. Santoni (INRA, Montpellier, France), M.E. Soupe, J. Riss, C. Richert for their respective contributions and/orsuggestions. They are also thankful Mr. Pierrick Gailhbaud and Mr Antoine Leveau of Koniambo Nickel Society (KNS), Vavouto, Koné, New Caledonia. The authors thank the anonymous referees for their valuable comments on this study, and Krista L. McGuire and Caitlyn Gillikin for improving the language.

References

Alexander, E., R. Coleman, T. Keeler-Wolfe and S. Harrison. 2007. Serpentine Geoecology of Northern North America. Geology, Soils, and Vegetation. Oxford University Press, New York.

Aggangan, N., B. Dell and N. Malajczuk. 1998. Effects of chromium and nickel on growth of the ectomycorrhizal fungus *Pisolithus* and formation of ectomycorrhizas on *Eucalyptus urophylla* S.T. Blake. Geoderma 84: 15–27.

Altschul, S.F., W. Gish, W. Miller, E.W. Myers and D.J. Lipman. 1990. Basic local alignment search tool. J. Mol. Biol. 215: 403–10.

Amir, H. and M. Ducousso. 2010. Les bactéries et les champignons du sol sur roches ultramafiques. *In*: L. L'huillier, T. Jaffré and A. Wulff (eds.). Mines et environnement en Nouvelle-Calédonie: les milieux sur substrats ultramafiques et leur restauration. IAC, Nouméa, Nouvelle-Calédonie, pp. 129–145.

Anderson, I.C., S.M. Chambers and J.W.G. Cairney. 1998. Use of molecular methods to estimate the size and distribution of mycelial individuals of the ectomycorrhizal basidiomycete *Pisolithus tinctorius*. Mycol. Res. 102: 295–300.

Bellion, M., M. Courbot, C. Jacob, D. Blaudez and M. Chalot. 2006. Extracellular and cellular mechanisms sustaining metal tolerance in ectomycorrhizal fungi. FEMS Microbiol. Let. 254: 173–181.

Blaudez, D., C. Jacob, K. Turnau, J.V. Colpaert, U. Ahonen-Jonnarth, R. Finlay, B. Botton and M. Chalot. 2000. Differential responses of ectomycorrhizal fungi to heavy metals *in vitro*. Mycol. Res. 104: 1366–1371.

Bolchi, A., R. Ruotolo, G. Marchini, E. Vurro, L. Sanità di Toppi, A. Kohler, E. Tisserant, F. Martin and S. Ottonello. 2011. Genome-wide inventory of metal homeostasis-related gene product including a functional phytochelatin synthase in th hypogeous mycorrhizal fungus *Tuber melanosporum*. Fungal Genet. Biol. 48: 573–584.

Bougher, N.L., B.A. Fuhrer and E. Korak. 1994. Taxonomy and biogeography of Australian *Rozites* species mycorrhizal with *Nothofagus* and Myrtaceae. Aust. J. Bot. 7: 353–375.

Brady, K., A. Kruckberg and H. Bradshaw. 2005. Evolutionary ecology of plant adaptation to serpentine soils. Ann. Rev. Ecol. Evol. Syst. 36: 243–266.

Branco, S. 2010. Serpentine soils promote ectomycorrhizal fungal diversity. Mol. Ecol. 19: 5566–5576.

Branco, S. and R.H. Ree. 2010. Serpentine soils do not limit mycorrhizal fungal diversity. PLoS One 5:e11757.

Brearley, F.Q. 2006. Differences in the growth and ectomycorrhizal community of Dryobalanops lanceolata (Dipterocarpaceae) seedlings grown inultramafic and non-ultramafic soils. Soil Biol. Biochem. 38: 3407–3410.

Brooks, R. 1987. Serpentine and its vegetation. Dioscorides Press, Portland.

Brundrett, M., N. Bougher, B. Dell, T. Grove and N. Malajczuk. 1996. Working with mycorrhizas in forestry and agriculture. ACIAR Monogr. 32. Australian Centre for International Agricultural Research, Canberra, Australia, p. 374.

Chiarucci, A. and A.J.M. Baker. 2007. Advances in the ecology ofserpentine soils. Plant Soil 293: 1–217.

Coleman R.G. and C. Jove. 1992. Geological origin of serpentinites. *In*: A.J.M. Baker, J. Proctor and R.D. Reeves (eds.). The Vegetation of Ultramafic (Serpentine) Soils. Proceedings of the First International Conference on Serpentine Ecology. Intercept, Hampshire, UK, pp. 1–17.

Colpaert, J.V., L.A.H. Muller, M. Lambaerts, K. Adriaensen and J. Vangronsveld. 2004. Evolutionary adaptation to Zn toxicity in populations of Suilloid fungi. New Phytol. 162: 549–560.

Colpaert, J.V., J. Wevers, E. Krznaric and K. Adriaensen. 2011. How metal-tolerant ecotypes of ectomycorrhizal fungi protect plants from heavy metal pollution. Annals of Forest Science 68: 17–24.

Dickie, I.A., N. Bolstridge, J.A. Cooper and D.A. Peltzer. 2010. Co-invasion by *Pinus* and its mycorrhizal fungi. New Phytol. 187: 475–484.

Ducousso, M., C. Contesto, M. Cossegal and Y. Prin. 2004. *Cantharellus garnierii* sp. nov., une nouvelle chanterelle des maquis nickélifères de Nouvelle-Calédonie. Cryptogamie Mycol. 25: 135–145.

Ducousso, M., S. Proust, V. Denis and G. Eyssartier. 2009. *Podoserpula miranda* nom prov., une nouvelle espèce de champignon très spectaculaire découverte en Nouvelle-Calédonie. Bois et Forêts des Tropiques 302: 73–75.

Ducousso, M., F. Carriconde, E. Frischt, F. Juillot, L. Hannibal, C. Majorel and Jourand P. 2012. Diversité des champignons ectomycorhiziens d'*Acaciaspirorbis* dans des habitats très contrastés en Nouvelle-Calédonie. Journées Francophones Mycorhizes, troisième édition, 5-7 septembre 2012, Nancy, France.

Echevarria, G., S. Massoura, T. Sterckeman, T. Becquer, C. Schwartz and J.L. Morel. 2006. Assessment and control of the bioavailability of Ni in soils. Environ. Toxicol. Chem. 25: 643–651.

Finlay, R.D. 2004. Mycorrhizal fungi and their multifunctional roles. Mycol. 18: 91–96.

Fuchs, B.B. and E. Mylonakis. 2009. Our paths might cross: the role of the fungal cell wall integrity pathway in stress response and cross talk with other stress response pathways. Eukaryot. Cell 8: 1616–1625.

Garnica, S., M. Weib and F. Oberwinkler. 2003. Morphological and molecular phylogenetic studies inSouth American *Cortinarius* species. Mycol. Res. 107: 1143–1156.

Gonçalves, S.C., A. Portugal, M.T. Goncalves, R. Vieira, M.A. Martins-Loucao and H. Freitas. 2007. Genetic diversity and differential *in vitro* responses to Ni in *Cenococcum geophilum*isolates from serpentine soils in Portugal. Mycorrhiza 17: 677–686.

Gonçalves, S.C., M.A. Martins-Louçao and H. Freitas. 2009. Evidence of adaptative tolerance to nickel in isolates of *Cenococcum geophilum* from serpentine soils. Mycorrhiza 19: 221–230.

Harrington, T.J. and D.T. Mitchell. 2002. Colonisation of root systems of *Carex flacca* and *C. pilulifera* by *Cortinarius* (*Dermocybe*) *cinnamomeus*. Mycol. Res. 106: 452–459.

Harrison, S. and N. Rajakaruna. 2011. Serpentine: the Evolution and Ecology of a Model System. The University of California Press, Berkeley and Los Angeles, CA.

Hartley, J., J.W.G. Cairney and A.A. Meharg. 1997. Do ectomycorrhizal fungi exhibit adaptive tolerance to potentially toxic metals in the environment? Plant Soil 189: 303–319.

Hawksworth, D.L. 1991. The fungal dimension of biodiversity: magnitude, significance, and conservation. Mycol. Res. 95: 641–655.

Hawksworth, D.L. 2001. The magnitude of fungal diversity: the 1.5 million species estimate revisited. Mycol. Res. 105: 1422–1432.

Horak, E. and J. Mouchacca. 1998. Annoted checklist of New Caledonian Basidiomycota. I. Holobasidiomycetes. Mycotaxon 68: 75–129.

Horak, E. and A.E. Wood. 1990. *Cortinarius* Fr. (Agaricales) in Australasia. 1. Subgen. *Myxacium* and subgen. *Paramyxacium*. Sydowia 42: 88–168.

Izzo, A., J. Agbowo and T.D. Bruns. 2005. Detection of plot-level changes in ectomycorrhizal communities across years in a old-growth mixed-conifer forest. New Phytol. 166: 619–630.

Jacob, C., M. Courbot, F. Martin, A. Brun and M. Chalot. 2004. Transcriptomic responses to cadmium in the ectomycorrhizal fungus *Paxillus involutus*. FEBS Letters 576: 423–427.

Jaffre, T. 1992. Floristic and ecological diversity of the vegetation on ultramafic rocks in New Caledonia. *In*: A.J.M. Baker, J. Proctor and R.D. Reeves (eds.). The Vegetation of Ultramafic Soils. Intercept Ltd, Andover, UK, pp. 101–107.

Jaffre, T. and L. L'Huillier. 2010. La végétation des roches ultramafiques ou terrains miniers. *In*: L. L'huillier, T. Jaffré and A. Wulff (eds.). Mines et environnement en Nouvelle-Calédonie: les milieux sur substrats ultramafiques et leur restauration. IAC, Nouméa, Nouvelle-Calédonie, pp. 45–103.

Jaffré, T., P. Morat, J.M. Veillon, F. Rigault and G. Dagostini. 2001.Composition and characterisation of the native flora of New Caledonia. Documents Scientifiques et Techniques—IRD: II, 121 p.

Jentschke, G. and D.L. Godbold. 2000. Metal toxicity and ectomycorrhizas. Physiol. Plant 109: 107–116.

Jourand, P., M. Ducousso, C. Loulergue-Majorel, L. Hannibal, S. Santoni, Y. Prin and M. Lebrun. 2010a. Ultramafic soils from New Caledonia structure *Pisolithus albus* in ecotype. FEMS Microbiol. Ecol. 72: 238–249.

Jourand, P., M. Ducousso, R. Reid, C. Majorel, C. Richert, J. Riss and M. Lebrun. 2010b. Nickel-tolerant ectomycorrhizal *Pisolithus albus* ultramafic ecotype isolated from nickel mines in New Caledonia strongly enhance growth of the host plant *Eucalyptus globulus* at toxic nickel concentrations. Tree Physiol. 30: 1311–1319.

Kazakou, E., P.G. Dimitrakopoulos, R.D. Reeves, A.J.M. Baker and A.Y. Troumbis. 2008. Hypotheses, mechanisms, and trade-offs of tolerance and adaptation to serpentine soils: from species to ecosystem level. Biol. Rev. 83: 495–508.

Khosla, B. and M.S. Reddy. 2008. Response of ectomycorrhizal fungi on the growth and mineral nutrition of eucalyptus seedlings in bauxite mined soil. Amer-Eurasian J. Agri. Environ. Sci. 3: 123–126.

L'Huillier, L., T. Jaffré and A. Wulf. 2010. Mines et environnement en Nouvelle-Calédonie : les milieux sur substrats ultramafiques et leur restauration. IAC Ed, Noumea, New Caledonia.

Majorel, C., L. Hannibal, M.E. Soupe, F. Carriconde, M. Ducousso, M. Lebrun and P. Jourand. 2012. Tracking nickel adaptive biomarkers in *Pisolithus albus* from New Caledonia using a transcriptomic approach. Mol. Ecol. 21: 2208–2223.

Martin, F., C. Delaruelle and J.L. Hilbert. 1990. An improved ergsoterol assay to estimate fungal biomass in ectomycorrhizas. Mycol. Res. 94: 1059–1064.

Martin, F., J. Diez, B. Dell and C. Delaruelle. 2002. Phylogeography of the ectomycorrhizal *Pisolithus* species as inferred from ribosomal DNA ITS sequence. New Phytol. 153: 345–357.

Marschner, H. 1995. Mineral nutrition of higher plants 2nd edn. Academic Press, London, UK, p. 889.

Marx, D.H. 1977. Tree host range and world distribution of the ectomycorrhizal fungus *Pisolithus tinctorius*. Can. J. Microbiol. 23: 217–223.

McCoy, S.G. 1991. Edaphic controls influencing the distribution of *Nothofagus aequilateralis* on ultrabasic soils at the Col the Mouirange, New Caledonia, Australian National University.

Meharg, A.A. 2003. The mechanistic basis of interactions between mycorrhizal associations and toxic metal cations. Mycol. Res. 107: 1253–1265.

Morat, P., T. Jaffré, F. Tronchet, J. Munzinger, Y. Pillon, J.-M. Veillon and M. Chalopin. 2012. The taxonomic database "FLORICAL" and characteristics of the indigenous flora of New Caledonia. Adansonia 34, in press.

Moser, A.M., C.A Petersen, J.A. D'Allura and D. Southworth. 2005. Comparison of ectomycorrhizas of *Quercus garryana* (*Fagaceae*) on serpentine and non serpentine soils in southwestern Oregon. Am. J. Bot. 92: 224–230.

Moser, A.M., J. Frank, J.A. D'Allura and D. Southwood. 2009. Ectomycorrhizal communities of *Quercus garryana* are similar on serpentine and non serpentine soils. Plant Soil 315: 185–194.

Mouchacca, J. 1998. Ascomycetes described from New Caledonia, South Pacific region. Mycotaxon 67: 99–121.

Mouchacca, J. and E. Horak. 1998. Annotated cheklist of New Caledonia Basidiomycota. II. Rusts and Smuts. Mycotaxon 69 : 13–30.

Moyersoen, B., Beever, R.E. and F. Martin 2003. Genetic diversity of *Pisolithus* in New-Zealand indicates multiple long-distance dispersal from Australia. New Phytol. 160: 569–579.

Muller, L.A., J. Vangronsveld and J.V. Colpaert. 2007. Genetic structure of *Suillus luteus* populations in heavy metal polluted and non-polluted habitats. Mol. Ecol. 16: 4728–4737.

Myers, N., R.A. Mittermeier, C.G. Mittermeier, G.A.B. da Fonseca and J. Kent. 2000. Biodiversity hotspots for conservation priorities. Nature 403: 853–858.

Perrier, N. 2005. Bio-géodiversité fonctionelle des sols latéritiques miniers: application à la restauration écologique (Massif du Koniambo, Nouvelle-Calédonie). PhD thesis. University of New Caledonia, New Caledonia.

Perrier, N., H. Amir and F. Colin. 2006a. Occurrence of mycorrhizal symbioses in the metal-rich lateritic soils of the Koniambo Massif, New Caledonia. Mycorrhiza 16: 449–458.

Perrier, N., J.P. Ambrosi, F. Colin and R.J. Gilkes. 2006b. Biogeochemistry of a regolith: the New Caledonian Koniambo ultramafic massif. J. Geochem. Explor. 88: 54–58.

Proctor, J. 2003. Vegetation and soil and plant chemistry on ultramafic rocks in the tropical Far East. Perspect. Plant Ecol. 6: 105–124.

Rajkumar, M., M.N.V. Prasad, H. Freitas and N. Ae. 2009. Biotechnological applications of serpentine soil bacteria for phytoremediation of trace metals. Crit. Rev. Biotechnol. 2: 120–130.

Ray, P., R. Tiwari, G.U. Reddy and A. Adholeya. 2005. Detecting the heavy metal tolerance level in ectomycorrhizal fungi *in vitro*. World J. Microb. Biot. 21: 309–315.

Reddell, P., V. Gordon and M.S. Hopkins. 1999. Ectomycorrhizas in *Eucalyptus tetrodonta* and *E. miniata* forest communities in tropical Northern Australia and their role in the rehabilitation of these forests following mining. Aust. J. Bot. 47: 881–907.

Ruytinx, J., A.R. Craciun, K. Verstraelen, J. Vangronsveld, J.V. Colpaert and N. Verbruggen. 2011. Transcriptome analysis by cDNA-AFLP of *Suillus luteus* Cd-tolerant and Cd-sensitive isolates. Mycorrhiza 21: 145–154.

Schoch, C.L., K.A. Seifert, S. Huhndorf, V. Robert, J.L. Spouge, C.A. Levesque, W. Chen and F.B. Consortium. 2012. Nuclear ribosomal internal transcribed spacer (ITS) region as a universal DNA barcode marker for Fungi. P. Natl. Acad. SCI USA 109: 6241–6246.

Smith, S. and D. Read. 2008. Mycorrhizal Symbiosis, 2nd edn. Academic Press, London.

Smith, M.E., G.W. Douhan and D.M. Rizzo. 2007. Ectomycorrhizal community structure in a xeric *Quercus* woodland based on rDNA sequence analysis of sporocarps and pooled roots. New Phytol. 174: 847–863.

Tam, P.C.F. 1995. Heavy metal tolerance by ectomycorrhizal fungi and metal amelioration by Pisolithus tinctorius. Mycorrhiza 5: 181–187.

Tedersoo, L., U. Kõljalg, N. Hallenberg and K.-H. Larsson. 2003. Fine scale distribution of ectomycorrhizal fungi and roots across substrate layers including coarse woody debris in a mixed forest. New Phytol. 159: 153–165.

Tedersoo, L., T. Jairus, B.M. Horton, K. Abarenkov, T. Suvi, I. Saar and U. Koljalg. 2008. Strong host preference of ectomycorrhizal fungi in a Tasmanian wet sclerophyll forest as revealed by DNA barcoding and taxon specific primers. New Phytol. 180: 479–490.

Urban, A., M. Puschenreiter, J. Strauss and M. Gorfer. 2008. Diversity and structure of ectomycorrhizal and co-associated fungal communities in a serpentine soil. Mycorrhiza 18: 339–354.

Wang, B. and Y.-L. Qiu. 2006. Phylogenetic distribution and evolution of mycorrhizas in land plants. Mycorrhiza 16: 299–363.

Wenzel, W.W. and F. Jockwer. 1999. Accumulation of heavy metals in plants grown on mineralised soils of the Austrian Alps. Environ. Pollut. 104: 145–15.

CHAPTER 7

Diversity and Function of Ectomycorrhiza between *Scleroderma* and *Afzelia* Species in Burkina Faso (West Africa)

Kadidia Bibata Sanon,[1,*] *Amadou Mustapha Bâ*[2] and *Robin Duponnois*[3]

1. Introduction

Most forest trees draw water and nutrients from the soil through fungi associated with their root systems. This symbiotic association, called mycorrhizas, contributes to water and mineral nutrition, and root protection. In return, the fungus receives photosynthetic products necessary for its growth and development (Smith and Read 2008).

[1]Institut de l'Environnement et de Recherches Agricoles, Département Productions Forestières (INERA/DPF), Laboratoire de Microbiologie Forestière, BP 7047 Ouagadougou 03, Burkina Faso.
[2]Laboratoire Commun de Microbiologie (UCAD-IRD-ISRA) BP 1386 Bel Air, Dakar, Sénégal.
[3]Laboratoire Ecologie & Environnement (Unité associée au CNRST, URAC 32). Faculté des Sciences Semalia. Université Cadi Ayyad. Marrakech. Maroc.
*Corresponding author: sbkady@gmail.com

Two main types of mycorrhizal association have been described following their morphological characteristics: the endomycorrhizas and ectomycorrhizas. The endomycorrhizas are widespread and associated with approximately 80% of terrestrial plants (Fortin et al. 2008). They are found in cultivated plants and most tropical trees. Ectomycorrhizas (ECM), are frequent and widespread in forests and woodlands of temperate and boreal regions (Alexander 2006). However, a minority of tropical trees, ecologically and economically important such as Dipterocarpaceae, Fagaceae, Myrtaceae, Phyllanthaceae and Caesalpinioideae form EcM. In tropical Africa, surveys of mycorrhizal symbiosis and its recognition in the technical itineraries of agricultural production and production of reforestation plants remain limited (Bâ et al. 2011, Bâ et al. 2012). In Burkina Faso, the first studies on mycorrhizal symbiosis dates back to 1989 with *Afzelia africana* Sm and *Pisolithus tinctorius* (Pers.) Coker. & Couch. (Sary 1989). A few years later, works have been carried out on both types of symbiosis. The studies on endomycorrhizal symbiosis include among others: (i) The diversity of Glomales associated with some Australian *Acacia* introduced in Burkina Faso (Bâ et al. 1996a); (ii) The effect of inoculation with mycorrhizal fungal strains on growth, nutrition and water stress resistance of some wild or domesticated fruit-tree species in tropical Africa [*Ziziphus mauritiana* Lam., *Balanites aegyptiaca* (L.) Delile, *Parkia biglobosa* (Jacq.) R. Br. (Guissou et al. 1998, Guissou et al. 2001, Bâ et al. 2001)]; (iii) The response to inoculation of Shea tree (*Vitellaria paradoxa* Kotchy) and *Faidherbia albida* (Delile) A. Chev. with or without fertilizers or rock phosphate (Bâ and Guissou 1996, Bâ et al. 1996b, Dianda et al. 2009).

Burkina Faso is a dry tropical country with 400 mm rainfall in the north to 1100 mm in the south-west. EcM species are very scarce, confined to the south and south-western woodlands and gallery forests. Pioneering EcM investigations (Sanon et al. 1997, Sanon 1999) addressed trees species and putative fungal partner. These works have allowed isolating and making pure cultures of *Scleroderma* species. *Scleroderma* species appear to be good candidates for controlled mycorrhization due to their relatively easy cultivation and their early stage, and have been the subject of detailed studies (Sanon et al. 1997, Bâ et al. 1999, 2002, Sanon et al. 2002, Sanon et al. 2009a, b).

2. Partners of Ectomycorrhizal Symbiosis in Burkina Faso

2.1 Some features of the climate and vegetation in Burkina Faso

Burkina Faso is a landlocked country located in the heart of West Africa between latitudes N 009° 20′ and 15° 05′, longitudes E 002° 03′ to W 5° 20′. It is a relatively flat country especially in its northern part and covers 274,000 km^2.

The climate is characterized by two strongly contrasted seasons: a dry season and a rainy season. The dry season varies from 8 to 9 months in the North and 5–6 months in the Southwestern part. The rainy season extends from May to October in the South, Southwest and June-July to September in the north. Precipitations are irregularly distributed in space and time. The average annual rainfall decreases from Southwest (1100 mm) to the north (less than 400 mm) (Fig. 1). Rainfall is concentrated during the single rainy season with a maximum in August. Since the drought of 1973–1974, the original isohyets lost between 100 and 200 mm (Kessler and Geerling 1994).

Based on rainfall distribution, the country is divided into three climatic regions (Laclavère 1993) (Fig. 1). The Sudanian zone which occupies the entire south of the country with about 6 months of rainfall (>900 mm), the Sudano-Sahelian zone, located in the center, accounts for half of the country

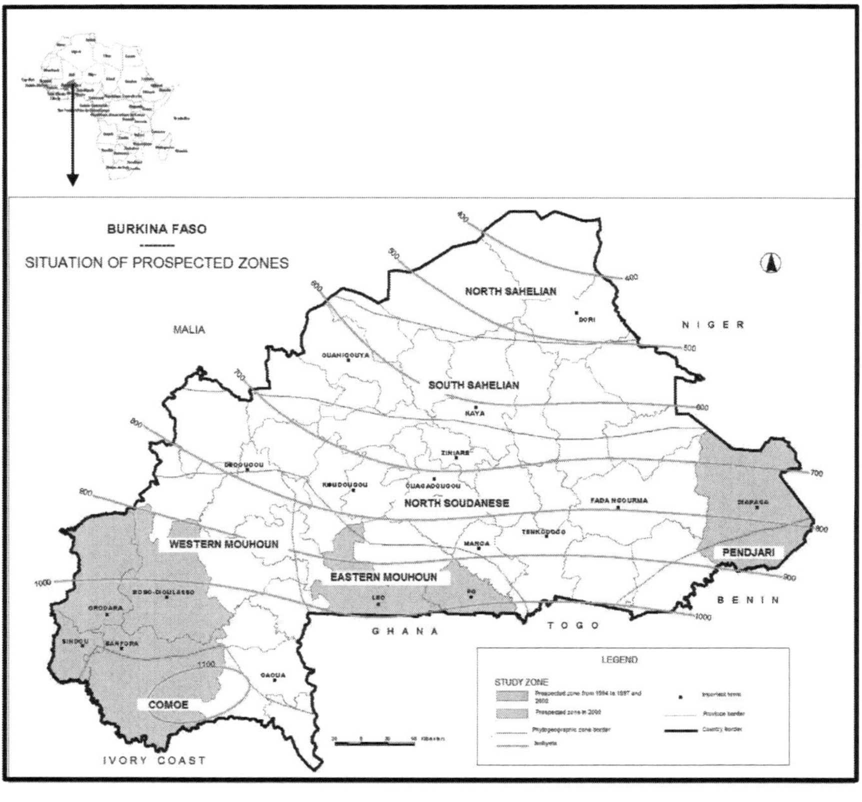

Fig. 1. Climatic regions of Burkina Faso and localization of prospected areas. Data from Sanon et al. (2009a).

Color image of this figure appears in the color plate section at the end of the book.

with 4–5 months of rainfall between 900 and 600 mm, the Sahelian zone in the north with about 2–3 months of rain and an annual rainfall less than 600 mm. According to Boussim (2010), two phytogeograpical zones can be distinguished: the Soudanian zone extending from N 009°20 to N 13° making 2/3 of the land area, and the Sahelian zone (N 13° to 15°) that represent one third of the country. The vegetation is characterized by the predominance of mixed woody and herbaceous formations patchily distributed mostly in the northern part (steppes, savannas, woodlands). The natural vegetation covers 60% of the country (FAO 1987). Gallery forests (1%) and woodlands (1%) occur in the southern part. Tiger bush (1%) and steppe (4%) are common in the North, whilst wooded savannas and shrublands (53%) can be found in the rest the country.

Ectomycorrhizal-rich vegetations are located in the wetter southern areas. They are mostly Caesalpinioid- and Phyllanthioid-dominated woodlands of the Sudanian zone in the Southwestern, Southern and Eastern parts of the country (Fig. 1). In many such woodlands, *Anogeissus leiocarpus* (DC.) Guill. & Perr. dominates with other species such as *Pterocarpus erinaceus* Poir., *Burkea africana* Hook., *Afzelia africana* Sm., *Albizia chevalieri* Harms. and mainly *Isoberlinia doka* Craib. & Stapf., *I. tomentosa* (Harms.) Craib. & Stapf. and *Detarium microcarpum* Guill. & Perr. (Thombiano et al. 2012). The presence of *Isoberlinia doka* as monospecific planting or associated with *Isoberlinia tomentosa* also characterizes this area. Gallery forests, mainly those of Kou and Mouhoun Reserve Forests are exclusively dominated by EcM tree species such as *Berlinia grandiflora* (Vahl) Hutch. & Dalziel.

2.2 Ectomycorrhizal trees species with their myco-symbionts

Different tropical plant families known to form EcM symbiosis have been reported (Ducousso et al. 2008, Bâ et al. 2011, Bechem and Alexander 2012). Compared to these data, an inventory was first performed in the field in the Sudanian zone. It allowed identifying seven ectomycorrhizal species sorted into three families or subfamily and five genera (Table 1).

Assessment of mycorrhizal status of these species was performed on the roots of young plants, mature trees and natural regeneration. Arbuscular mycorrhizal were not observed on roots of *Isoberlinia* spp. Some studies have reported the presence of arbuscular mycorrhizal fungi on *A. africana* roots (Thoen and Ducousso 1989), however we did not find this type of symbiosis in our samples. For *B. grandiflora*, the presence or not of AM has not been evaluated. Sites containing these EcM species were identified mainly in the south-west and to a lesser extent in the south and east of the country (Fig. 1). These sites can be grouped as shown in Table 2.

Sites contained *U. guineensis* and *B. grandiflora* (3 sites in Southwest), and the site of *A. africana* in Southwest are located in gallery forests. Apart

Table 1. Ectomycorrhizal tree species so far in Burkina Faso.

Family or Subfamily	Genera	Species	Symbiotic statut[a]
Caesalpinioideae	*Afzelia*	*A.africana*	ECM
	Berlinia	*B. grandiflora*	ECM
	Isoberlinia	*I. doka*	ECM
	Isoberlinia	*I. tomentosa*	ECM
Dipterocarpaceae	*Monotes*	*M. kerstingii*	ECM & AM
Phyllanthaceae	*Uapaca*	*U. guineensis*	ECM & AM
	Uapaca	*U. somon*	ECM & AM

Sanon et al. (1997). [a] ECM : Ectomycorrhiza ; MA ; Arbuscular Mycorrhiza.

Table 2. The sites prospected with the dominant host plants.

Regions	Sites number	Dominant host plants
South-west	3	*U. guineensis* and *B. grandiflora*
	2	*Isoberlinia* spp.
	1	*A. africana*
	1	*I. dalziellii* and *M. kerstingii*
	1	*Isoberlinia* spp. and *M. kerstingii*
	1	*M. kerstingii*
	1	*U. somon* and *Isoberlinia* spp.
South	2	*I. doka*
	1	*A. africana*
	1	*I. doka* and *A. africana*
East	2	*I. doka*
	1	*A. africana*
	1	*I. doka* and *A. africana*

from the site of *Isoberlinia* spp. and *M. kerstingii* in Southwest which were identified in 2000 and explored, all other sites were surveyed in 1994, 1995, 2000, 2005 and 2006. Sites in South and East were surveyed in 2000 and fungi frequently encountered were *Scleroderma* species. Except gallery forests, the other trees species are found in woodlands and savannas. Surveys were conducted during the rainy season, the period of fungi fruiting, particularly in June, July and August. All putative ECM fungi fruitbodies near these trees were harvested. They were cleaned, photographed and described morphologically. They were transported to the laboratory for isolation in pure culture and/or dried in an oven at 50°C and stored in herbarium of the Laboratory of Microbiology INERA/DPF.

The sampled specimens are sorted into 18 genera (Fig. 2) and 8 orders (Aphyllophorales, Agaricales, Boletales, Cantharellales, Gautierales, Hymenogastrales, Russulales and Sclerodermatales) and identified according to the herbarium of the Microbiology Laboratory of IRD Senegal

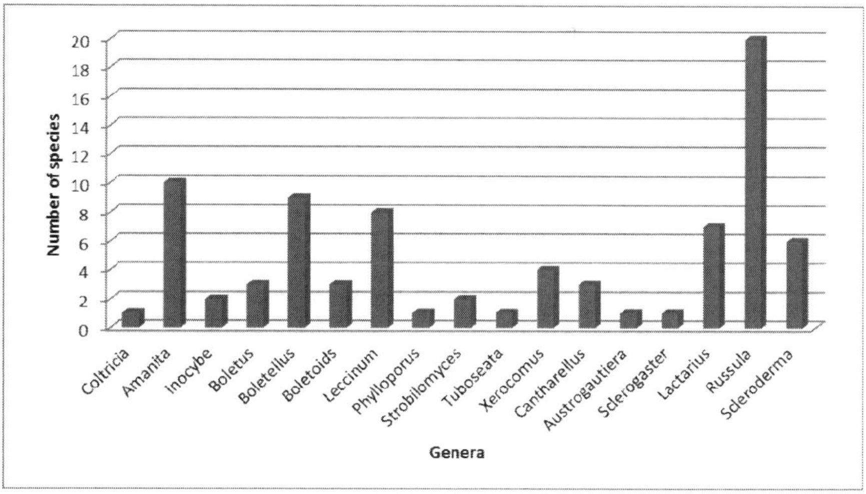

Fig. 2. Genera of ectomycorrhizal fungi identified by their morphology. (From Sanon et al. 1997, Sanon 1999, Sanon et al. 2012, in press).

(formerly ORSTOM), Ducousso (personal communication) and Verbeken (personal communication). The most represented genera are *Russula*, *Amanita*, *Boletelus*, *Leccinum* and *Scleroderma*.

From the above mentioned EcM taxa, isolation and pure culture were made from some *Scleroderma* species as highlighted in Sanon et al. (1997) and Sanon (1999). The mycorrhizal status of these isolates was confirmed *in vitro* with certain host plants (*Afzelia africana*, *A. quanzensis* Welw., *Isoberlina doka*, *I. tomentosa*, *Brachystegia speciformis*). An analysis of the genetic diversity of different species of *Scleroderma* and testing for controlled mycorrhization in nurseries were then carried out (Bâ et al. 1999, 2002, Sanon et al. 2009a).

3. *Scleroderma* species

Scleroderma belongs to Gasteromycete and to the Sclerodermataceae family. This genus is widespread in tropical and temperate zone and is a symbiont of many families of temperate and tropical plants of economic importance such as Pinaceae, Myrtaceae, Fagaceae, Dipterocarpaceae, Gnetaceae and subfamily Caesalpinioideae (Munyanziza and Kuyper 1995, Sims et al. 1997, Bechem and Alexander 2012). The genus *Scleroderma* encompasses more than 25 accepted species (Sims et al. 1995, Sims et al. 1997, Guzman et al. 2004). However, many species remain unidentified especially in tropical Africa and Asia (Sanon et al. 1997, David and David 1998, Sims et al. 1999). A revised key to the species of the genus *Scleroderma* was developed by

Sims et al. (1995). This key is based primarily on the characteristics of the peridium and spores. Thus, three morphotypes or sections are accepted with the genus *Scleroderma* depending on the morphology of their spores: Section *Aculeatispora* with spiny or warty spores; section *Sclerangium* with sub-reticulate spores and the section *Scleroderma* with spores totally covered with reticulum (Sims et al. 1995).

3.1 Morphological diversity of *Scleroderma*

Scleroderma sporocarps occur most frequently at the beginning of the rainy season (June-early July) and are almost absent when the rains are well established and abundant. The fruiting bodies collected from 1994 to 1997 and 2000 have been described morphologically. They are sub-spherical or spherical and contracted to the base to form a pseudo-stem structure (Fig. 3). The outer surface of the peridium can be smooth or scaly. Based on colour and ornamentation of spores and peridium, six morphological species have been identified belonging to the sections *Aculeatispora* and *Scleroderma*, and a new group not yet described in the literature (Table 3, Fig. 4).

Fig. 3. Sporocarps of *S. verrucosum*.

3.4 Genetic Diversity of *Scleroderma*

Inter- and intraspecies differences in the ability to stimulate host plant growth were identified among ectomycorrhizal fungi in different studies performed in the laboratory, nursery and in the field (Malajczuk et al. 1990, Lei et al. 1990, Le Tacon et al. 1992, Thomson et al. 1994, Bâ et al.

Table 3. Characteristics of sporocarps and basidiospores of *Scleroderma* species.

Species	**Sporocarps color**	**Spores diameter (µm)**	**Spores ornamentation**	**Sections**
Scleroderma dictyosporum Pat.	Brown-black	7–9	reticulated	*Scleroderma*
Scleroderma verrucosum Pers.	Brown	5–9	warty 1–2 µm	*Aculeatispora*
Scleroderma sp1	Pinkish-brown	9–13	Spiny 2–3 µm	*Aculeatispora*
Scleroderma sp2	Brown	5–7	-	New
Scleroderma sp3	Brown	5–10	Very small spines	*Aculeatispora*
Scleroderma sp4	Yellowish	7–9	-	New

Sanon et al. (2009a).

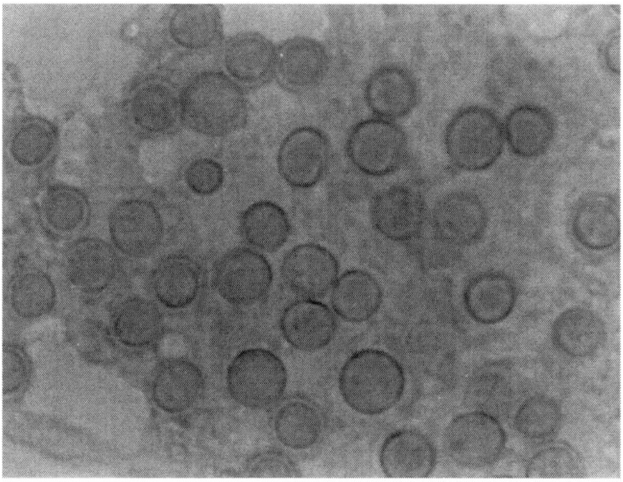

Fig. 4. Spores of *Scleroderma* sp2 without ornamentations (G. 100x, photonic microscope).

1999, 2002, Diédhiou et al. 2005). To attest whether the increased growth of host plants resulted from inoculated symbionts, it was essential to assess the persistence of these symbionts. In the absence of fructification, study of the persistence of mycosymbionts implied that it was possible to identify them from the mycorrhizas formed *in situ*. However, tracing the identity of a fungal species from the EcM root tips was quite tricky, due to poorly variable morphological and structural characteristics (Agerer 1987, Gardes et al. 1991, Sanon 1999, Sanon et al. 2002). To overcome these difficulties, molecular biology techniques are increasingly used to enable the identification of ectomycorrhizal fungi from fruiting bodies, cultured mycelia and mycorrhizas (Gardes et al. 1991, Liang et al. 2004, Matsushita et al. 2005, Ruiz-Diez et al. 2006, Sica et al. 2007). Also, although the morphological characteristics allowed for the identification of a relatively large number of species of the genus *Scleroderma*, molecular tools provided more detail in the identification of ectomycorrhizal fungi (Hansen et al. 2002, Rivière et al. 2007, Tedersoo et al. 2007). Various studies have shown that *in vitro* DNA amplification (PCR) followed by the analysis of Polymorphism Restriction Fragment Length (RFLP), or sequencing the internal transcribed spacer (ITS) and inter-genic spacer (IGS) are among the most important tools for analyzing inter and intra-specific fungal symbionts (Liang et al. 2004, Matsushita et al. 2005, Ruiz-Diez et al. 2006, Sica et al. 2007). The ribosomal DNA spacers (ITS and IGS) are known to be inter-and intra-specific variable and used as markers to distinguish multiple species or isolates of ectomycorrhizal fungi (Karen et al. 1997, Peter et al. 2001, Horton 2002, Gomes et al. 2002).

To assess the inter- and intraspecific diversity of *Scleroderma* harvested in Burkina Faso, the ITS and IGS1 regions were amplified and digested with two restriction enzymes *HinfI* and *MboI*. DNA was extracted from isolates and fruiting bodies using the methods described by Grube et al. (1995) and Martin et al. (1997) or the DNeasy kit according to the manufacturer's instructions (Qiagen, France). The universal primers and ITS1/ITS4 and CNL12/5SA were used to amplify ITS and IGS1, respectively. The amplification conditions, digestion and electrophoresis are described by Sanon et al. (2009a).

Among the six morphological species identified (Table 3), the analysis of length polymorphism restriction fragments revealed eight ribotypes noted, A, B, C, D, E, F, G and H with 1-3 ribotypes within each morphological species (Table 4). Two ribotypes were identified within the morphotype *S. dictyosporum* (A, B), three for *S. verrucosum* (C, D, E), one for *Scleroderma* sp1 (F) and two for *Scleroderma* sp2 (G, H). Some ribotypes were more represented than others; this was the case of ribotype A of *S. dictyosporum*, C of *S. verrucosum* and H of *Scleroderma* sp2. However, no relation was established between the species origin (host plant) and ribotypes observed. RFLP profiles of *Scleroderma* sp3 and *Scleroderma* sp4 were identical to those of ribotype D of *S. verrucosum* and A of *S. dictyosporum* respectively (Sanon et al. 2009a). This suggested that these two species are rather ecotypes of *S. verrucosum* and *S. dictyosporum*, respectively. Thus, the six morphospecies identified should be grouped into four species: *S. dictyosporum*, *S. verrucosum*, *Scleroderma* sp1 and *Scleroderma* sp2.

Except ribotype G of *Scleroderma* sp2, ITS of at least two ribotypes of each species have been sequenced. ITS sequences were obtained from 9 fruiting bodies and 16 isolates in culture. They were submitted to GenBank database and similar sequences were identified using the algorithm Blastn.

Table 4. Size of restriction fragments (in base pairs) of the ITS and IGS1 regions digested with MboI and HinfI. The fruiting bodies are noted "SD, SV, SP1, SP2, SP3 or SP4" and isolates "IR". Fragments less than 50 bp are not shown on the table. Data from Sanon et al. (2009a).

Species	ITS		IGS1		Ribotypes
	HinfI	*MboI*	*HinfI*	*MboI*	
S. dictyosporum					
(4IR et 4SD)	174, 127, 113, 87, 60	208, 159, 135, 119	143, 135, 128	228, 153, 86	A
(1IR et 1SD)	287, 240	218, 198, 167	160, 151, 82	280, 240	B
S. verrucosum					
(5IR et 10SV)	311, 246, 93	171, 162, 153, 119	182, 167, 119	238, 151, 54	C
(2IR)	285, 146, 110, 66	254, 146, 124, 98, 77	182, 147, 63	228, 178	D
(2SV)	264, 114, 104, 94	208, 159, 135, 92	182, 139, 119	238, 210	E
Scleroderma sp1					
(all 4 samples)	322, 149, 110, 66	241, 153, 140, 119, 102	286, 143	228, 196, 54	F
Scleroderma sp2					
(1SP2 et 2IR)	331, 294, 87	284, 227, 91	232, 143, 88	228, 153	G
(10SP2 et 2IR)	206, 186, 99	152, 137, 113, 106	173, 139, 124	231, 209	H

The sequences of *Scleroderma* from Burkina showed low similarity with other *Scleroderma* in GenBank (89% maximum similarity, Sanon et al. 2009a). Seven genotypes were identified for all 25 samples analyzed and sequences were deposited in GenBank (Table 5).

The phylogenetic tree (Neighbor-joining method, MEGA software) (Kumar et al. 2004) from these sequences can distinguish three different branches (Fig. 5). Branch 1 summarizes the morphological species *S. verrucosum* and *Scleroderma* sp1, branch 2 summarizes the morphotypes of *S. dictyosporum* and the third branch includes the morphotype *Scleroderma* sp2. For each branch, the ribotypes were supported by high bootstrap values, suggesting that the ribotypes identified within each morphological species can be considered as different phylogenetic species as for *Pisolithus* (Martin et al. 2002). Thus, samples identified as *S. verrucosum* include three species, those identified as *S. dictyosporum* include two species and one species sequenced for *Scleroderma* sp2 (Fig. 5). *Scleroderma* sp1 is closer to *S. verrucosum* than any other species, from which it differs by a few deletions and insertions at ITS2 (sequence alignment matrix, data not shown), thus belonging to the *S. verrucosum* morphotype (species 4). Both morphological species *Scleroderma* sp3 (IR252) and *Scleroderma* sp4 (IR408) were supported by high bootstrap values (100%) with species 2 of *S. verrucosum* and *S. dictyosporum* morphotypes, respectively (Fig. 5). This confirms the results obtained by PCR/RFLP and shows that *Scleroderma* sp3 (IR252) and *Scleroderma* sp4 (IR408) are respectively morphotypes of *S. verrucosum* and *S. dictyosporum*. *Scleroderma* sp2 was genetically distant

Table 5. Sporocarps and isolates of *Scleroderma* sequenced with their GenBank accession numbers. Isolates are noted "IR" and sporocarps "SD, SV, SP1, SP2, SP3 or SP4". Data from Sanon et al. (2009a).

Species/Ribotypes/Voucher number	GenBank accession number
S. dictyosporum	
Ribotype A	
IR109	FJ840442
IR250	FJ840444
IR412	FJ84044
IR602	FJ840447
IR408	FJ840445
ORS7731	FJ840448
Ribotype B	
IR215	FJ840443
SD-4901	FJ840449
S. verrucosum	
Ribotype C	
IR256	FJ840454
IR114	FJ840451
IR110	FJ840450
IR118	FJ840452
SV1803	FJ840458
SV2802	FJ840460
SV5602	FJ840461
Ribotype D	
IR500	FJ840455
IR600	FJ840456
IR252	FJ840453
Ribotype E	
SV1801	FJ840457
SV1804	FJ840459
Scleroderma sp1	
Ribotype F	
IR406	FJ840462
IR410	FJ840463
Scleroderma sp2	
Ribotype H	
SSP2-1806	FJ840464
SSP2-2803	FJ840465
SSP2-3901	FJ840466

from other species analyzed (52% bootstrap value) but was grouped with the genus *Scleroderma*. The morphology of spores of this species suggested that it might be a new species specific to Africa. However, further studies are needed on a larger number of samples and also the sequencing of the second ribotype of this species.

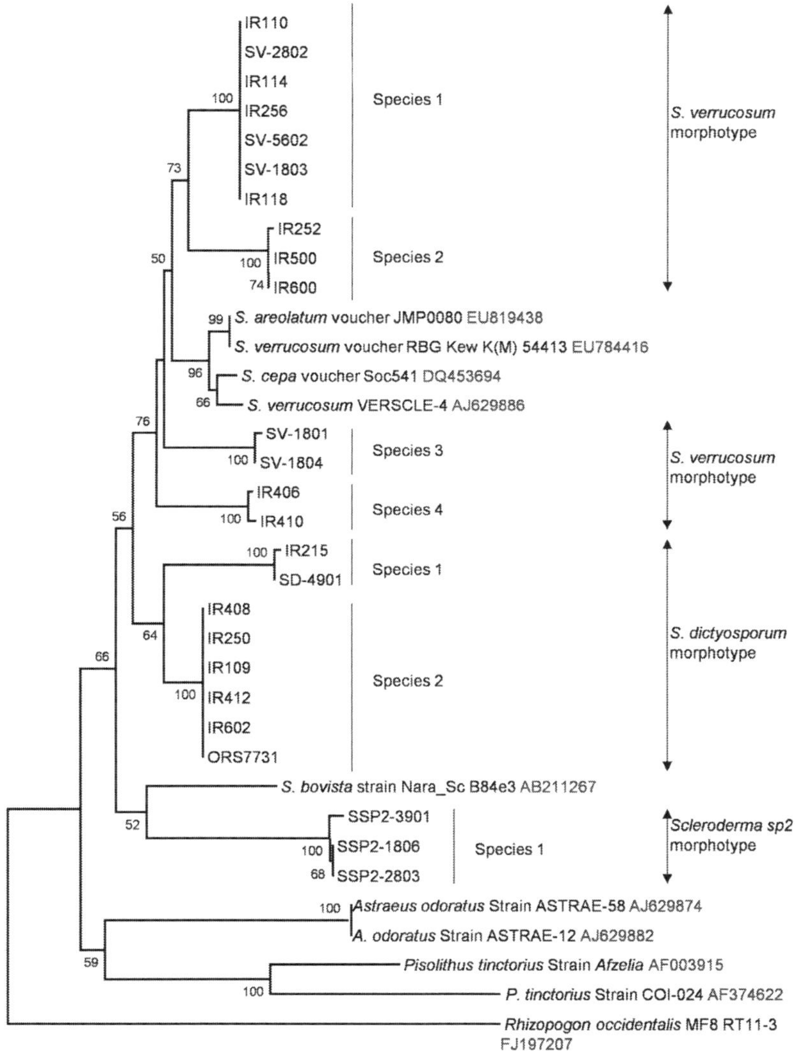

Fig. 5. Neighbor joining phylogenetic tree of *Scleroderma* species based on ITS sequences. *Rhizopogon occidentalis* is used as outgroup. Numerical values on the branches are the bootstrap values of 1000 replications. Data from Sanon et al. (2009a).

4. Controlled Mycorrhization of *Afzelia africana*

The inoculation of a host plant with a well-defined ectomycorrhizal fungus strain under controlled conditions and the evaluation of the effect of the fungus on host growth is called "controlled mycorrhization." It allows the selection of fungi compatible and efficient for the production of seedlings for reforestation. The criteria for the selection of the best strains include the ability to form mycorrhizas with seedlings, the beneficial effect on growth of inoculated plants and the mineral content (especially phosphorus) of various organs of the plants. Also, the length of the extramatrical network developed in the soil by the mycosymbionts, or the proportion of metabolically active hyphae in the soil, associated with other parameters, have been proposed by some authors to characterize the efficacy of the fungus (Jones et al. 1990, Hamel et al. 1990). This technique is widely used in temperate regions (France, Spain, USA, Canada) (Le Tacon et al. 1992). However, there are few studies on tropical trees, especially in Africa. Nevertheless, in recent years, there is a renewed interest in the study of controlled mycorrhization of some local species. In Burkina Faso, works on controlled mycorrhization were carried out mainly on the genus *Scleroderma* and the host plant *A. africana*, one of the most important ectomycorrhizal species in Burkina Faso, and *A. quanzensis*, a species of East Africa (Bâ et al. 1999, 2002).

4.1 Inoculum production

The solid mycelial inoculum was used in two experiments on *A. africana* and *A. quanzensis* in the nursery. The inoculum was produced on a solid substrate composed of peat and vermiculite according to the method described by Duponnois and Garbaye (1991).

4.2 Fungal inoculation

Controlled mycorrhization of *A. africana* and *A. quanzensis* was carried out on a sandy soil from Sangalkam (Senegal) which was free of ectomycorrhizal propagules and displayed the following physico-chemical characteristics: clay (5.4%), silt (5.8%), sand (88.8%), pH in water (6.03), 0.39% total C, total N (0.027%), total MO (0.68%), C/N 14, P total (41.92%), available P, Bray 1 (2.15%). The substrate was sieved (2 mm) and mixed with the inoculum (10: 1, v/v). The mixture was dispensed into polyethylene bags of 2 liters and watered with tap water. Two germinated seeds were planted at 2 seeds per bag. Ten days after planting, the number of seedlings was reduced to one per bag.

For *A. africana*, two provenances (Nazinga, Burkina Faso (AaBF) and Diatock, Senegal (AaSN)) were inoculated with two isolates of *S. dictyosporum* (IR109 and IR408), one isolate of *S. verrucosum* (IR406) and one isolate of *Thelephora* sp (ORSXM002) (Bâ and Thoen 1990, Sanon 1999, Diédhiou et al. 2004) or not inoculated. The experiment was factorial (2 provenances x (4 fungal strains + 1 control). Each treatment was replicated 10 times and the treatments were arranged into randomized blocks in a screen house under the climatic conditions of Dakar (average daily temperature 25–30° C, photoperiod approximately 12 h). Watering was done once a day with tap water. The experiment duration was-4 months (Bâ et al. 1999).

For *A. quanzensis*, a species from East Africa, one provenance was inoculated with *S. dictyosporum* (IR408) and *Thelephora* sp. (ORSXM002). The experiment design was single randomization with 3 treatments and 10 replicates per treatment. The experimental conditions were the same as above and the duration of the test was also 4 months (Bâ et al. 2002).

4.3 Growth and nutrition of two provenances of A. africana

The two provenances of *A. africana* differed in their response to inoculation with the four fungal strains. Mycorrhization rate varied in roughly the same proportions for the two provenances, 32% to 69% for AaBF and 22% to 64% for AaSN (Table 6). For the Burkina Faso provenance (AaBF), the highest rates of mycorrhization were obtained with *S. dictyosporum*, IR109 (69%) and IR408 (64%). *Thelephora* sp., ORSXM002 formed mycorrhiza with AaBF with medium rate of 46%. *S. verrucosum* IR406 has a lower rate of 32%. Isolates IR408 and IR406 gave the lowest rates of mycorrhization (22% and 32%) with the provenance AaSN (Bâ et al. 1999).

Growth (total dry weight) of AaSN was not significantly improved by inoculation despite mycorrhization rate of 58% and 64% with IR109 and ORSXM002 respectively, while that of AaBF was improved by inoculation with IR408 and ORSXM002 (Table 6). The increased growth observed for this provenance could be attributed to the development of the root system as reflected by the total biomass of inoculated plants. These results confirmed those obtained by Bâ (1990) on the positive effect of inoculation with ORSXM002 on growth of *A. africana* (from Bayottes) due to an increase in root biomass. They were also consistent with those obtained by Diédhiou et al. (2005) from a Senegalese *A. africana* inoculated for 7 months in the nursery with different mycorrhizal fungi with IR408 and ORSXM002.

The mycorrhizal dependency of the two provenances of *A. africana* was relatively low 0.15% to 32% compared to the values obtained with tropical species introduced in West Africa such as the hybrid *Eucalyptus urophylla* x *E. kirtonia* and *Acacia mangium* (Garbaye et al. 1988, Duponnois and Bâ 1999). However, the comparison was difficult because of different culture conditions. A mycorrhizal dependency of at most 30% was also obtained with another provenance of *A. africana* inoculated with different isolates of *Scleroderma*, *Pisolithus* and *Thelephora* after 7 months of culture (Diédhiou et al. 2005).

Table 6. Effect of inoculation with different isolates of ectomycorrhizal fungi on growth parameters and nutrition of two provenances of *A. africana* seedlings. MD, Mycorrhizal dependency. Data from Bâ et al. (1999).

Provenances of *A. africana* and fungal isolates	**Height (cm)**	**Roots dry weight (g)**	**Total dry weight (g)**	**Colonization (%)**	**MD (%)**	**N (%)**	**P (%)**	**K (%)**
AaBF								
IR109	41.97 b	2.83 abc	7.81 ab	69.00 a	21.25 ab	1.96 d	0.11 bc	1.41 cde
IR406	44.96 ab	2.53 bc	7.73 ab	32.00 abc	20.43 ab	2.33 ab	0.09 c	2.12 a
IR408	47.78 ab	3.29 ab	9.00 a	64.00 a	31.66 a	1.94 d	0.14 a	1.66 bc
ORSXM002	42.75 b	3.98 a	9.10 a	46.00 ab	32.24 a	1.85 d	0.10 bc	1.45 cd
Control	38.98 b	1.76 c	6.15 b	0.00 c	-	2.02 cd	0.10 bc	1.02 e
AaSN								
IR109	46.85 ab	2.07 bc	6.34 b	58.00 ab	0.15 b	2.03 cd	0.13 ab	1.55 c
IR406	47.80 ab	2.71 abc	6.95 ab	38.00 ab	8.92 ab	2.28 abc	0.10 bc	2.05 ab
IR408	52.40 a	2.17 bc	7.03 ab	22.00 bc	9.95 ab	2.54 abc	0.11 bc	2.11 a
ORSXM002	52.78 a	2.92 abc	8.18 ab	64.00a	22.61 ab	2.08 bcd	0.12 abc	1.80 abc
Control	52.60 a	1;73 c	6.33 b	0.00 c	-	1.90 d	0.10 bc	1.13 de

Values followed by the same letter within columns are not significantly different using ANOVA (SAS, Bonferroni test, $P<0.05$).

One major effect of mycorrhization is generally to improve the phosphorus nutrition of host plants. Here, however, when comparing the control treatments and those inoculated, the increased P, N, Mg and Ca contents of plants inoculated with some strains was not related to the stimulation of plant growth in terms of biomass produced, except for P content of AaBF inoculated with IR408 (Table 6). Only the leaf K concentration was higher in inoculated plants compared to controls but unrelated to the stimulation of plant growth (Bâ et al. 1999). This suggested that these minerals were not limiting factors for the growth of these two *A. africana* provenances. Thus, the high rate of mycorrhization do not necessarily translate into a high concentration of minerals in leaves of inoculated plants, there was no clear link between the level of colonization of inoculated plants and the potential benefit for plants. In conclusion, two fungal species, *S. dictyosporum* IR408 and *Thelephora* sp. ORSXM002 seemed to be the most suitable species to improve seedling growth of *A. africana* in nursery.

4.4 Growth and nutrition of A. quanzensis

Based on the results of the previous study, the two promising isolates identified were tested on *A. quanzensis* (Bâ et al. 2002). *A. quanzensis* responded differently to inoculation depending on the isolate. *Thelephora* sp. ORSXM002 efficiently colonized the root system of inoculated plants and stimulated root biomass compared to *S. dictyosporum* IR408 (Table 7). The mycorrhization rate obtained with ORSXM002 (66.40%) was similar to that obtained with the Senegalese *A. africana* provenance (64%) (Bâ et al. 1999). However, for IR408, this rate was very low (19.40%). Stimulation of root biomass by ORSXM002 was not reflected in the total biomass of plants. Conversely, this strain had stimulated the total biomass of *A. africana* in the same conditions. Therefore, mycorrhizal dependency of *A. quanzensis* inoculated with these two species does not exceed 16%, which was quite

Table 7. Effect of inoculation with IR408 and ORSXM002 on growth parameters and nutrition of *A. quanzensis* seedlings. MD, Mycorrhizal dependency. Data from Bâ et al. (2002).

A. quanzensis/ Isolates inoculated	Height (cm)	Shoot dry weight (g)	Root dry weight (g)	Colonization (%)	MD (%)	N (%)	P (%)	K (%)
IR408	69.86 b	8.84 a	2.26 a	19.40 b	3.50 a	2.09 a	0.13 a	1.63 b
ORSXM002	68.22 ab	9.14 a	3.60 b	66.40 a	16.00 a	2.14 a	0.14 a	1.56 b
Control	64.28 a	8.80 a	1.91 a	0.00 c	-	2.15 a	0.12 a	0.81 a

Values followed by the same letter within columns are not significantly different using ANOVA (SAS, Bonferroni test, P<0.05).

low compared to *A. africana* (Bâ et al. 1999). These data suggested that the different origin of strains and host plant may influence the response to inoculation with *A. quanzensis* compared to *A. africana*. As in the case of *A. africana*, only the K content of the shoot was greater in plants inoculated with both strains (Table 7).

5. Conclusion

This work allowed the identification of ectomycorrhizal fungi associated with seven host plants in Southwest, South and East of Burkina Faso (*A. africana, B. grandiflora, I. doka, I. dalzeillii, M. kerstingii, U. guineensis, U. somon*). All harvested fungi belonged to genera already identified as ectomycorrhizal and some were identified at the species level. An unexpected diversity of ectomycorrhizal fungi (about 78 morphological species) could be detected despite the geo-climatic conditions of the regions surveyed and phenological fluctuation of the fungi from one year to another in low humidity areas.

Our inventory revealed the dominance of species of the genera *Russula, Boletelus, Leccinum, Amanita and Scleroderma*. For the latter, ITS sequencing of ribotypes showed that they belonged to at least seven phylogenetic species: 4 species for *S. verrucosum* morphotype, 2 for *S. dictyosporum* and a single ribotype sequenced for *Scleroderma* sp2. The particular morphology of spores of *Scleroderma* sp2 and its position on the phylogenetic tree compared with other species of *Scleroderma* suggested a new species unique to Africa, especially in West Africa.

We tested the effectiveness of four fungal isolates, *S. dictyosporum*, IR109 and IR408, *S. verrucosum*, IR406 and *Thelephora* sp, ORSXM002, on the growth and mineral nutrition on two provenances of *A. africana* in the nursery. We have shown that the inoculation induced a wide range of plant growth stimulation (0.17% to 48% of biomass increase) and improved mineral nutrition, especially K. The use of two provenances of *A. africana* has highlighted the strong influence of the genetic background of the host plant on the beneficial effect of mycorrhization. Mycorrhizal dependency indicated that the Senegalese origin was less dependent on mycorrhiza for growth than that of Burkina Faso. The use of fungal strains effective to improve *A. africana* plant growth in controlled mycorrhization programs can be oriented towards *S. dictyosporum* IR408 and *Thelephora* sp. ORSXM002 for both tree provenances. These two isolates tested on *A. quanzensis* have shown a limited significant effect compared to *A. africana* suggesting that the origin of the host plant in relation to the fungal isolates tested may influence the response to inoculation.

However, these strains should also be tested under other soil conditions because our experiment was conducted in a sandy soil unrepresentative of *A. africana* forests in West Africa. Any selection of strain must therefore be supplemented by tests under various conditions and respecting local conditions of plantations.

Acknowledgements

We thank the anonymous referees for their valuable comments on this study, Krista McGuire and Jean Garbaye for improving the language.

References

Agerer, R. 1987–1996. Color Atlas of ectomycorrhizae with glossary. Pp??? *In*: R. Agerer (ed.). Einhorn-Verlag Eduard Dietenberger.

Alexander, I.J. 2006. Ectomycorrhizas—out of Africa? New Phytol. 172: 589–591.

Bâ, A.M. 1990. Contribution à l'étude de la symbiose ectomycorhizienne chez deux essences forestières d'Afrique intertropicale: *Afzelia africana* Sm. et *Uapaca guineensis* Müll. Arg. Thèse de l'Université des Sciences et Techniques du Languedoc, Montpellier, pp. 193.

Bâ, A.M. and D. Thoen. 1990. First syntheses of ectomycorrhizas between *Afzelia africana* Sm. (Caesalpinioideae) and native fungi from West Africa. New Phytol. 103: 441–448.

Bâ, A.M. and T. Guissou. 1996. Rock phosphate and vesicular-arbuscular mycorrhiza effects on growth and nutrient uptake of *Acacia albida* (Del.) seedlings in an alkaline sandy soil. Agrofor. Syst. 34: 129–137.

Bâ, A.M., Y. Dalpé and T. Guissou. 1996a. Les Glomales d'*Acacia holosericea* Cunn. ex G. Don. et d'*Acacia mangium* Willd.: diversité et abondance relative des champignons endomycorhiziens à arbuscules dans deux types de sols de plantations au Burkina Faso. Bois For. Trop. 250: 5–18.

Bâ, A.M., M. Bazié and T. Guissou. 1996b. Effet du phosphate naturel sur de jeunes *Acacia albida* Del. en présence ou non de mycorhizes. Cahiers Scientifiques du CIRAD 12: 237–244.

Bâ, A.M., B.K. Sanon, R. Duponnois and J. Dexheimer. 1999. Growth responses of *Afzelia africana* Sm. seedlings to ectomycorrhizal inoculation in nutrient-deficient soil. Mycorrhiza 9: 91–95.

Bâ, A.M., T. Guissou, R. Duponnois, C. Plenchette, O. Sacko, D. Sidibé, K. Sylla and B. Windou. 2001. Mycorhization contrôlée et fertilisation phosphatée: applications à la domestication du jujubier (*Ziziphus mauritiana* Lam.). Fruits 56: 261–269.

Bâ, A.M., B.K. Sanon and R. Duponnois. 2002. Influence of ectomycorrhizal inoculation on *Afzelia quanzensis* Welw. seedlings in a nutrient-deficient soil. For. Ecol. Manag. 161: 215–219.

Bâ, A.M., R. Duponnois, M. Diabaté and B. Dreyfus. 2011. Les champignons ectomycorhiziens des arbres forestiers en Afrique de l'Ouest: méthodes d'étude, diversité, écologie, utilisation en foresterie et comestibilité. Editions IRD pp. 264.

Bâ, A.M., R. Duponnois, B. Moyersoen and A. Diédhiou. 2012. Ectomycorrhizal symbiosis of tropical African trees. Mycorrhiza 22: 1–29.

Bechem, E.E. and I.J. Alexander. 2012. Mycorrhizal status of *Gnetum africanum* in Cameroon: Evaluating diversity with a view to ameliorating domestication efforts. Mycorrhiza 22: 99–108.

Boussim J.I. 2010. Les territoires phytogéographiques. *In:* A. Thiombiano and D. Kampmann (eds.). Atlas de la Biodiversité de l'Afrique de l´Ouest, Tome II : Burkina Faso. BIOTA, Ouagadougou/Frankfurt am Main, pp. 152–155.

David, J.F. and M.S. David. 1998. Variation in the ribosomal DNA internal transcribed spacer of a diverse collection of ectomycorrhizal fungi. Mycol. Res. 102: 859–865.

Dianda, M., J. Bayala, T. Diop and S.J. Ouedraogo. 2009. Improving growth of shea butter tree (*Vitellaria paradoxa* C.F. Gaertn.) seedlings using mineral N, P and arbuscular mycorrhizal (AM) fungi. Biotechnol. Agron. Soc. Environ. 13: 93–102.

Diédhiou, A.G., A.M. Bâ, S.N. Sylla, B. Dreyfus, M. Neyra and I. Ndoye. 2004. The early-stage ectomycorrhizal Telephoroid fungal sp. is competitive and effective on *Afzelia africana* Sm. in nursery conditions in Senegal. Mycorrhiza 14: 313–322.

Diédhiou, A.G., O. Guèye, M. Diabaté, Y. Prin, R. Duponnois, B. Dreyfus and A.M. Bâ. 2005. Contrasting responses to ectomycorrhizal inoculation in seedlings of six tropical African tree species. Mycorrhiza 16: 11–17.

Ducousso, M., H. Ramanankierana, R. Duponnois, R. Rabévohitra, L. Randrihasipara, M. Vincelette, B. Dreyfus and Y. Prin. 2008. Mycorrhizal status of native trees and shrubs from eastern Madagascar littoral forests with special emphasis on one new ectomycorrhizal endemic family, the Asteropeiaceae. New Phytol. 178: 233–238.

Duponnois, R. and J. Garbaye. 1991. Techniques for controlled synthesis of the Douglas-fir-*Laccaria laccata* ectomycorrhizal symbiosis. Ann. Sci. For. 48: 641–650.

Duponnois, R. and A.M. Bâ. 1999. Growth stimulation of *Acacia mangium* Willd. by *Pisolithus* sp. in some Senegalese soils. For. Ecol. Manag. 119: 209–215.

FAO. 1987. Etude sur la contribution du secteur forestier à l'économie du Burkina Faso. Rapport de synthèse du projet préparé par le gouvernement du Burkina Faso et la FAO. Rome.

Fortin, J.A., C. Plenchette and Y. Piché. 2008. Les mycorhizes: La nouvelle révolution verte. Edition Multimondes. Québec, Canada, pp. 131.

Garbaye, J., J.C. Delwaulle and D. Diangana. 1988. Growth response of eucalyptus in the Congo to ectomycorrhizal inoculation. For. Ecol. Manag. 24: 151–157.

Gardes, M., T.J. White, J.A. Fortin, T.D. Bruns and J.W. Taylor. 1991. Identification of indigenous and introduced symbiotic fungi in ectomycorrhizae by amplification of nuclear and mitochondrial ribosomal DNA. Can. J. Bot. 69: 180–190.

Gomes, E.A., M.C.M. Kasuya, E.G. de Barros, A.C. Borges and E.F. Araujo. 2002. Polymorphism in the internal transcribed spacer (ITS) of the ribosomal DNA of 26 isolates of ectomycorrhizal fungi. Genetics and Mol. Biol. 25: 477–483.

Grube, M., P.T. Depriest, A. Gargas and J. Hafellner. 1995. DNA isolation from lichen ascomata. Mycol. Res. 99: 1321–1324.

Guissou T., A.M. Bâ, J.M. Ouadba, S. Guinko and R. Duponnois. 1998. Responses of *Parkia biglobosa* (Jacq.) Benth., *Tamarindus indica* L. and *Ziziphus mauritiana* Lam. to arbuscular mycorrhizal fungi in a phosphorus-deficient sandy soil. Biol. Fert. Soils 26: 194–198.

Guissou, T., A.M. Bâ, C. Plenchette, S. Guinko and R. Duponnois. 2001. Effets des mycorhizes à arbuscules sur la tolérance à un stress hydrique chez quatre arbres fruitiers: *Balanites aegyptiaca* (L.) Del., *Parkia biglobosa* (Jacq.) Benth., *Tamarindus indica* L. et *Ziziphus mauritiana* Lam. Sécheresse 12: 121–127.

Guzman, G., F. Ramirez-Guillém, O.K. Miller and D.J. Lodge. 2004. *Scleroderma stellatum* versus *Scleroderma bermudense*: the status of *Scleroderma echinatum* and the first record of *Veligaster nitidum* from the Virgin Islands. Mycologia 96: 1370–1379.

Hamel, C., H. Fyles and D.L. Smith. 1990. Measurement of development of endomycorrhizal mycelium using three different vital strains. New Phytol. 115: 297–302.

Hansen, K., T. Laessoe and D. Pfister. 2002. Phylogenetic diversity in the core group of *Peziza* inferred from ITS sequences and morphology. Mycol. Res. 106: 879–902.

Horton, T. 2002. Molecular approaches to ectomycorrhizal diversity studies: variation in ITS at a local scale. Plant Soil 244: 29–39.

Jones, M.D., D.M. Durall and P.B. Tinker. 1990. Phosphorus relationships and production of extramatrical hyphae by two types of willow ectomycorrhizas at different soil phosphorus levels. New Phytol. 115: 259–267.

Karen, O., N. Högberg, A. Dahlberg, L. Jonsson and J.E. Nylund. 1997. Inter- and intraspecific variation in the ITS region of rDNA of ectomycorrhizal fungi in *Fennoscandia* as detected by endonuclease analysis. New Phytol. 136: 313–325.

Kessler, J.J. and C. Geerling. 1994. Profil environnemental du Burkina Faso. Université Agronomique, Département de l'Aménagement de la Nature. Wageningen, les Pays Bas, pp. 63.

Kumar, S., K. Tamura and M. Nei. 2004. MEGA3: Integrated software for Molecular Evolutionary Genetics Analysis and Sequence Alignment. Briefings in Bioinform. 5: 150–163.

Laclavère, G. 1993. Atlas du Burkina Faso. Jeune Afrique, pp. 53.

Le Tacon, F., I.F. Alvarez, D. Bouchard, B. Henrion, R.M. Jackson, S. Luff, J.I. Parlade, J. Pera, E. Stenström, N. Villeneuve and C. Walker. 1992. Variations in field response of forest trees to nursery ectomycorrihzal inoculation in Europe. *In*: D.J. Read, D.H. Lewis, A.H. Fitter and I.J. Alexander (eds.). Mycorrhizas in Ecosystems pp. 119–134.

Lei, J., F. Lapeyrie, N. Malajczuk and J. Dexheimer. 1990. Infectivity of pine and eucalypt isolates of *Pisolithus tinctorius* (Pers.) Coker & Couch on roots of *Eucalyptus urophylla* S. T. Blake *in vitro*. II. Ultrastructural and biochemical changes at the early stage of mycorrhiza formation. New Phytol. 116: 115–122.

Liang, Y., L.D. Guo LD and K.P. Ma. 2004. Genetic structure of a population of the ectomycorrhizal fungus *Russula vinosa* in subtropical woodlands in southwest China. Mycorrhiza 14: 235–240.

Malajczuk, N., F. Lapeyrie and J. Garbaye. 1990. Infectivity of pine and eucalypt isolates of *Pisolithus tinctorius* on roots of *Eucalyptus urophylla in vitro*. 1. Mycorrhiza formation in model systems. New Phytol. 114: 627–631.

Martin, F., G. Costa, C. Delaruelle and J. Diez. 1997. Genomic fingerprinting of ectomycorrhizal fungi by microsatellite-primed PCR. *In*: A. Varma and B. Hock (eds.). Mycorrhiza Manual. Springer Lab Manuel Berlin, Springer-Verlag, pp. 463–474.

Martin, F., J. Diez, B. Dell and C. Delaruelle. 2002. Phylogeography of the ectomycorrhizal *Pisolithus* species as inferred from the nuclear ribosomal DNA ITS sequences. New Phytol. 153: 345–358.

Matsushita, N., K. Kikuchi, Y. Sasaki, A. Guerin-Laguette, F. Lapeyrie, L.M. Vaario, M. Intini and K. Suzuki. 2005. Genetic relationship of *Tricholoma matsutake* and *T. nauseosum* from the Northern Hemisphere based on analyses of ribosomal DNA spacer regions. Mycoscience 46: 90–96.

Munyanziza, E. and T.W. Kuyper. 1995. Ectomycorrhizal synthesis on seedlings of *Afzelia quanzensis* Welw. using various types of inoculum. Mycorrhiza 5: 283–287.

Peter, M., U. Büchler, F. Ayer and S. Egli. 2001. Ectomycorrhizas and molecular phylogeny of the hypogeous russuloid fungus *Arcangeliella borziana*. Mycol. Res. 105: 1231–1238.

Rivière, T., A.G. Diedhiou, M. Diabaté, G. Senthilarasu, K. Natarajan, M. Ducousso, A. Verbeken, B. Buyck, B. Dreyfus, G. Bena and A.M. Bâ. 2007. Diversity of ectomycorrhizal Basidiomycetes in West African and Indian tropical rain forests. Mycorrhiza 17: 415–428.

Ruiz-Diez, B., A.M. Rincon, M.R. de Felipe and M. Fernandez-Pascual. 2006. Molecular characterisation and evaluation of mycorrhizal capacity of *Suillus* isolates from Central Spain for the selection of fungal inoculants. Mycorrhiza 16: 465–474.

Sanon, K.B., A.M. Bâ and J. Dexheimer. 1997. Mycorrhizal status of some fungi fruiting beneath indigenous trees in Burkina Faso. For. Ecol. Manag. 98: 61–69.

Sanon, B.K. 1999. La symbiose ectomycorhizienne chez quelques Césalpiniacées et Euphorbiacées des forêts du Sud-Ouest du Burkina Faso. Etude morphologique et cytologique, mycorhization contrôlée et étude de la diversité inter- et intraspécifique de Sclérodermes ectomycorhiziens. Thèse de Doctorat de l'Université Henri Poincaré Nancy I pp. 119.

Sanon, B.K., J. Dexheimer, A.M. Bâ, M. Dianda and J. Gerard. 2002. Structure comparée des ectomycorhizes de *Afzelia africana* Sm. et *Scleroderma* sp. Science et Technique, série Sci. Nat. Agro. 1: 17–28.

Sanon, B.K., A.M. Bâ, C. Delaruelle, R. Duponnois and F. Martin. 2009a. Morphological and molecular analyses in *Scleroderma* species associated with some Caesalpinioid legumes, Dipterocarpaceae and Phyllanthaceae trees in southern Burkina Faso. Mycorrhiza 19: 571–584.

Sanon, B.K., M. Dianda, T. Guissou and A.M. Bâ. 2009b. Description des champignons ectomycorhiziens du genre *Scleroderma* de quelques formations forestières du Sud du Burkina Faso. Cameroon J. Experiment. Biol. 2: 69–78.

Sanon, B.K., R. Duponnois, A.M. Bâ, M. Doucousso and B. Dreyfus. 2012. Ectomycorrhizal fungi diversity from south-west Burkina Faso using part of mitochondrial rDNA sequence analysis. In press.

Sary, H. 1989. Essai de mycorrhization de *Afzelia africana* avec *Pisolithus tinctorius* (Pers.) Coker et Couch. *In*: Trees for development in Sub-Saharan Africa, Proceedings of a regional seminar held by the International Foundation for Science (IFS), ICRAF House, Nairobi, Kenya.

Sica, M., L. Gaudio and S. Aceto. 2007. Genetic structure of *Tuber mesentericum* Vitt. Based on polymorphism at the ribosomal DNA ITS. Mycorrhiza 17: 405–414.

Sims, K.P., R. Watling and P. Jeffries. 1995. A revised key to the genus *Scleroderma*. Mycotaxon: 403–420.

Sims, K.P., R. Watling, R. De La Cruz and P. Jeffries. 1997. Ectomycorrhizal fungi of the Philippines: a preliminary survey and notes on the geographic biodiversity of the Sclerodermatales. Biod. Conserv. 6: 45–58.

Sims, K.P., R. Sen, R. Watling and P. Jeffries. 1999. Species and population structures of *Pisolithus* and *Scleroderma* identified by combined phenotypic and genomic marker analysis. Mycol. Res. 103: 449–458.

Smith, S.E. and D.J. Read. 2008. Mycorrhizal symbiosis. Third edition pp. 800.

Tedersoo, L., T. Suvi, K. Beaver and U. Kõljalg. 2007. Ectomycorrhizal fungi of the Seychelles: diversity patterns and host shifts from the native *Vateriopsis seychellarum* (Dipterocarpaceae) and *Intsia bijuga* (Caesalpiniaceae) to the introduced *Eucalyptus robusta* (Myrtaceae), but not *Pinus caribea* (Pinaceae). New Phytol. 175: 321–333.

Thombiano, A., M. Schmidt, S. Dressler, A. Ouédraogo, K. Hahnand and G. Ziska. 2012. Catalogue des plantes vasculaires du Burkina Faso. Conservatoire et Jardin Botanique de la Ville de Genève pp. 396.

Thoen, D. and M. Ducousso. 1989. Champignons et ectomycorhizes du Fouta Djalon. Bois For. Trop. 221: 45–63.

Thomson, B.D., T.S. Grove, N. Malajczuk and G.E. St. J. Hardy. 1994. The effectiveness of ectomycorrhizal fungi in increasing the growth of *Eucalyotus globulus* Labill. in relation to root colonisation and hyphal development in soil. New Phytol. 126: 517–524.

CHAPTER 8

The Physiology of *Scleroderma sinnamariense* Mont. (Sclerodermaceae), an Ectomycorrhizal Fungus Associated with *Gnetum* spp. (Gnetaceae)

Eneke Esoeyang Tambe Bechem

1. Introduction

Scleroderma sinnamariense Mont. is a bright yellow ectomycorrhizal fungus with a pantropical distribution (Guzman and Ovrebo 2000) which was originally described in French Guiana. This fungus is a gasteromycete and produces macroscopic sporocarps called "puffballs" (Fig. 1). These yellowish leathery, rounded structures become dry and crack at maturity, releasing dry powdery basidiospores which are dispersed by wind. The basidiomes are not eaten in Cameroon by humans but it is possible that they are eaten by some animals including small mammals, birds and insects. They are, however, eaten by humans in Asia. *S. sinnamariense* is commonly recorded

Department of Botany and Plant Physiology, Faculty of Science, University of Buea, P. O. Box 63 Buea, South West Region, Republic of Cameroon.
Email: tamenekeso@yahoo.co.uk

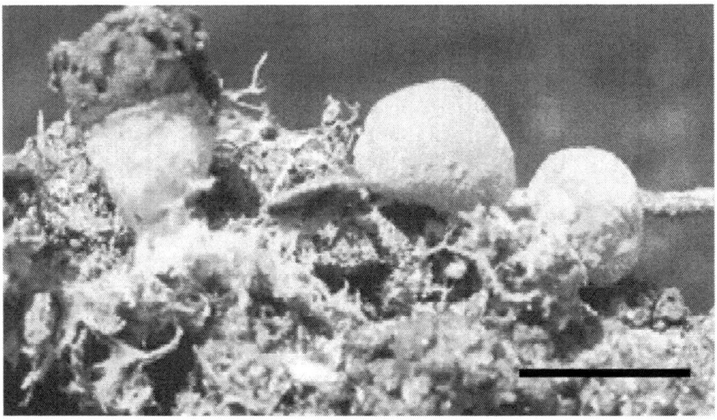

Fig. 1. Sporocarps of *S. sinnamariense*. Bar is 2.5 mm.
Color image of this figure appears in the color plate section at the end of the book.

in SE Asia (Sims et al. 1995) where it is reported to form ectomycorrhizas with members of the Dipterocarpaceae and Gnetaceae (Watling 1994, Cadiz and Florido 2001). It has also been recorded in Cameroon (Bechem and Alexander 2009, 2012a, Onguene and Kuyper 2001, Ingleby 1999) where it forms ectomycorrhiza with *Gnetum* spp. The ectomycorrhizal status of this fungus is proven, but unlike other species of the genus *Scleroderma*, such as *S. citrinum* Pers., *S. aurantium* Viall.: Pers., *S. vulgare* Horn., which have also been shown to occur as saprotrophs on rotting wood (Jeffries 1999), *S. sinnamariense* has been demonstrated to form typical ECM (Bechem and Alexander 2009). There are no records to show that this species is capable of a free-living saprotrophic existence. Richter and Bruhn (1989) suggested that the genus was wholly mycorrhizal and not partly saprotrophic. This suggestion was based on data collected from enzyme tests.

The difficulty encountered with the growth of this fungus *in vitro* as well as its relatively slow growth rate are some of the reasons why few studies exist on its physiology as well as on its fungus-host interaction.

2. Taxonomy and Ecology

S. sinnamariense is a variable species. It belongs to the Sclerodermaceae which also contains *Pisolithus*, a genus that has been widely studied. This species is believed to have been described using several different names such as *S. aureum* Massee from New Guinea, *S. chrysastrum* Martin from Panama, *S. luteum* Lloyd from Malaya and *S. pantherinum* Matt. from Africa (Guzman 1970). Demoulin and Dring (1971) described two related species, *S. congolense* and *S. schmitzii* from the Congo basin in Africa which also have a yellow endoperidium. However, their spores and basidiomata are smaller than those of *S. sinnamariense*. A molecular phylogenetic analysis (Fig. 2) carried out by Bechem and Alexander (2012a), placed a sequence from *S. sinnamariense* associated with *Gnetum* spp. in Cameroon in the same group as that obtained from a basidiome collected beneath *G. gnemon* in Malaysia. Another group included a sequence from *S. sinnamariense* associated with *Gnetum* spp. as well as that obtained from a culture isolated from a basidiome growing with *Hopea odorata* in Malaysia. From the said phylogenetic analysis (Fig. 2), it was concluded that *S. sinnamariense* forming EM with *Gnetum* spp. in Cameroon was not distinct and therefore was the same as that forming EM with dipterocarps and *Gnetum* in Asia. This conclusion was however based on a single "non-*Gnetum*' sequence.

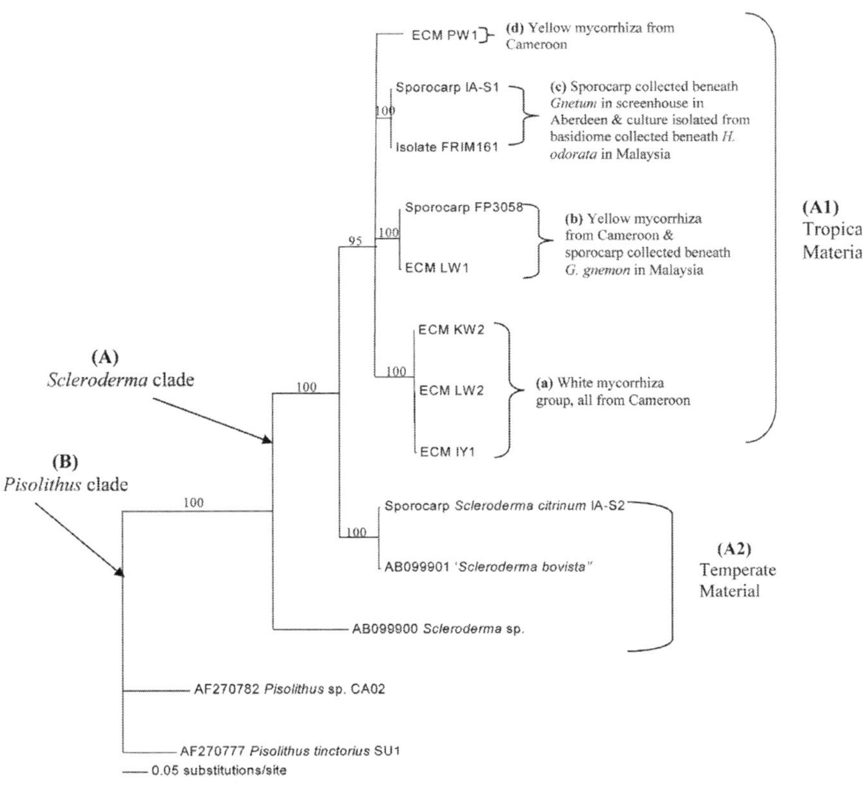

Fig. 2. Molecular phylogenetic tree-diagram of single linkage analysis of RFLP patterns from digestion of the ITS region of rDNA amplified from EM root tips from *Gnetum*, mycelium, and EM fruiting bodies. Numbers at nodes indicate bootstrap indices over 50% obtained after 1,000 replicates. Data from Bechem and Alexander (2012a).

These authors also observed variations within the ITS sequences obtained from yellow mycorrhizas and yellow basidiomes, and even between yellow mycorrhizas from different provenances in Cameroon. *S. sinnamariense* isolated from *Gnetum* spp. in Cameroon could not form EM with *Hopea odorata* from Malaysia (Bechem and Alexander 2009), in spite of the fact that they were not quite distinct based on molecular studies.

Generally, *Scleroderma* has a wide and inconsistent distribution of species. Tropical species such as *S. sinnamariense* are able to withstand high temperatures and can grow at 30°C in axenic cultures. This species has been recorded principally in forest habitats in association with *Gnetum* spp. (Bechem and Alexander 2009, 2012a, Cadiz and Florido 2001), *Shorea* sp. and *Hopea* sp. (Lee et al. 1997). This fungus like most basidiomycetes produces extensive extramatrical mycelia (Fig. 3), which may aggregate into rhizomorphs (Bechem and Alexander 2012a), used in channeling nutrients and water to and from the host (Hasselquist et al. 2010).

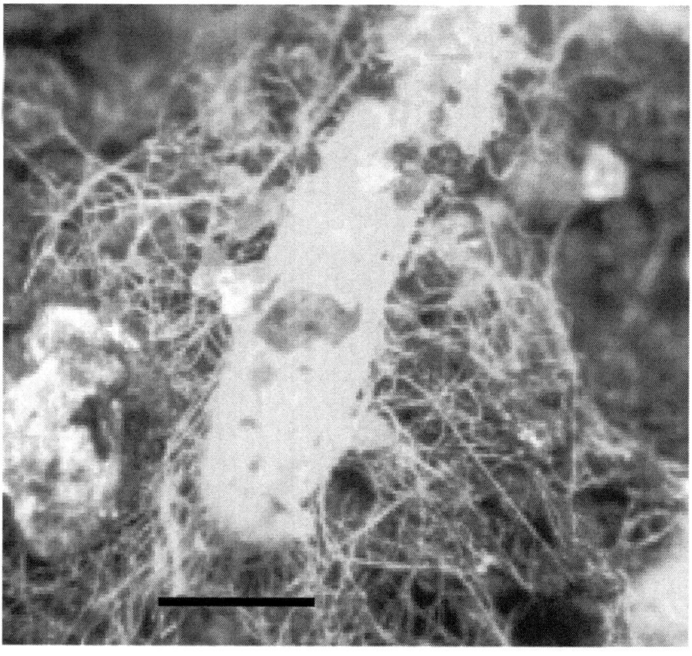

Fig. 3. Yellow *Gnetum* EM root tip formed with *S. sinnamariense*, showing extensive extramatrical mycelia. Bar is 0.3 mm.

Color image of this figure appears in the color plate section at the end of the book.

3. Growth in Culture

Just like most mycobionts, *S. sinnamariense* can be isolated from EM root tips but recovery is often less than 20% and sometimes can be as low as 5% or less of the number of plated tips (Molina and Palmer 1982). The low success rates could be attributed to many factors including the inability of fungi to grow on the media used, insufficient surface sterilization, and requirement of special nutrients. The mycobiont can also be isolated from spores and sporocarps. Several media types have been used for the isolation of EM fungi from root tips. These include modified Melin-Norkran (MMN) agar (Marx 1969), Hagems agar (Modess 1941), Pachlewski agar (Pachlewski and Pachlewska 1974), modified woody plant medium, potato dextrose agar, etc.

S. sinnamariense was successfully isolated from *Gnetum* EM root tips on MMN agar containing benomyl (1 ug/ml), chlortetracycline (30 ug/ml) and streptomycin (10 ug/ml) (Bechem 2004). More than 350 individual EM root tips were plated with a fungal symbiont recovery rate of 38%. This was obtained after surface sterilization using 6% Cl_2 for 4 min. In the said study, only one fungus type was isolated from the yellow morphotype unlike other studies such as Zak and Marx (1964), who isolated two or more fungi from the same pine EM roots.

Scleroderma sinnamarense was a yellow culture with prominent hyphae which had numerous clamp connections. The arrangement of hyphae in culture was similar to that observed on the mycorrhiza tip. Growth rate of the fungus was very slow in comparison to some isolates of *Pisolithus* sp., an observation similar to that of Sims et al. (1999). Growth rarely exceeded 35–40 mm in ten weeks, giving a growth rate of 0.5–0.57 mm d^{-1}. This growth rate was much slower than that observed on the same fungus species studied by Sims et al. (1999) where growth rate was as high as 1.5 mm d^{-1} on some isolates. Growth was slower on Malt extract agar (MEA), Potato dextrose agar (PDA), and on Hagem's agar, in comparison to MMN and Pachlewski agar (Fig. 4) (Bechem 2004).

S. sinnamariense species possessed colonies with matted mycelial growth which sometimes had a floccose periphery, with an irregular to occasionally feathery outline. Cultures were bright yellow in color, with the reverse showing the same coloration throughout. Observation of this fungal isolate was quite similar to that of Sims et al. (1999). Color was impermanent, with cultures occasionally changing to white after prolonged storage. The reverse side of these white cultures was whitish in the extremities but yellowish in the centre. The fungus exuded a yellowish-green pigment and sometimes produced faint greenish droplets into the agar. Putra et al. (2011) isolated two known coloring constituents, methyl 4, 4′-dimethoxyvulpinate and

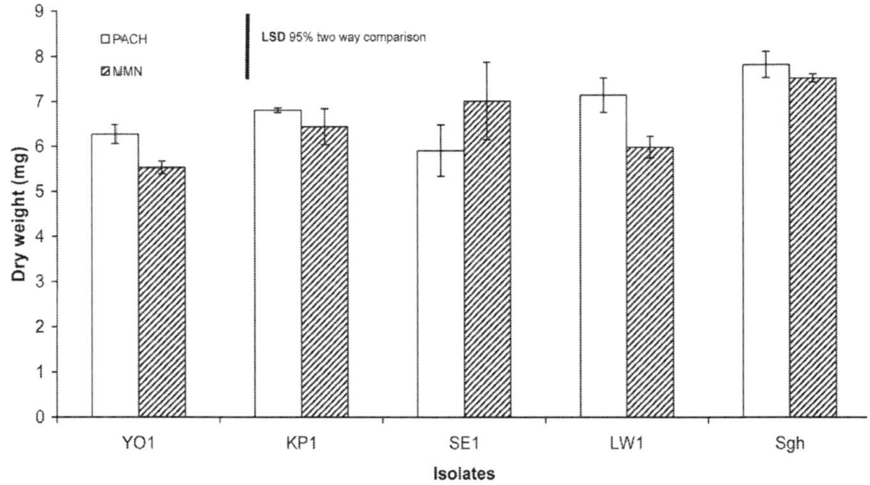

Fig. 4. Dry weight yields (mg) of five isolates of *S. sinnamariense* following growth for 30 days on liquid nutrient media, modified Melin Norkran (MMN) and Pachlewski (PACH). Vertical bars represent standard errors of mean. Each point is a mean of three replicates.

4, 4'-dimethoxyvulpinic acid, from the fruiting body of this fungus. The second compound was shown to have inhibitory effects on the growth of some pathogenic microbes. This fungus could tolerate temperatures of up to 30°C.

4. Nitrogen Nutrition

Nitrogen is present in soil in diverse forms such as ammonium, nitrate, proteins, peptides and amino acids. Studies using different fungal species have shown that EM fungi have the ability to take up and transfer N to the partner plant. Studies on the use of different N sources by *Scleroderma* spp. are lacking but some fungi species such as *Hebeloma* spp. and *Paxillus* species have been widely studied in this respect (Abuzinadah and Read 1986).

S. sinnamariense was studied to evaluate its ability to use different N sources for growth (Bechem 2012a). This fungus produced measurable biomass (Table 1) following growth on MMN liquid broth containing a range of nitrogen sources including ammonium, nitrate, glutamic acid, arginine, alanine, glycine, peptone, and bovine serum albumin (BSA) (Bechem 2012a).

This *Scleroderma* produced greater biomass on ammonium-N than that produced on nitrate-N. Growth in ammonium was accompanied by a fall in pH of the medium, while an increase in pH was observed with

Table 1. Dry weight yields (mg) after 30 days of growth of *Scleroderma sinnamariense* on liquid MMN media containing different nitrogen sources.

N source	*S. sinnamariense*
	Mean ± s.e.m
Calcium nitrate	4.2 ± 0.1
Ammonium sulphate	7.2 ± 1.1
Glutamic acid	8.8 ± 1.2
Arginine	12.9 ± 1.9
Alanine	11.2 ± 0.6
Glycine	2.8 ± 0.5
Bovine Serum Albumin	3.5 ± 0.2
Peptone	16.7 ± 1.4
No Nitrogen	2.5 ± 0.1

Values are means ± standard errors of mean. Number of replicates (n) = 3. Least significant difference at 95 % (one-way comparison) was 3.324.

growth in nitrate. This inability to use nitrate had been observed in other ectomycorrhizal fungi species (Rangel-Castro et al. 2002, Sawyer et al. 2003).

The fungus grew best on peptone-N, producing a biomass which was more than twice (131.9%) the yield observed on ammonium-N. Growth on BSA was poor, when compared to ammonium. However, utilization of this N source was better at very low pH (3.0). This low pH optimum for proteolytic activity suggested that broad spectrum acid proteases were involved in the catalytic process.

Glycine could not provide the N needed for growth by the fungus. Its utilization was therefore limited. But arginine, alanine and glutamic acid were good sources of nitrogen for the fungus. Its growth on these N sources was higher than that observed in ammonium. It is possible to conclude from this observation that free amino acid pools in soil as well as those released directly by the action of fungal proteases might help to supplement the supply of mineral N.

Ammonium has always been thought to be the N source for the growth of microorganisms. The results of the study carried out by Bechem (2012a) showed that organic N sources like peptone, arginine, and alanine were better for the growth of *Scleroderma sinnamariense*.

In Cameroon where this fungus was collected, the pH of the soils is mostly acidic and rarely exceeds 4.5. It is therefore possible that under such conditions the activity of the proteinases necessary for protein uptake and utilization would be optimal, enabling the endophyte to access organic N from the rhizosphere and bulk soil. The phytobiont can therefore access these N sources via the mycobiont.

5. Carbon Nutrition

There has been no published study on the growth of *Scleroderma* spp. on different carbon sources, except for the study by Bechem (2012b). She evaluated the ability of *S. sinnamariense* to grow on different C sources in the presence of either ammonium-N or peptone-N. The C sources used in the study included; glucose, sucrose, soluble starch, cellulose, cellulose plus 'starter' glucose. She observed that glucose was the best carbon source for this fungus and that growth on glucose was significantly greater when peptone was the N source. Growth on the other C sources was relatively poor, implying that the fungus could hardly utilize soluble starch, cellulose, and cellulose with 'starter' glucose in peptone or ammonium-N. The ability of the fungus to use sucrose only in ammonium-N might have been due to changes in pH which resulted following ammonium uptake. Hampp and Schaeffer (1994) had observed that sucrose hydrolysis was significant at pH levels below 4.0. In the study by Bechem (2012b) the media were not buffered and uptake of ammonium would have caused acidification of the medium. The drop in pH may have favored the hydrolysis of sucrose to its constituent glucose and fructose, which might have been used by the fungus for growth. In this study, 'starter' glucose caused adaptive growth of the fungus in cellulose in the presence of ammonium and peptone-N. There is a possibility that the limited growth of *Scleroderma* on sucrose and cellulose could restrict the fungus to the root of the host plant due to unavailability of nutrients.

6. Phosphorus Nutrition

Some researchers had suggested that the release of organic acids from EM fungi may solubilize recalcitrant mineral phosphates (Leyval and Berthelin, 1986). Others believed that before insoluble organic P can be broken down by fungal phosphatases and the resulting phosphate absorbed by the roots, they would first have to be solubilized (Lapeyrie et al. 1991).

In the experiment reported by Bechem (2011), the ability of *S. sinnamariense* to use calcium tetra hydrogen di-orthophosphate $CaH_4(PO_4)_2$, hydroxyapatite $Ca_5(PO_4)_3OH$, calcium phytate $C_6H_6Ca_6O_{24}P_6$ and amorphous iron phosphate $FePO_4$ as P sources for growth was evaluated. The effect of the internal P status of the inoculum, the medium N source and the presence/absence of buffer on growth was also determined.

In the said study, P-sufficient *S. sinnamariense* showed growth on all P sources with peptone as sole N source except on $FePO_4$ in unbuffered media. The source of medium P and the presence of buffer had significant effects ($P< 0.001$) on colony diameter. Maximum growth of 40 mm was observed on $FePO_4$ in buffered conditions. Growth of fungus in buffered medium was superior to that observed in unbuffered medium. The difference was greatest for $FePO_4$. The interaction between the P source and buffer had a significant effect ($P<0.001$) on growth.

Fungal growth on phytate in unbuffered medium was comparable to that on $CaH_4(PO_4)_2$ and hydroxyapatite but better than that on $FePO_4$. The fungus grew on all P sources with ammonium as the N source, but growth on $CaH_4(PO_4)_2$ and hydroxyapatite was better than growth on $FePO_4$. P-sufficient inoculum grew better in comparison to P-starved inoculum on $CaH_4(PO_4)_2$ and hydroxyapatite but not on $FePO_4$.

P-starved fungus demonstrated an ability to solubilize hydroxyapatite and $CaH_4(PO_4)_2$ which was superior to that shown by P-sufficient fungus. This difference was greatest for hydroxyapatite. Solubilization of $CaH_4(PO_4)_2$ by P-sufficient fungus was comparable to solubilization of hydroxyapatite. This was probably due to proton expulsion. The resulting pH change in the medium might have led to the observed solubilization. This phenomenon is confirmed by studies which showed that during ammonium transport by the hyphae, there is H^+ secretion into the medium (Jennings 1997).

In an earlier assay, this author had observed that growth of *Scleroderma* on peptone was followed by fall in the pH of medium. However, the resulting decrease in pH was not significant enough to bring about solubilization of hydroxyapatite and $CaH_4(PO_4)_2$ in peptone-N as observed on ammonium-N. Nonetheless, the fact that *Scleroderma* grew on these P sources implied that there was some solubilization to permit growth. It is probable that an unknown proportion of the insoluble P provided must have dissolved, thus supporting growth without any apparent indication of solubilization. It is unfortunate that other similar studies such as Van Leerdam et al. (2001) did not indicate whether growth was recorded in the absence of solubilization or not.

The inability of *Scleroderma* to solubilize some of the P sources assayed does not necessarily imply that they would not be able to solubilize these P sources under natural conditions, with the help of some bacteria. Villegas and Fortin (2002) showed that the extramatrical mycelium of *Glomus intraradices* could solubilize sparingly soluble calcium phosphate *in vitro* only when grown in association with *Pseudomonas aeroginosa* and *P. putida*. These bacteria were also unable to solubilize this P source without the help of the fungus. There is therefore a possibility that in association with some mycorrhiza helping bacteria, many more P sources could be accessed *in situ*.

It was also interesting to note that solubilization of a P source did not necessarily mean better growth on that P source. Bechem (2011) noted that within the same P source, in situations where solubilization occurred, the fungal growth diameter was inferior to that observed in the absence of solubilization. A probable reason for such an observation may be that the fall in pH which is thought to have assisted in the solubilization process demonstrated adverse effects on the fungal culture, causing a reduction in growth in comparison to the situation where solubilization did not occur. Alternatively, the observed solubilization of the P sources led to an increase in the amount of available P in the growth medium. The attained levels may eventually become toxic to the culture.

Cairney and Smith (1992) found that inoculum P status does influence hyphal absorption of soluble phosphate. The internal P status of the fungal inoculum had an effect on the solubilization of hydroxyapatite but not of $CaH_4(PO_4)_2$. Van Leerdam et al. (2001) did not observe any effect of P status of inoculum on solubilization in their study. Although they concluded that solubilization activities were not dependent on internal phosphorus concentrations, the author of this chapter believes from her assay that this may vary with P source and with the fungal species.

The fact that *Scleroderma* could solubilize hydroxyapatite and calcium tetrahydrogen di-orthophosphate *in vitro* does not necessarily imply that its plant partner can utilize these P sources *in situ*. It would be necessary to demonstrate that solubilization of a P source leads to uptake and eventual translocation to the host plant. Such a demonstration would confirm that the host plant can utilize these P sources via their fungal partners.

Scleroderma exhibited cell-bound phosphatase activity. The greatest activity was shown by P-starved mycelium assayed in medium lacking inorganic P. It was apparent that the absence of N from the assay medium, just like the presence of P, suppressed activity.

Extracellular phosphatase activity was less than cell-bound activity by a factor of 100. Again the greatest activity was shown by P-starved mycelia assayed in medium lacking inorganic P. The absence of P from the assay medium appeared to not affect extracellular activity to the same extent as cell-bound activity.

Cell-bound activity was almost completely suppressed by the presence of P in the assay medium but there was no effect of P on the low levels of extracellular phosphatase detected. The results showed that levels of *p*-nitrophenol phosphatase (*p*NPPase) activity of some species of ectomycorrhizal fungi grown in pure culture are partially influenced by the nitrogen and phosphorus source as well as internal P status of the inocula and assay medium. Cell-bound phosphatase activity of *Scleroderma* was greatly influenced by the presence of inorganic P and peptone-N in the inoculum which was demonstrated in the absence of inorganic P and presence of peptone-N in assay solution.

This study showed that a source of nitrogen in the assay solution is vital for the activation of cell-bound acid phosphatase activity. Overall, activity was slightly higher when a phosphorus source was absent in the assay medium. This observation goes to confirm findings by others that the production of acid phosphatase is enhanced in the absence of P in ectomycorrhiza fungi (Kropp 1990, Antibus et al. 1992, Tibbett et al. 1998). The highest expression of both cell-bound and extracellular acid phosphatase activity was observed following growth with peptone as the sole N source but in the absence of P. This indicated that inorganic P is an important regulator of both cell-bound and extracellular phosphatase activity. Such activity was also dependent on the nitrogen source. Growth was not proportional to acid phosphatase production. From such an observation it is possible that P may not be the only limiting factor of mycelia growth in this experiment.

The ability to restrict secretion when a product is plentiful and to increase production where product is limited suggests that ectomycorrhizal fungi have an economic regulation of phosphatase production brought into harmony by environmental P concentration and therefore aimed at external (soil) substrates (Tibbett et al. 1998a).

With phosphorus being one of the least available plant nutrients found in soil, phosphatases are believed to be very important in their uptake. Phosphatases produced by plants and microbes are presumed to convert organic P into available inorganic P, which is then absorbed by plants. Therefore the observations from this experiment would shed some light into the possible mechanisms by which these fungi in mycorrhizal associations facilitate their host plants with P uptake in a P-limiting environment.

In this study, cell-bound phosphatase was the most active enzyme, an observation similar to that of McElhinney and Mitchell (1993). This was probably because cell-bound phosphatases are thought to be most important in cleavage and procurement of P (Calleja et al. 1980). Nonetheless, in the study carried out by Tibbett et al. (1998a) it was observed that extracellular phosphatases accounted for the largest fraction of pNPPase. They put forward an argument that intimated that contact between substrate and enzyme was not the only factor required for enzyme activity. They emphasised the fact that proper alignment of both substrate and enzyme was also vital for substrate to be transformed to product.

Acid phosphatases are thought to be vital in the uptake and eventual transfer of phosphate to the plant host. The fact that *S. sinnamariense* could express acid phosphatase activity in pure culture does not necessarily mean that it could access unavailable P sources when in symbiosis. There is a possibility that the phosphatases produced in symbiosis might be slightly different from those produced in pure culture. But some researchers such as Gianinazzi-Pearson and Gianinazzi (1989) showed that acid phosphatases produced by ectomycorrhizal hyphae were the same as those produced by the fungus in pure culture. Since phosphorus acquisition is one of the most important functions of the ectomycorrhizal symbiosis, acid phosphatase activity is of considerable interest in selecting strains for inoculation experiments.

In view of the association of *Gnetum* plants with organic soils of low phosphorus status, the demonstration that *Scleroderma sinnamariense* (the ectomycorrhizal endophyte of *Gnetum* spp. in Cameroon) can produce high acid phosphatase activity in organic environments with low P may be of considerable importance in the establishment and growth of *Gnetum* plants in tropical soil communities. In such an environment, absorption of phosphorus would be enhanced by the mycorrhizal system. The phosphatase released by the mycorrhizal fungal mantle and attached mycelium ramifying through the soil would probably catalyze the hydrolysis of complex phosphorus compounds found in the organic layer into more readily absorbed forms. Organic acids produced by the fungus may lead to the solubilization of insoluble P sources, which would be taken up by the fungus and eventually translocated to the plant host. This improved P nutritional status would make the plant more fit so as to better compete for other limiting factors in its environment.

The results from Bechem (2011) also indicated that the nitrogen and phosphorus status of the mycorrhizosphere is an important determinant of acid phosphatase activity. For there to be any meaningful and concrete conclusion, more of such studies are needed on tropical mycorrhizal fungi species.

7. Ectomycorrhiza Formation

Scleroderma spp. have been reported generally to form white mycorrhizas from clamped hyphae (Jeffries 1999). *S. sinnamariense*, however, has been reported to form bright yellow mycorrhizas with *Gnetum* spp. (Bechem and Alexander 2012a, Onguene and Kuyper 2001, Ingleby 1999). The same fungus was reported to form dark brown EM with *Shorea leprosula* Miq. but the hyphae in the rhizomorphs linking the sporocarps to the EM were said to be bright yellow (Lee et al. 1997). This fungus seems to have a narrow host range and has been associated only with *Gnetum*, *Parashorea*, and *Shorea*. EM synthesis with rooted *Gnetum* cuttings under controlled conditions was possible with the use of spores, soil, and mycorrhizal tips as inoculum (Bechem and Alexander 2009).

8. Growth Responses and Nutritional Benefits to the Host Plant

Species of *Scleroderma* have been used to improve on the early growth of some tree species and growth responses have been positive both in screen house and field trials (Jeffries 1999, Castellano 1996, Bâ et al. 2002). Most of the studies reported used species other than *S. sinnamariense* except for Dodd et al. (1996) who used this fungus to inoculate seedlings of *Parashorea malanonan*. Height increments of over 22% were observed on inoculated seedlings when compared to uninoculated plants.

Bechem and Alexander (2012b) carried out an experiment in which they evaluated the effects of mycorrhization and the addition of P (30mg/kg and 300mg/kg) on the growth of rooted *Gnetum* cuttings. It was noted that the number of new EM tips decreased with an increase in P level implying that the rate of colonization decreased with increasing P concentrations. This finding demonstrated that the level of fertilization that can be applied to *Gnetum* plants in farms and plantations is of critical importance and the type of fertilizer must be carefully chosen so as to allow for the continuous development and survival of the mycorrhizas. Ectomycorrhization led to an increase in shoot, root and total plant dry weights, root: shoot ratio, and shoot elongation as well as the total number of new leaves. Ectomycorrhization of *Gnetum* in this study therefore led to improved access to added P. Non-ectomycorrhizal plants in the said study responded to phosphorus addition by increased root and total plant dry weights, shoot elongation and total number of new leaves. An increase in foliar P concentration was also demonstrated, thus suggesting that P availability may have been one of the limiting factors under these experimental conditions. In a previous study, Lee and Alexander (1994) had demonstrated that non-mycorrhizal *Hopea odorata* responded to P addition by showing an increase in shoot dry weight and P concentration. They also observed that ectomycorrhizal colonization increased shoot P concentration and dry weight by about 100% as compared to those of uncolonized plants growing on P amended soil.

Both P addition and EM colonization led to an increase in total foliar P. Mycorrhizal colonization increased foliar N concentrations at all P concentrations. Foliar N concentrations of mycorrhizal plants were higher when P was added, but this was not the case for non-mycorrhizal plants. There was a linear relationship between percent ectomycorrhizal colonization and total foliar N at low P concentrations but this was significant ($P < 0.05$) only in absence of added P. However, there was no evidence of an association between total foliar P and percent colonization.

The relationship between ectomycorrhizal colonization and total foliar N at low P concentration in the study by Bechem and Alexander (2012b) gives an indication of the functional relationship between EM colonization and nutrient (N, P) acquisition by plant species of the tropical rain forests. Although ectomycorrhization and P addition had significant effects on total foliar P in the said study, the analysis of data did not show any linear relationship between percent colonization and foliar P. This observation was inconsistent with that of Moyersoen et al. (1998) in which a clear relationship between P uptake and mycorrhizal colonization was observed in *Tetraberlinia moreliana*, an ectomycorrhizal tree species from the lowland tropical forests of Cameroon.

The linear relationship between total foliar N and ectomycorrhizal colonization at low P concentration suggested that *Gnetum* would be partially dependent on EM for the uptake of N in limiting situations. Nonetheless, N is more available for uptake by trees in most lowland tropical rain forests (Vitousek and Sandford 1986, Martinelli et al. 1999). It is probable that in situations where there is a seasonal fluctuation of nutrients, mycorrhizally-assisted N capture would be important (Hodge 2003). EM colonization increased total foliar P at both low and high P availability, an observation similar to that of Moyersoen et al. (1998) in which EM colonization was observed to increase P uptake of *T. moreliana* at both low and high P availability.

The increase in shoot elongation observed on mycorrhizal plants was probably not due to improved P nutrition only. This was because there was a response at all P concentration. It was therefore possible that there were additional factors brought about by EM colonization besides improved P nutrition.

This experiment showed that *Gnetum* plants do respond positively to ectomycorrhizal colonization and that the response may not be attributed solely to increased uptake of nutrients including phosphorus and nitrogen. It also demonstrated that mycorrhization and fertilization would be necessary for the establishment of *Gnetum* farms and plantations.

9. Conclusion

From the findings of this researcher and her collaborators, there is further evidence that some tropical ectomycorrhiza fungi are able to use amino acids and some proteins as a sole N source. The stimulatory effect of peptone was interesting, although the implication of this observation in nature is to be determined. There was evidence that the fungus was able to utilize organic P and N. It expressed both extracellular and cell-bound phosphatase activity, with activity being higher in the absence of inorganic P. There is therefore a possibility that the fungus would be able to access some organic P sources in nature when inorganic P is limiting, an ability which would be beneficial to the host plant. It is conclusive that ectomycorrhizas are an integral part of *Gnetum*'s physiology. Any useful and effective conservation strategy therefore would have to incorporate the diversity and biology of the fungal partner.

Acknowledgements

I appreciate the assistance from colleagues at the Department of Botany and Plant Physiology, University of Buea for their support. I am indebted to Dr Egbe Enow Andrew for reading this manuscript. I thank the anonymous referees for their valuable comments on this study, and Krista L. McGuire and Caitlyn Gillikin for improving the language.

References

Abuzinadah, R.A. and D.J. Read. 1986. The role of proteins in the nitrogen nutrition of ectomycorrhizal plants .I. Utilisation of peptides and proteins by ectomycorrhizal fungi. New Phytol. 103: 481–493.

Antibus, R.K., R.L. Sinsabaugh and A.E. Linkins. 1992. Phosphatase activities and phosphorus uptake from inositol phosphate by ectomycorrhizal fungi. Can. J. Bot. 70: 794–801.

Bâ, A.M., K.B. Sanon and R. Duponnois. 2002. Influence of ectomycorrhizal inoculation on *Afzelia quanzensis* Welw. seedlings in a nutrient-deficient soil. For. Ecol. Manag. 161: 215–219.

Bechem, E.E. 2004. Mycorrhizal status of *Gnetum* spp. in Cameroon: Evaluating diversity with a view to ameliorating domestication efforts. PhD Thesis, University of Aberdeen, Aberdeen, Scotland.

Bechem, E.E. 2011. Growth and *in vitro* phosphate solubilizing ability of *Scleroderma sinnamariense*: A tropical mycorrhiza fungus isolated from *Gnetum africanum* ectomycorrhiza root tips. J. Yeast Fungal Res. 2: 132–142.

Bechem, E.E. 2012a. Utilisation of organic and inorganic nitrogen sources by *Scleroderma sinnamariense* Mont. isolated from *Gnetum africanum* Welw. Afr. J. Biotechnol. 11: 9205–9213.

Bechem, E.E. 2012b. Effect of carbon and nitrogen sources on *in vitro* growth of *Scleroderma sinnamariense* Mont., a pantropical ectomycorrhizal fungus. Int. J. Biol. Chem. Sci. 6: 1192–1201.

Bechem, E.E. and I.J. Alexander. 2009. Inoculum production and inoculation of *Gnetum africanum* rooted cuttings using a range of mycorrhizal fungi. Int. J. Biol. Chem. Sci. 3: 578–586.

Bechem, E.E. and I.J. Alexander. 2012a. Mycorrhizal status of *Gnetum africanum* in Cameroon: Evaluating diversity with a view to ameliorating domestication efforts. Mycorrhiza 22: 99–108.

Bechem, E.E. and I.J. Alexander. 2012b. Phosphorus nutrition of ectomycorrhizal *Gnetum africanum* plantlets from Cameroon. Plant Soil 353: 379–393.

Cadiz, R.T. and H.B. Florido. 2001. *Gnetum gnemon* Linn. BAGO. Res. Info. Series Ecosysts. 13: 1–6.

Calleja, M., D. Mousain, B. Lecouvreur and J. d'Auzac. 1980. Influence de la carence phosphatée sur les activitées phophatases acides de trois champignons mycorrhizens: *Hebeloma edurum* Metrod., *Suillus granulutus* (L. ex Fr.) O. Kuntze et *Pisolithus tinctorius* (Pers.) Coker et Couch. Physiol. Vég. 18: 489–504.

Cairney, J.W.G. and S.E. Smith. 1992. Influence of intracellular phosphorus concentration on phosphate absorption by the ectomycorrhizal basidiomycete *Pisolithus tinctorius*. Mycol. Res. 96(8): 673–676.

Castellano, M.A. 1996. Outplanting performance of mycorrhizal inoculated seedlings. *In*: K.G. Mukerji (ed.). Concepts in Mycorrhizal Research. Kluwer, Dordrecht, pp. 223–301.

Demoulin, V. and D.M. Dring. 1971. Two new species of *Scleroderma* from tropical Africa. Trans. Br. Mycol. Soc. 56: 163–165.

Dodd, J.C., P. Jeffries and R. De La Cruz. 1996. The use of mycorrhiza fungi in reforestation programmes in the Philippines. Final Report EU Contract C II*-CT91-0904, STD3 Programme, DG XII, Brussels.

Gianinazzi-Pearson, V. and S. Gianinazzi. 1989. Cellular and genetical aspects of the interactions between host and fungal symbionts in mycorrhizae. Genome 31: 341–366.

Guzman, G. 1970. Monografia del genero *Scleroderma* Pers. Amend. Fr. Darwinania 16: 233–407.

Guzman, G. and C.L. Ovrebo. 2000. New observations in sclerodermataceous fungi. Mycologia 92: 174–179.

Hampp, R. and C. Schaeffer. 1994. Mycorrhiza-carbohydrate and energy metabolism. *In*: A. Varma and B. Hock (eds.). Mycorrhiza-Structure, Function, Molecular biology and Biotechnology. Springer-Verlag, Germany.

Hasselquist, N.J., L.S. Santiago and M.F. Allen. 2010. Belowground nitrogen dynamics in relation to hurricane damage along a tropical dry forest chronosequence. Biogeochemistry 98: 89–100.

Hodge, A. 2003. N capture by *Plantago lanceolata* and *Brassica napus* from organic material: the influence of spatial dispersion, plant competition and an arbuscular mycorrhizal fungus. J. Exp. Bot. 54: 2331–2342.

Ingleby, K. 1999. *Scleroderma sinnamariense* Mont. + *Gnetum africanum*. *In*: R. Agerer, R.M. Danielson, K. Ingleby, K. Luoma and R. Treu (eds.). Descriptions of Ectomycorrhizae 4: 127–133.

Jennings, D.H. 1997. The Physiology of Fungal Nutrition. Cambridge University Press, Cambridge.

Jeffries, M. 1999. *Scleroderma*. *In*: J.W.G Cairney and S.M Chambers (eds.). Ectomycorrhizal Fungi: Key Genera in Profile 7: 187–200.

Kropp, B.R. 1990. Variation in acid phosphatase activity among progeny from controlled crosses in the ectomycorrhizal fungus *Laccaria bicolor*. Can. J. Bot. 68: 864–866.

Lapeyrie, F., J. Ranger and D. Vairelles. 1991. Phosphate-solubilizing activity of ectomycorrhiza fungi *in vitro*. Can. J. Bot. 69: 342–346.

Lee, S.S. and I.J. Alexander. 1994. The response of seedlings of two dipterocarp species to nutrient additions and ectomycorrhizal infection. Plant Soil 163: 299–306.

Lee, S.S., I.J. Alexander and R. Watling. 1997. Ectomycorrhizae and putative ectomycorrhizal fungi of *Shorea leprosula* Miq. (Dipterocarpaceae). Mycorrhiza 7: 63–81.

Leyval, C. and J. Berthelin. 1986. Comparison between the utilization of phosphorus from insoluble mineral phosphates by ectomycorrhizal fungi and rhizobacteria. *In*: V. Gianinazzi-Pearson and S. Gianinazzi (eds.). Physiological and Genetical Aspects of Mycorrhizae INRA: Paris, France, pp. 340–345.

Marx, D.H. 1969. The influence of ectotrophic mycorrhizal fungi on resistance of pine roots to pathogenic infections. I. Antagonism of mycorrhizal fungi to root pathogenic fungi and soil bacteria. Phytopathology 59: 153–163.

Martinelli, L.A., M.C. Piccolo, A.R. Townsend, P.M. Vitousek, E. Cuevas, W. Mcdowell, G.P. Robertson, O.C. Santos and K. Treseder. 1999. Nitrogen stable isotopic composition of leaves and soil: tropical versus temperate forests. Biogeochemistry 46: 45–65.

McElhinney, C and D.T. Mitchell. 1993. Phosphate activity of four ectomycorrhizal fungi in a Sitka spruce-Japanese larch plantation in Ireland. Mycol. Res. 97(6): 725–732.

Modess, 0. 1941. Zur Kenntnis der Mykorrhizabildner von Kiefer and Fichte. Symb. Bot. Upsal. 5: 1–146.

Molina, R. and J.G. Palmer. 1982. Isolation, maintenance and pure culture manipulation of ectomycorrhizal fungi. *In*: N. Schenck (ed.). Methods and Principles of Mycorrhizal Research. The American Phytopathological Society, pp. 115–129.

Moyersoen, B., I.J. Alexander and A.H. Fitter. 1998. Phosphorus nutrition of ectomycorrhiza and arbuscular mycorrhizal tree seedlings from a lowland tropical forest in the Korup National Park, Cameroon. J. Trop. Ecol. 14: 47–61.

Onguene, N.A. and T.W. Kuyper. 2001. Mycorrhiza associations in the rain forest of South Cameroon. For. Ecol. Man. 140: 277–287.

Pachlewski, R. and J. Pachlewska. 1974. Studies on symbiotic properties of mycorrhizal fungi of pine (*Pinus sylvestris*) with the aid of the method of mycorrhizal synthesis in pure culture on agar, Warsaw, Poland: Forest Research Institut, pp. 139.

Putra, D.P., I. Nurmilasari, Y. Komala, Asakawa and D. Arbain. 2011. The coloring constituents of *Scleroderma sinnamariense* (Sclerodermaceae). Nat. Prod. Commun. 6: 357–360.

Rangel-Castro, J.I., E. Danell and A.F.S. Taylor. 2002. Use of different nitrogen sources by the edible ectomycorrhizal mushroom *Cantharellus cibarius*. Mycorrhiza 12: 131–137.

Richter, D.L. and J.N. Bruhn. 1989. *Pinus resinosa* ectomycorrhizae: seven host-fungus combinations synthesized in pure culture. Symbiosis 7: 211–228.

Sawyer, N.A., S.M. Chambers and J.W.G. Cairney. 2003. Utilisation of inorganic and organic nitrogen sources by *Amanita* species native to temperate eastern Australia. Mycol. Res. 107: 413–420.

Sims, K., R. Watling and P. Jeffries. 1995. A revised key to the genus *Scleroderma*. Mycotaxon 56: 403–420.

Sims, K., R. Sen, R. Watling and P. Jeffries. 1999. Species and population structures of *Pisolithus* and *Scleroderma* identified by combined phenotypic and genomic marker analysis. Mycol. Res. 103: 449–458.

Tibbett, M., F.E. Sanders and J.W.G Cairney. 1998. The effect of temperature and inorganic phosphorus supply on growth and acid phosphatase production in arctic and temperate strains of ectomycorrhizal *Hebeloma* spp. in axenic culture. Mycol. Res. 102: 129–135.

Van Leerdam, D.M., P.A. Williams and J.W.G. Cairney. 2001. Phosphate-solubilizing abilities of ericoid mycorrhizal endophytes of *Woollsia pungens* (Epacridaceae). Aust. J. Bot. 49: 75–80.

Villegas, J. and J.A. Fortin. 2002. Phosphorus solubilization and pH changes as a result of the interactions between soil bacteria and arbuscular mycorrhizal fungi on a medium containing NO_3^- as nitrogen source. Can. J. Bot. 80: 571–576.

Vitousek, P.M. and R.L. Sandford. 1986. Nutrient cycling in moist tropical forest. An. Rev. Ecol. and Syst. 17: 137–167.

Watling, R. 1994. Taxonomic and Floristic notes on some Malaysian Larger Fungi- I. Malayan Nat. J. 48: 67–78.

Zak, B. and D.H. Marx. 1964. Isolation of individual fungi from roots of individual slash pines. For. Sci. 10: 214.

CHAPTER 9

Alleviation of Salt Stress by *Scleroderma bermudense* in *Coccoloba uvifera* Seedlings in the French West Indies

Amadou Mustapha Bâ,[1,2,3,*] *Raymond Avril,*[1] *Eric Bandou,*[1] *Seynabou Sène,*[2] *Robin Duponnois,*[3] *Régis Courtecuisse,*[4] *Samba Sylla*[2] and *Abdala Diédhiou*[2]

1. Introduction

More than 800 million ha of land throughout the world is affected by salt levels. The soil types most affected by excess salt are found in 30% of Australian soils (Flowers and Colmer 2008). Excess salt in soils can have two origins: (i) natural causes, i.e., weathering of parent rocks, deposition of oceanic salts carried in wind and rain (10 mg/kg of NaCl would deposit

[1]Laboratoire de Biologie et Physiologie Végétales (LBPV), Faculté des Sciences Exactes et Naturelles, Université des Antilles et de la Guyane, BP. 592, 97159, Pointe-à-Pitre, Guadeloupe, France.
[2]Laboratoire Commun de Microbiologie (LCM) IRD/UCAD/ISRA, Centre de Recherche de Bel Air, BP. 1386, CP. 18524 Dakar, Sénégal.
[3]Laboratoire des Symbioses Tropicales et Méditerranéennes (LSTM), UMR113, Campus de Baillarguet, A10/J, 34398 Montpellier, Cedex 5, France.
[4]Département de Mycologie et de Botanique, Université de Lille, BP. 83, F-59006 Lille cedex, France.
*Corresponding author: amadou.ba@ird.fr, amadou.ba@univ-ag.fr

10 kg/ha of salt for each 100 mm of rainfall per year); (ii) human causes, i.e., excess amounts of fertilizers, irrigation with brackish water (20% of irrigated lands in the world are affected), and (iii) increased sea level caused by global climate changes. Subsequently, two main groups of salted soils exist: saline soils (soluble salts) and alkaline sodic soils (adsorbed sodium). Saline soils affect plant growth without changing the physical properties of soil. Soil is classified as saline when electrical conductivity ECe is ≥ 4 dS m^{-1} (~40 mM or 2.4‰ NaCl) and generates an osmotic pressure of 0.2 MPa. Saline soils cover 397 million ha worldwide (FAO 2005). Sodic soils affect plant growth and soil structure by saturating the adsorption complex by Na^+, which causes soil structure to evolve to an unstable state of dispersion. This can be overcome through the flocculant power of Ca^{++} from gypsum ($CaSO_4 2H_2O$). Soil is classified as sodic when Exchangeable Sodium Percentage (ESP) is ≥ 15% of cation exchange capacity. Sodic soils cover 434 million ha worldwide (FAO 2005).

Plants respond to salinity stress at two levels: (i) osmotic stress (rapid response) due to the increase in external osmotic pressure (reduction of shoot growth rate); this is the osmotic or water-deficit effect of salinity; (ii) Salt-specific effect (slow response) due to the accumulation of toxic ions (Na^+, Cl^-) in leaves (decrease of shoot growth rate only for sensitive plant). This is the toxic effect of salinity (Munns 2005). The osmotic stress not only has an immediate effect on growth, but also has a greater effect on growth rates than the ionic stress. To overcome salt-stress problems, it is possible to select salt-tolerance plants, to use biological processes such as mycorrhizal interactions, or to desalinate soil by leaching excessive salts (Parida and Das 2005). The desalinisation of soils is not economically viable for sustainable agriculture. Although the results are somewhat variable, salt tolerant plants develop mechanisms such as osmotic and ionic homeostasis. Osmotic homeostasis consists of osmotic adjustment through the active accumulation of compatible solutes (e.g., K+, sucrose, proline, glycine betaine) in the cytosol to maintain turgor of cells (e.g., stomatal opening) that would otherwise dehydrate. Ionic homeostasis consists of Na^+ exclusion by roots through transporters (e.g., type SOS1) to ensure that Na^+ does not accumulate to toxic concentrations in leaves, or Na^+ and Cl^- compartmentation through transporters (e.g., type NHX) in the vacuoles to avoid toxic concentrations within the cytoplasm.

The majority of terrestrial plants form symbioses with mycorrhizal fungi (Smith and Read 2008). There is considerable evidence that mycorrhizal fungi can improve plant growth and nutrition in soils subject to a range of saline stress (Giri and Mukerji 2004, Parida and Das 2005, Bandou et al. 2006, Bois et al. 2006). Mycorrhiza can help moderately salt-tolerant plants prevent Na^+ and Cl^- translocation to shoot and leaf tissues, and improved salt tolerance after mycorrhizal colonization may also result from more efficient P uptake by mycorrhizal plants in P-deficient soils, leading to increased growth and the subsequent dilution of toxic ions effects (Juniper and Abbott 1993, Giri and Mukerji 2004). In this respect, ectomycorrhizal (ECM) fungi contribute to render trees more resistant to salt stress than non-ECM plants. The identification of salt-tolerant ECM fungi could help enhance tree survival and growth in salted soils (Bois et al. 2006, Chen et al. 2001, Tang et al. 2009). Many studies have been conducted to test salt tolerance of ECM fungi in axenic cultures and little is known about the salt tolerance of the ECM association (Kernaghan et al. 2002). Chen et al. (2001) screened eighteen isolates of *Pisolithus* for resistance to NaCl during growth in axenic cultures and found that three of them are broadly resistant to salinity and may represent useful ECM inoculants for the outplanting of compatible, salt-resistant host trees at saline sites. Levels of mannitol, glycerol, and trehalose, and some other metabolites, such as the proline amino acid, were observed to increase in fungi under osmotice stress (Shen et al. 1999). Bois et al. (2006) tested the effects of three selected ECM fungi for their *in vitro* tolerance to excess NaCl, on the growth of *Picea glauca* and *Pinus banksiana* seedlings over a gradient of NaCl concentration. Although the different ECM fungi altered the physiological response of the host in different ways (e.g., organic osmolyte accumulation, water content, Na^+ accumulation and allocation), inoculation with salt-stress tolerant ECM fungi increased growth and reduced the negative effects of excess NaCl (Bois et al. 2006). *Paxillus involutus* was salt tolerant, showing biomass increases in media containing up to 500 mM NaCl after 4 wk of growth (Langenfeld-Heyser et al. 2007). Whole salt-sensitive poplar hybrid *Populus* x *canescens* performance was positively affected by inoculation with *P. involutus* because total biomass was greater and its leaves accumulated less Na^+ than non-mycorrhizal plants (Langenfeld-Heyser et al. 2007).

Coastal vegetation usually consists of low numbers of plant species that are specifically adapted to environments that experience frequent stress such as intermittent salinity, drought, intense insolation, instability of sand, low fertility, etc. In this context, *Coccoloba uvifera* (L.) L. (Polygonaceae), also named seagrape, often grows in pure stands within well-drained sandy soils that are slightly to moderately alkaline (Parrota 1994). It is considered a drought-hardy and non-halophyte woody plant relatively tolerant to salt. This tree is naturally associated with ECM fungi and one of them, *Scleroderma bermudense*, alleviates saline stress and stimulates growth and improves nutrition of seagrape seedlings (Bandou et al. 2006). Here, we examined (i) how the salinity affected the ECM communities associated with three species of *Coccoloba* (*C. uvifera*, *C. swartzii* and *C. pubescens*) along a salinity gradient in Guadeloupe and Martinique (Lesser Antilles); (ii) the ability of ECM fungi to grow in pure culture and to alleviate salt stress in seagrape seedlings over a range of NaCl levels.

2. Distribution of *Coccoloba uvifera*, *C. swartzii* and *C. pubescens*, and their Putative Ectomycorrhizal Fungi along a Salinity Gradient

The genus *Coccoloba* belongs to the family of Polygonaceae. This plant family has two ECM genera, *Polygonum* and *Coccoloba*, including 300 and 400 species, respectively (Howard 1988). *Coccoloba* includes several species that predominate in South America, the largest number of species being in Brazil. Of the *Coccoloba* species found in the Caribbean, ten species have been identified, including six species in Guadeloupe and Martinique (Table 1). Of these species of *Coccoloba*, three have been particularly encountered: *C. uvifera*, *C. pubescens* and *C. swartzii*. *C. uvifera* is the most known and widespread species in the Caribbean basin. It is an important tree for edible fruits, ornamental plantings, and coastal windbreak along Caribbean beaches and roadsides. Seagrape is the dominant tree species in beach canopies. It is encountered in rocky, calcareous and volcanic sandy soils which are alkaline and poor in terms of P, N and K. *C. pubescens*, also named grandleaf (diameter 2.5–45 cm) seagrape, is a tree distributed in the xero-mesophytic forests in Guadeloupe and Martinique (Lesser Antilles). *C. swartzii* is a tree encountered in the degraded xerophytic forests in Guadeloupe and Martinique. *C. pubescens* and *C. swartzii* are often found together in Martinique. Given the natural distribution of the three species of *Coccoloba* along a salinity gradient, it could be inferred that *C. uvifera* is more adapted to salinity than the other two *Coccoloba* species (Fig. 1). In fact, only mature trees of *C. uvifera* with abundant seedling recruitment were well developed within areas far from the sea where salt levels were low (0–2‰), whereas mature trees near the sea (2–15‰) were shunted and seedlings were almost absent (Fig. 1). We observed that mature trees of seagrape adjacent to the sea were smaller than mature trees growing far from the sea. Measurements were not made to estimate their difference in terms of sizes. Nevertheless, we noted that the number of mycorrhizal and non-mycorrhizal root tips was nearly three times higher on mature trees growing in salted soil at 0–2‰ than those at 2–15‰, suggesting a negative impact of salinity on root development of mature trees of *C. uvifera*. Naturally regenerating seedlings under the canopy of mature trees were high in low salinity levels, whereas very little occurred under the mature trees growing near the sea. This suggests a negative impact of salinity on naturally regenerating *C. uvifera* seedlings. Salinity is considered to be one of the most significant environmental factors limiting plant germination, growth, and productivity (Flowers and Colmer 2008). Salinity reduces the availability of water and increases toxic ion concentrations in soils. This could affect imbibition, germination, and root elongation. However, the

Table 1. Diversity of *Coccoloba* species in the Caribbean basin (nd= not determined).

Species	Synonymous	Distribution	Country	Habitat	Reference
Coccolobaascendens	*Uviferuuscendens*	Fairly common	Guadeloupe, Martinique, South America, Trinidad, Dominica, St. Lucia, Grenade	Basaltic hills and forests	Fournet (2002)
C. caravellae	none	rare	Endemic to Martinique	nd	Fournet (2002)
C. coronata	*C. caribaea*	nd	Central America, Trinidad and Tobago, South America, Grenadines, Grenade	nd	Howard (1988)
C. diversifolia	*C. laurifolia*	nd	Florida, Mexico, Central America, Greater Antilles, Antigua	nd	Howard (1988)
C. dussii	none	Fairly common	Guadeloupe, Trinidad, St. Lucia, St. Vincent, Grenade	nd	Fournet (2002) Howard (1988)
C. krugii	*Uviferakrugii*	rare	St. Martin, Bahamas, Greater Antilles, Anguilla, Guadeloupe, St. Christophe, St. Kitts	nd	Fournet (2002)
C. pubescens	*C. grandifolia, C. antiguensis, C. bonfilsiana*	Fairly common	Guadeloupe, Desirade, Marie-Galante, Martinique, Hispaniola, Porto Rico, Barbuda, Antigua, Nevis, Montserrat, Dominica, St. Lucia, Barbados	Coastal mesophytic, Basaltic or calcareous hills	Fournet (2002) Howard (1988)
C. swartzii	*C. diversifolia, C. barbadensis, C. punctata, Coccolobensisdiversifolia, C. neglecta*	Common	Bahamas, Greater Antilles, Central America, South America, Antigua, Saba, St. Eustache, St. Christophe, St. Kitts, Nieves, Montserrat, Guadeloupe, Dominica, Marie-Galante, Martinique, St. Lucia, St. Vincent, Grenadines, Grenade, Barbados	Basaltic hills. Degraded xerophytic forest.	Fournet (2002) Howard (1988)
C. uvifera	*Polygonumuvifera*	Very common	Bahamas, Greater Antilles, Mexico, Central America, South America, Anguilla, Antigua, Saba, St. Eustache, St. Bartelemy, St. Christophe, St. Kitts, Nieves, Montserrat, Guadeloupe, Dominica, Marie-Galante, Martinique, St. Lucia, St. Vincent, Grenadines, Grenade, Barbados	Sandy coast and sometimes rocky	Fournet (2002) Howard (1988)
C. venosa	*C. nivea*	Quite rare	St. Martin, St. Bartelemy, Guadeloupe, Les Saintes, Marie-Galante, Martinique, Hispaniola, Porto Rico, Trinidad and Tobago, Venezuela, Antigua, Saba, St. Eustache, Montserrat, Dominica, St. Lucia, Grenadines, Grenade, Barbados	Coastal dry sand, tuff or basalt. Calcareous ills in *Pterocarpus* forest behind the mangrove swamp	Fournet (2002) Howard (1988)

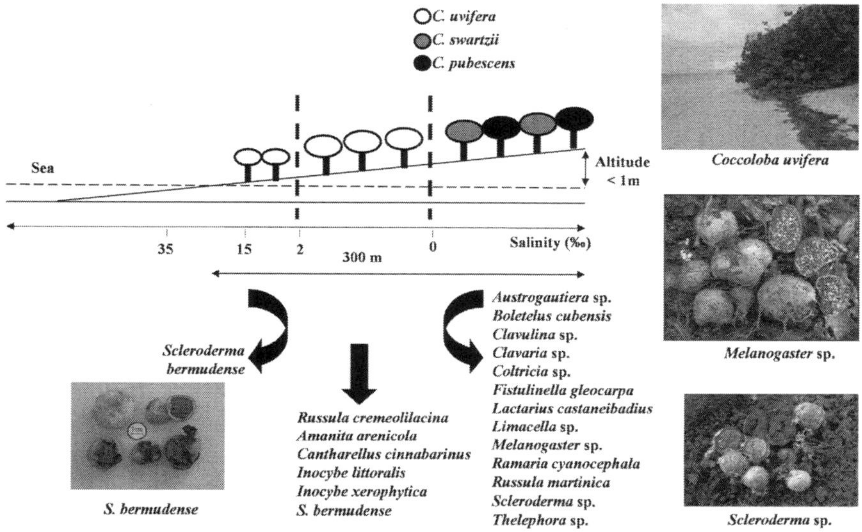

Fig. 1. Distribution of three *Coccoloba* species (*Coccoloba uvifera, C. swartzii* and *C. pubescens*) and their putative ectomycorrhizal fungi along a salinity gradient in Martinique and Guadeloupe (Lesser Antilles).

Color image of this figure appears in the color plate section at the end of the book.

way in which salinity exerts its influence on these vital processes through an osmotic effect or a specific ion toxicity is still not resolved (Katembe et al. 1998).

Table 2 shows the diversity of putative ECM fungi associated with different *Coccoloba* species in the Greater and Lesser Antilles. Some of them were common with those we collected in Martinique and Guadeloupe. We collected 19 putative ECM fungi, of which 6 were found under *C. uvifera* and 13 under *C. pubescens* and *C. swartzii* along a gradient of salinity (Fig. 1) (Bessard et al. 2008). The 11 epigeous (*Russula, Lactarius, Fistulinella, Scleroderma, Boletelus, Clavaria, Clavulina, Coltricia, Limacella, Ramaria* and *Thelephora*) and 2 hypogeous (*Austrogautieria* and *Melanogaster*) fungal species were collected under *C. pubescens* and *C. swartzii* in areas without salt, whereas 8 different fungal species were collected in *C. uvifera* in salted sandy soils (0–2‰ of salinity) (Fig. 1). Diversity of ECM fungi increased when the salinity was low. It could be structured by salinity, host preference, and/or charcteristics of soils along the salinity gradient. Also, abundance of sporocarps fruiting from under mature trees and seedlings in *C. uvifera* coastal forest were assessed over the salinity gradient (Fig. 2). Six species of sporocarps were identified in low salinity levels (0–2‰) and only one of them (*S. bermudense*) fructified in high salinity levels (2–15‰). Thus, sporocarps of *S. bermudense* were more abundant in high salt levels than

Table 2. Putative ectomycorrhizal fungi associated with different *Coccoloba* species in the Greater and Lesser Antilles.

Fungal family	Fungal species	Host plant	Country	Reference
Amanitaceae	*Amanitacystidiosa*	*C. uvifera*	Anguilla, Guana (Virgin Islands), PuertoRico	Miller et al. 2000
	A.antillana	*C. diversifolia, C. pubescens*	Martinique, Trinidad	Pegler 1983
	A .arenicola	*C. uvifera*	PuertoRico, Guadeloupe	Miller et al. 2000, Bandou 2005
	A .microspora	*C. uvifera*	PuertoRico	Miller et al. 2000
	A. craseoderma	*C. pubescens*	Martinique	Pegler 1983
Boletaceae	*Austrogautieria* sp.	*C. pubescens, C. swartzii*	Martinique	Bessard et al. 2008
	Boletusroborculus	*Coccoloba* sp.	Porto Rico	Miller et al. 2000
	Boleteluscubensis	*C. pubescens, C. swartzii*	Martinique	Bessard et al. 2008
	Fistulinellagloeocarpa	*C. diversifolia*	Martinique	Pegler 1983, Bessard et al. 2008
	Limacella sp.	*C. pubescens, C. swartzii*	Martinique	Pegler 1983, Bessard et al. 2008
	Melanogaster sp.	*C. pubescens, C. swartzii*	Martinique	Bessard et al. 2008
	Xerocomushypoxanthus	*C. uvifera*	Martinique	Pegler 1983
	X. coccolobae	*C. uvifera, C. pubescens, C. swartzii*	Martinique	Pegler 1983
	X.guadelupae	*Coccoloba* sp.	Martinique, Guadeloupe, Dominica	Pegler 1983
	X.cuneipes	*C. uvifera*	Martinique	Pegler 1983
	Sclerodermabermudense		PuertoRico, Guadeloupe, Martinique	Guzman et al. 2004, Bandou et al. 2006
	S. stellatum	*Coccoloba* sp., *C. uvifera*	Cuba, Puerto Rico	Kreisel 1971
	Scleroderma sp.	*C. swartzii, C. pubescens*	Martinique	Bessard et al. 2008

Cantharellaceae	*Cantharelluscinnabarinus*	*C. uvifera*	Martinique, Guadeloupe, PuertoRico, Cuba	Pegler 1983, Bandou 2005
Cortinariaceae	*Inocybe littoralis*	*C. uvifera*	Martinique, Guadeloupe	Pegler 1983, Bandou 2005
	I.xerophytica	*C. uvifera*	Guadeloupe	Pegler 1983, Bandou 2005
Gomphaceae	*Ramariacyanocephala*	*C. swartzii, C. pubescens*	Martinique	Bessard et al. 2008
Hymenochaetaceae	*Coltricia* sp.	*C. swartzii, C. pubescens*	Martinique	Bessard et al. 2008
Inocybaceae	*Inocybe littoralis*	*C. uvifera*	Martinique, Guadeloupe	Pegler 1983, Bandou 2005, Bessard et al. 2008
	I. xerophytica	*C. uvifera*	Martinique, Guadeloupe	Pegler 1983, Bandou 2005, Bessard et al. 2008
Russulaceae	*Lactariuscoccolobae*	*C. uvifera*	Anguilla, Guana (Virgin Islands), Puerto Rico	Miller et al. 2000
	L. caribaeus	*C. pubescens, C. diversifolia*	Martinique	Pegler 1983
	L. castaneibadius	*C. diversifolia, C. swartzii, C. pubescens*	Martinique	Pegler 1983, Bessard et al. 2008
	L. hygrophoroides	*C. pubescens*	Martinique	Pegler 1983
	L. murinipes	*Coccoloba* sp.	Martinique	Pegler 1983
	L. nebulosus	*C. pubescens, C. diversifolia*	Martinique	Pegler 1983
	Russulacremeolilacina	*C. diversifolia, C. uvifera*	Martinique, Guadeloupe	Pegler 1983, Bandou 2005
	R. brevipes	*C. swartzii*	Martinique	Pegler 1983
	R. littoralis	*C. uvifera*	Martinique, Virgin Islands, PuertoRico	
	R. martinica	*C. diversifolia*	Martinique	Pegler 1983
	R. metachromaticae	*C. uvifera*	Martinique	Pegler 1983
Thelephoraceae	*Thelephora* sp.	*C. swartzii, C. pubescens*	Martinique	Bessard et al. 2008

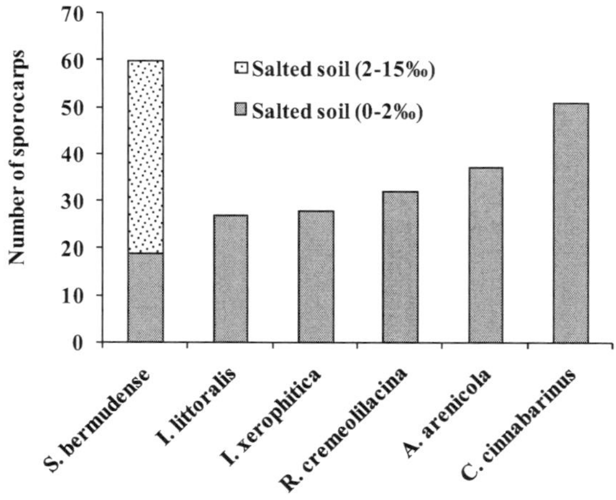

Fig. 2. Number of fungal species described by examining epigeous sporocarps collected in an ECM sampling plot within a period of 8 months in a *C. uvifera* coastal forest in Guadeloupe.

in low salt levels. Only *Scleroderma bermudense* fructified at salinity levels above 15‰. This suggests that fruiting of *S. bermudense* could tolerate salinity stress. The productivity of *S. bermudense* was greater than of any other species of ECM fungi, which permitted it to occupy soils with high salt levels in coastal forests (Fig. 2). It is well known that fungi require localized outbreaks of rain to provide the soil with sufficient water for them to develop and fructify. Their fruiting depends on rainfall distribution and intensity. The lack of available soil water due to salinity could have a detrimental effect on the fruiting of all the fungal species except *S. bermudense*. The species accumulation curves for sporocarps in both salt levels tended to level off, indicating that most ECM fungal taxa were detected in our sampling effort (Fig. 3).

3. *In vitro* Selection of ECM Fungi for use in Saline Soils

Among the sporocarps collected from under the three *Coccoloba* species, only 3 of them (*S. bermudense* isolate GUA 09.06.06, *Scleroderma* sp. isolate CLMART07.157 and *Melanogaster* sp. isolate CLMART06.004) were successfully isolated and cultivated in pure culture. Two fungal species (*Scleroderma* sp. and *Melanogaster* sp.) were from a non-salt-affected soil, and one (*S. bermudense*) was from salt-affected soil along the salinity gradient (Fig. 1). *S. bermudense* grew better in pure culture than *Melanogaster* sp. and *Scleroderma* sp. whatever the levels of NaCl (Fig. 4). *S. bermudense* showed significant peaks in radial growth at two concentrations of NaCl (100–200 mM). Other fungi showed an apparent general decline in radial growth with

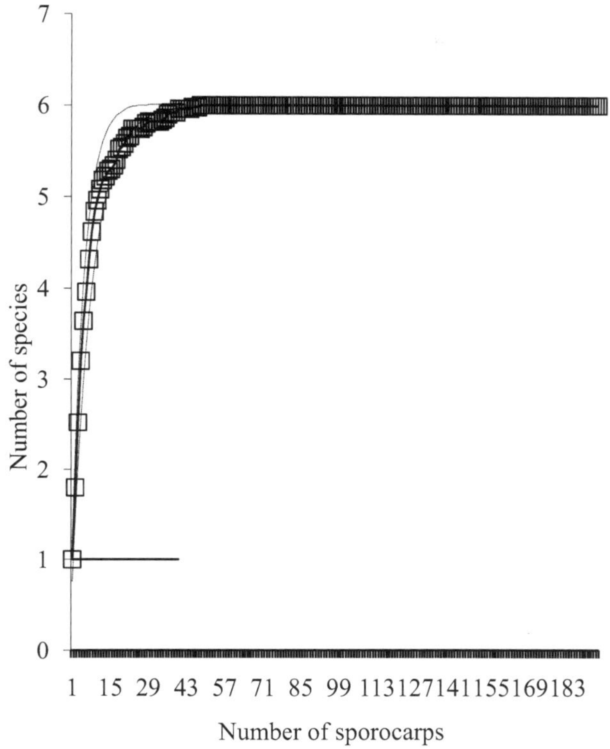

Fig. 3. Species accumulation curve for sporocarps in a *C. uvifera* coastal forest in Guadeloupe. Dotted lines indicate 95% CI.

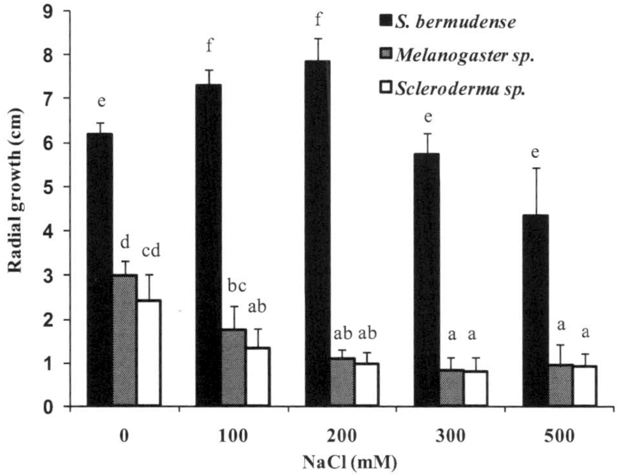

Fig. 4. Radial growth of *S. bermudense*, *Melanogaster* sp. and *Scleroderma* sp. in the presence of NaCl at a range of concentrations in axenic culture for 1 mon (Two-ways ANOVA, $P<5\%$).

increasing NaCl concentration, although the decline was not significant with incresasing NaCl up to 100 mM. Similar results were found for biomass yield for all parameters (Fig. 5). Selection of the salt concentration ranges used in the present study was based on field measurements of soil salinity (Bandou 2005, Bandou et al. 2006). Our data indicated that *S. bermudense* were broadly resistant to NaCl at concentrations encountered in the saline soils of littoral forests. This fungus could also require Na^+ for growth as their radial growth and biomass yield increased at two concentrations (100–200 mM) relative to the zero NaCl treatment. Chen et al. (2001) also found that some isolates of *Pisolithus* showed clear peaks in biomass yield at concentrations in the range of 25–100 mM NaCl. Much of the information on fungal salt tolerance comes from studies on aquatic fungi (Clipson and Jennings 1992). While some marine fungi require Na^+ for growth i.e. one fungal species growsat 12.8 M NaCl in the Great Salt Lake, terrestrial fungi appear to have less of a requirement. The observed peaks could also reflect responses to lowered external water potential generated by NaCl in the medium. This hypothesis is supported by similar peaks observed in experiments in which the growth of ECM fungi was measured at different

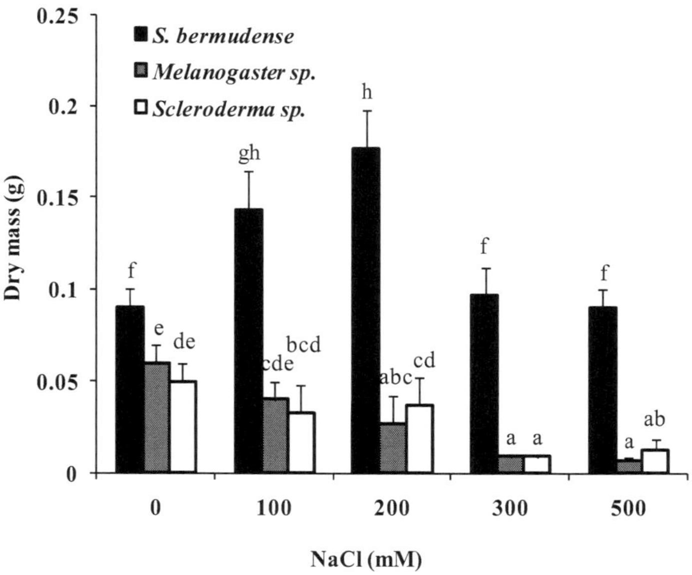

Fig. 5. Biomass yield of *S. bermudense, Melanogaster* sp. and *Scleroderma* sp. in the presence of NaCl at a range of concentrations in axenic culture for 1 mon (Two-ways ANOVA, P<5%).

external water potentials through application of the organic osmotica such as polyethylene glycol (Jennings 1995).

It is well known that ECM fungi tolerate high osmotic potential and toxic metal ions by combining compartmentalization of toxic ions and production of polyols that acts as nontoxic osmoregulators (Clipson and Jennings 1992, Chen et al. 2001, Bois et al. 2006). Such an ability may explain the observed growth of ECM fungi under conditions of high external osmotic stress, and may in part explain the resistance to salt that we have identified in *S. bermudense*. Interestingly, *S. bermudense* produced more fungal biomass and accumulated more Na^+ than *Melanogaster* sp. and *Scleroderma* sp. under salt stress, showing that *S. bermudense* was more tolerant to high NaCl concentrations (Figs. 5 and 6). Also, compatible osmolytes (e.g., proline) accumulated in the cytoplasm of *S. bermudense* to counter excess ions in the vacuole (Fig. 7). The results of this study showed that *S. bermudense* accumulated more Na^+ than the other ECM fungi, which means a physical barrier for plant development supported by *S. bermudense* was stronger than those supported by the other fungi. *S. bermudense* has been selected to alleviate salt stress in seagrape seedlings.

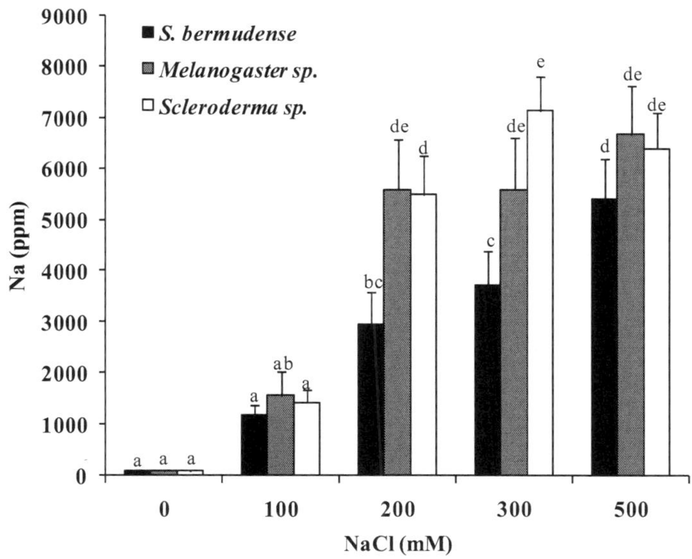

Fig. 6. Na contents of *S. bermudense, Melanogaster* sp. and *Scleroderma* sp. in the presence of NaCl at a range of concentrations in axenic culture for 1 mon (Two-ways ANOVA, P<5%).

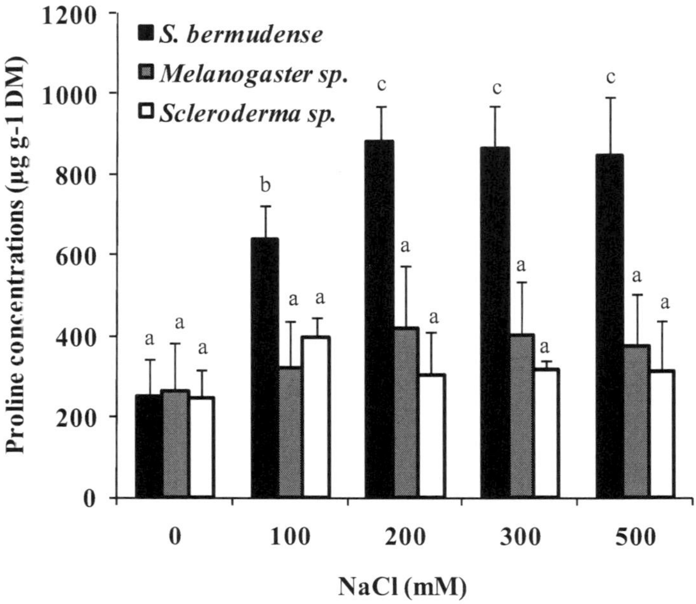

Fig. 7. Proline contents of *S. bermudense, Melanogaster* sp. and *Scleroderma* sp. in the presence of NaCl at a range of concentrations in axenic culture for 1 mon (Two-ways ANOVA, P<5%).

4. *Scleroderma bermudense* Alleviates Salt Stress in Seagrape Seedlings

The beneficial effects of *S. bermudense* on seagrape seedlings occurred not only during NaCl stress but also under non-stressed conditions, suggesting that the ECM effect in improving plant growth was not a specific process induced by saline stress (Table 3). These results recall observations on the beneficial effect of AM fungi on plant growth under salt stress conditions (Al-Karaki et al. 2001, Yano-Melo et al. 2003, Giri and Mukerji 2004, Tian et al. 2004). Although ECM colonization by *S. bermudense* was reduced with increasing NaCl levels, ECM dependency of seagrape seedlings was increased. This suggests that the activity of the ECM symbiosis between seagrape seedlings and *S. bermudense* was strengthened under saline stress conditions once the symbiosis was established.

In our study, ECM plants showed considerably reduced Na and Cl uptake into roots as compared to non-inoculated plants (Figs. 8 and 9). Moreover, ECM inoculation of plants lowered Na and Cl translocation to shoot and leaf tissues. The marked differences in Na and Cl concentrations between ECM and non-ECM plant tissues suggest that the mechanisms of enhanced tolerance to salt stress in ECM plants were related to exclusion of

Table 3. Effects of inoculation with *Scleroderma bermudense* and NaCl levels on growth variables and ectomycorrhizal (ECM) colonization in seedlings of *Coccoloba uvifera* at 3 mon (p<1%) (Bandou et al. 2006).

ECM status	NaCl levels (mM)	Height (cm)	Total biomass (g)	EM dependency (%)	EM colonization (%)
Non-inoculated	0	6.2 b	1.4 b	-	0.0 a
	200	5.0 a	0.7 a	-	0.0 a
	350	5.5 a	0.6 a	-	0.0 a
	500	5.1 a	0.6 a	-	0.0 a
Inoculated	0	7.2 d	3.1 f	55.8 a	89.4 d
	200	6.8 cd	2.6 e	71.7 c	66.6 c
	350	7.0 d	2.3 d	75.2 c	46.3 b
	500	6.2 bc	1.9 c	69.5 c	41.6 b

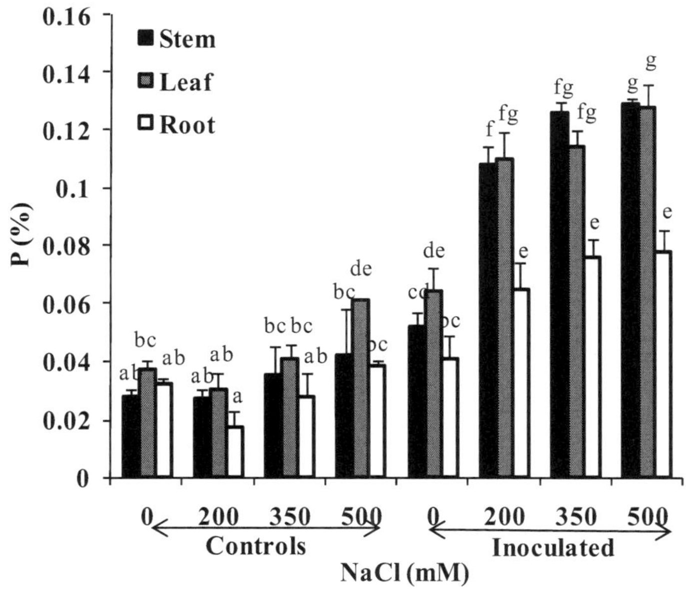

Fig. 8. Changes in phosphorus (P) concentrations in the leaves, shoots, and roots of *C. uvifera* seedlings in response to NaCl levels and inoculation with the ectomycorrhizal (ECM) fungus *S. bermudense* at 3 mon (Bandou et al. 2006).

Na and Cl by the ECM plants growing in the NaCl treatments. One of the main mechanisms for enhanced salinity tolerance in ECM plant has also been proposed to be the exclusion of Na and Cl (Giri and Mukerji 2004, Tian et al. 2004).

ECM plants had higher concentrations of P than non-inoculated plants particularly under salt stress conditions. This could be due to the extended network of hyphae beyond the depletion zone around roots that acquire more P and therefore suppress the adverse effect of salinity stress. These results suggest that *S. bermudense* can alleviate the deleterious effects of NaCl stress on plants by mechanisms which may also be related to improvement of Puptake (Fig. 8).

If Na and Cl are sequestered into plant vacuoles, K and organic solutes (e.g., proline) should accumulate in the cytoplasm and organelles to balance the osmotic pressure of intravacuolar ions (Munns 2002). Na and Cl were not accumulated under salt stress and found at low concentrations in ECM seedlings of *C. uvifera* when compared to controls (Figs. 9 and 10).

It was also noteworthy that K concentrations were higher in leaves of ECM seagrape seedlings than in controls at all salinity levels. *S. bermudense* seemed to protect seedlings by increasing the K/Na ratio to maintain higher turgor under salt stress (Figs. 11 and 12). It is well known that high concentrations of K and low concentrations of Na can play a major role in salt stress tolerance by regulating stomatal openings and osmotic potential in the vacuoles (Zhu 2003, Munns 2005). Some studies indicated that plants subjected to increased salinity accumulated less K because Na uptake competed with the uptake of K, leading to decreased K levels (Larcher 1995, Parida and Das 2005). Reductions in Na and Cl uptake and

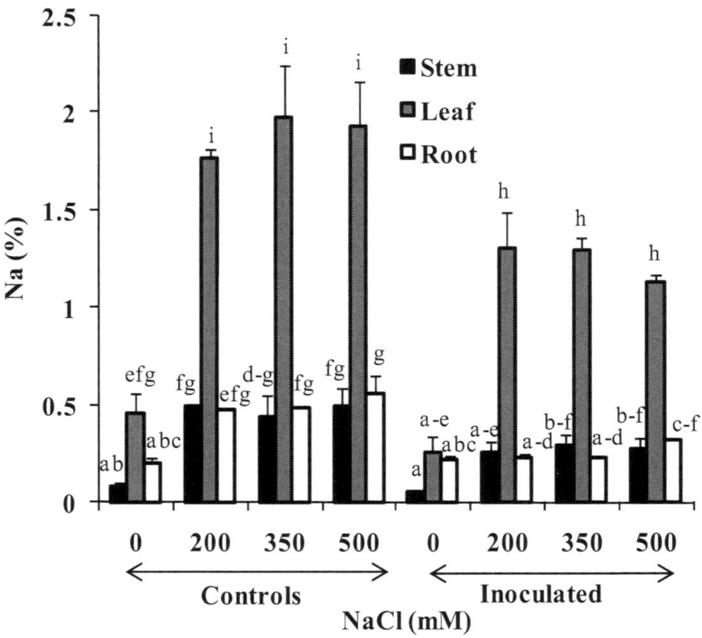

Fig. 9. Changes in sodium (Na) concentrations in the leaves, shoots, and roots of *C. uvifera* seedlings in response to NaCl levels and inoculation with the ectomycorrhizal (ECM) fungus *S. bermudense* at 3 mon (Bandou et al. 2006).

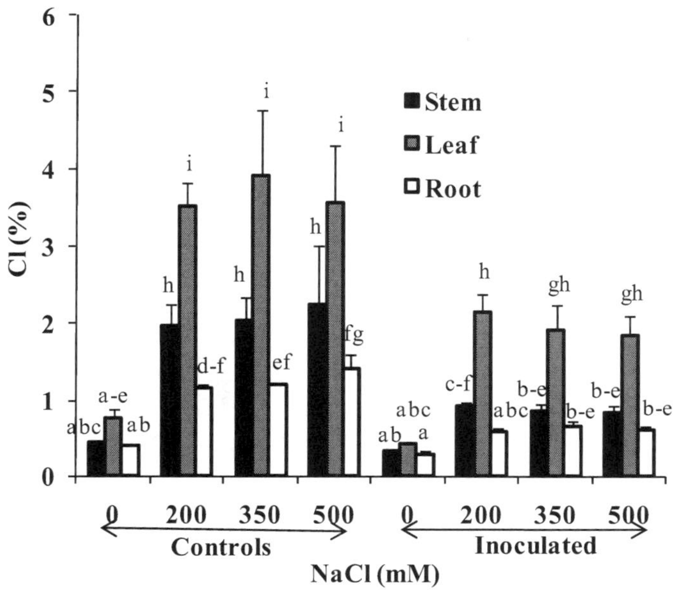

Fig. 10. Changes in chloride (Cl) concentrations in the leaves, shoots, and roots of *C. uvifera* seedlings in response to NaCl levels and inoculation with the ectomycorrhizal (ECM) fungus *S. bermudense* at 3 mon (Bandou et al. 2006).

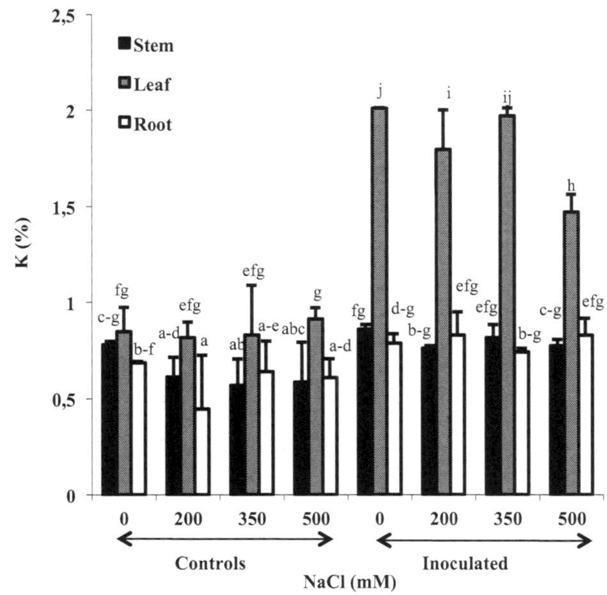

Fig. 11. Changes in potassium (K) concentrations in the leaves, shoots, and roots of *C. uvifera* seedlings in response to NaCl levels and inoculation with the ectomycorrhizal (ECM) fungus *S. bermudense* at 3 mon (Bandou et al. 2006).

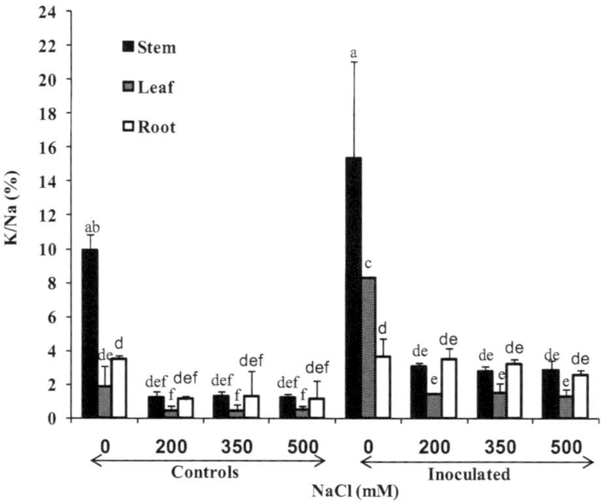

Fig. 12. Changes in ration K/Na concentrations in the leaves, shoots, and roots of *C. uvifera* seedlings in response to NaCl levels and inoculation with the ectomycorrhizal (ECM) fungus *S. bermudense* at 3 mon (Bandou et al. 2006).

enhanced P and K nutrition may be significant in helping the ECM plants to grow under saline conditions (Table 4). There were significant negative correlations between Na and Cl concentrations in the leaves, shoots, and roots, and production of biomass (Table 4). The exclusion of Na and Cl could be the main mechanism whereby mycorrhiza protected seagrape seedlings against salinity.

The contribution of ECM fungi to plant growth in stressed soil is not limited to the improvement in mineral nutrition but also to the changes in the morphological characters of the leaves and the water status of plants (Augé 2001). Measurements of minimal leaf water potentials are considered informative on evapotranspiration to assess the water status of plants (Larcher 1995). *S. bermudense* improved the water status of seagrape

Table 4. Correlation coefficients (r) between K, P, Na, Cl, and proline, and total biomass production (Bandou et al. 2006).

	Inoculated[a]	Non-inoculated[a]	Inoculated plus non-inoculated[b]
K	0,65*	0.50	0.91*
P	0.85*	0.25	0.55*
Na	–0.70*	–0.94*	–0.74*
Cl	–0.71*	–0.91*	–0.79*
Proline	0.95*	–0.93*	0.94*

* Significant (p<1%); [a] 10 *df*; [b] 22 *df*

seedlings despite their higher evaporative leaf surface (Table 5). Also, it is noteworthy that leaf proline content was higher in leaves of ECM seagrape seedlings than in the controls at all salinity levels (Fig. 13). *S. bermudense* seemed to also protect seagrape seedlings by increasing the proline content to maintain higher turgor under salt stress.

Table 5. Effects of inoculation with *Scleroderma bermudense* and NaCl levels on the number of leaves, leaf area, and minimal leaf water potential in seedlings of *Coccoloba uvifera* at 3 mon ($p<1\%$) (Bandou et al. 2006).

ECM status	NaCl levels (mM)	Number of leaves	Leaf area (cm^2)	Minimal leaf water potential (-bars)
Non-inoculated	0	3.60 d	33.7 b	9.2 e
	200	2.90 c	18.7 a	13.4 bc
	350	2.30 b	15.6 a	14.0 b
	500	1.90 a	15.7 a	15.0 a
Inoculated	0	4.10 e	80.9 e	5.9 f
	200	4.20 e	76.4 d	11.1 d
	350	3.80 de	68.9 d	12.2 cd
	500	3.60 d	56.05 c	15.50 a

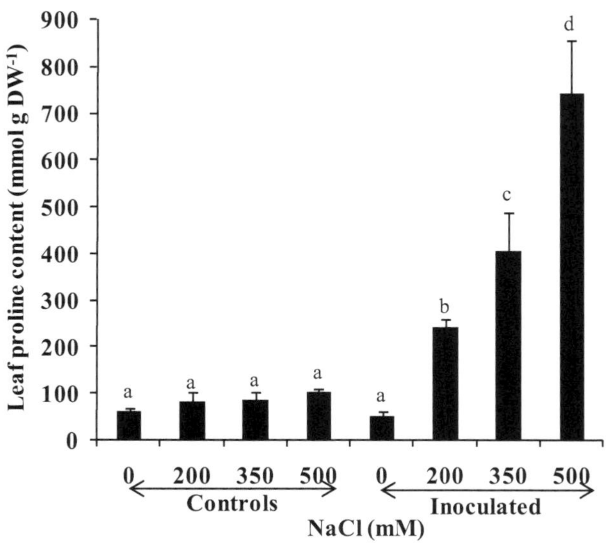

Fig. 13. Changes in proline concentrations in the leaves of *C. uvifera* seedlings in response to NaCl levels and inoculation with the ectomycorrhizal (ECM) fungus *S. bermudense* at 3 mon.

5. Conclusion

Salinity had a negative impact on growth of mature trees and naturally regenerating seedlings of *C. uvifera*. Salinity could contribute to the structuration of the three *Coccoloba* species as well as their associated ECM fungi. The data from the present study also suggest that *S. bermudense* possess considerable resistance to salinity. Also, seagrape tolerance to salt stress was considerably enhanced by *S. bermudense*. Reductions in Na and Cl uptake and enhanced P nutrition may be significant in helping the ECM plants to grow under saline conditions. Furthermore, the higher accumulation of K in the leaves of ECM plants could render ECM plants more tolerant to osmotic stress induced by exposure to salt. The overall improved water status of seagrape seedlings related to ECM inoculation with *S. bermudense* may also be a contributing factor. Further studies should be done to elucidate the physiological mechanisms (e.g., stomatal conductance, water use efficiency, transpiration, photosynthetic rate, osmotic adjustment) by which *S. bermudense* improved the water status of seagrape seedlings. Our results should be taken cautiously as the experiment was a greenhouse study with 3 month-old seedlings. From an ecological point of view, the ECM symbiosis may be an important factor for the establishment and survival of seagrape stands under saline stress along the beaches of the Caribbean. Moreover, transplanting ECM-inoculated seagrape to such degraded sites not only benefits the individual plant but, more importantly, may result in the development of ornamental plantings and coastal windbreaks along Caribbean beaches and roadsides.

Acknowledgements

We thank the anonymous referees for their valuable comments on this study, and Krista L. McGuire and Caitlyn Gillikin for improving the language.

References

Al-Karaki, G.N., R. Hammad and M. Rusan. 2001. Response of two tomato cultivars differing in salt tolerance to inoculation with mycorrhizal fungi under salt stress. Mycorrhiza 11: 43–47.

Augé, R.M. 2001. Water relations, drougth and vesicular-arbuscularmycorrhizal symbiosis. Mycorrhiza 11: 3–42.

Bandou, E. 2005. Diversité et fonctionnement des symbioses ectomycorhiziennes de *Coccoloba uvifera* (L.) L. en situation de stress salin et hydrique. Master of Science thesis, UAG, p. 36.

Bandou, E., F. Lebailly, F. Muller, M. Dulormne, A. Toribio, J. Chabrol, R. Courtecuisse, C. Plenchette, Y. Prin, R. Duponnois, M. Thiao, S. Sylla, B. Dreyfus and A.M. Bâ. 2006. The ectomycorrhizal fungus *Scleroderma bermudense* alleviates salt stress in seagrape (*Coccolobauvifera* L.) seedlings. Mycorrhiza 16: 559–565.

Bessard, S., I. Boulogne, J. Chabrol, J.P. Fiard, C. Lécuru, R. Courtecuisse and A.M. Bâ. 2008. Diversity of ectomyorrhizal fungi and ectomycorrhizas of three *Coccoloba* species in Martinique. Poster presented at the Workshop « Mycorrhizas in tropical forests », Universidad Téchnica Particular de Loja, Ecuador.
Bois, G., A. Bertrand, Y. Piché, M. Fung and D.P. Khasa. 2006. Growth, compatible solute and salt accumulation of five mycorrhizal fungal species grown over a range of NaCl concentrations. Mycorrhiza 16: 99–109.
Chen, D.M., S. Ellul, K. Herdman and J.W.G. Cairnay. 2001. Influence of salinity on biomass production by Australian *Pisolithus* spp. isolates. Mycorrhiza 11: 231–236.
Clipson, N.J.W. and D.H. Jennings. 1992. *Dendryphiellasalina* and *Debaryomyceshansenii*: models for ecophysical adaptation to salinity by fungi that grow in the sea. Canadian Journal of Botany 70: 2097–2105.
FAO. 2005. ftp://ftp.fao.org/agl/iptrid/salinity_brochure_fr.pdf.
Fournet, J. 2002. Flore illustrée des phanérogames de Guadeloupe et de Martinique. Editions Quae, Tomes 1 and 2, p. 2538.
Flowers, T.J. and T.D. Colmer. 2008. Salinity tolerance in halophytes. New Phytol. 179: 945–963.
Giri, B. and K.G. Mukerji. 2004. Mycorrhizal inoculant alleviates salt stress in *Sesbaniaaegyptiaca* and *Sesbaniagrandiflora* under field conditions : evidence for reduced sodium and improved magnesium uptake. Mycorrhiza 14: 307–312.
Guzman, G., F. Ramirez-Guillén, O.K. Miller, D.J. Lodge and T.J. Baroni. 2004. *Sclerodermastellatum* versus *Scleroderma bermudense* : the status of *Scleroderma echinatum* and the first record of *Veligasternitidum* from the Virgin Islands. Mycologia 96: 1370–1379.
Howard, R. 1988. Flora of the lesser Antilles. (F LAnt) 4, pp. 132.
Jennings, D.H. 1995. The physiology of fungal nutrition. Cambridge University Press, Cambridge, England.
Juniper, S. and L. Abbott. 1993. Vesicular-arbuscularmycorrhizas and soil salinity. Mycorrhiza 4: 45–57.
Katembe, W.J., I.A. Ungar and J.P. Mitchell. 1998. Effect of salinity on germination and seedling growth of two *Atriplex*species (Chenopodiaceae). Ann. Bot. 82: 165–175.
Kernaghan, G., B. Hambling, M. Fung and D. Khasa. 2002. *In vitro* selection of Boreal ectomycorrhizal fungi for use in reclamation of saline-alkaline habitats. Restor. Ecol. 10: 1–9.
Kreisel, K. 1971. Clave para la identificacion de los macromicetos de Cuba. La Habana: Ser. A, CiencasBiologicas 16, Universidad de la Habana, p. 101.
Langenfeld-Heyser, R., J. Gao, T. Dicic, P.H. Tachd, C.F. Lu, E. Fritz, A. Gafur and A. Polle. 2007. *Paxillusinvolutus* mycorrhiza attenuate NaCl-stress responses in the salt-sensitive hybrid poplar *Populus* x *canescens*. Mycorrhiza 17: 121–131.
Larcher, W. 1995. Physiological Plant Ecology: Ecophysiology and Stress Physiology of Functional Groups. Springer, Third edition.
Miller, O.K., D.J. Lodge and T.J. Baroni. 2000. New and interesting ectomycorrhizal fungi from Puerto Rico, Mona, and Guana Islands. Mycologia 92: 558–570.
Munns, R. 2002. Comparative physiology of salt and water stress. Plant Cell Env. 25: 239–250.
Munns, R. 2005. Genes and salt tolerance: bringing them together. New Phytol. 167: 645–663.
Parida, A.K. and A.B. Das. 2005. Salt tolerance and salinity effects on plants: a review. Ecotox. Env. Safety 60: 324–349.
Parrota, J.A. 1994. *Coccoloba uvifera* (L.) L., seagrape, uva de playa. Research note SOITF-SM-74. U.S. Department of Agriculture, Forest Service, Southern Forest Experiment Station, New Orleans, LA pp. 5.
Pegler, D.N. 1983. Agaric flora of the Lesser Antilles. Kew Bull Add Ser 9, pp. 668.
Smith, S. And J. Read. 2008. Mycorrhizal symbiosis. Hardcover, pp. 800.

Shen, B., S. Hohman, R.G. Jensen and H.J. Bohnert. 1999. Roles of sugar alcohols in osmotic stress adaptation. Replacement of glycerol by mannitol and sorbitol in yeast. Plant Physiol. 121: 45–52.
Tang, M., M. Sheng, H. Chen and F.F. Zhang. 2009. *In vitro* resistance of three ectomycorrhizal fungi. Soil Biol. Bioch. 41: 948–953.
Tian, C.Y., G. Feng, X.L. Li and F.S. Zhang. 2004. Different effects of arbuscularmycorrhizal fungal isolates from saline or non-saline soil on salinity tolerance of plants. Appl. Soil Ecol. 26: 143–148.
Yano-Melo, A.M., O.J. Saggin and L.C. Maia. 2003. Tolerance of mycorrhized banana (*Musa* sp. cv. Pacovan) plantlets to saline stress. Agr. Eco. Env. 95: 343–348.
Zhu, J.K. 2003. Regulation of ion homeostasis under salt stress. Curr. Opinion Plant Biol. 6: 441–445.

CHAPTER 10

The Contribution of Ectomycorrhizal Fungal Feedbacks to the Maintenance of Tropical Monodominant Rain Forests

Krista L. McGuire

1. Introduction

Tropical monodominant forests have been classically defined as areas of forest where one tree species comprises >60% of the stems or basal area of canopy trees (Torti et al. 2001). These forests occur throughout all tropical regions (Fig. 1) and have been noted as interesting deviations from the more extensive high-diversity rain forests (Leigh et al. 2004). A number of mechanisms have been suggested to explain their existence and persistence among the mosaics of highly diverse forest (Connell and Lowman 1989, Hart 1990, Torti and Coley 1999), however, a single explanation for tropical monodominance does not suffice (Torti et al. 2001, Peh et al. 2011). It is now widely accepted that multiple mechanisms must be operating simultaneously that lead to feedbacks, which enable a monodominant tree species to persist and maintain site dominance (McGuire 2008).

Columbia University, Barnard College Department of Biology, USA.
Email: kmcguire@barnard.edu

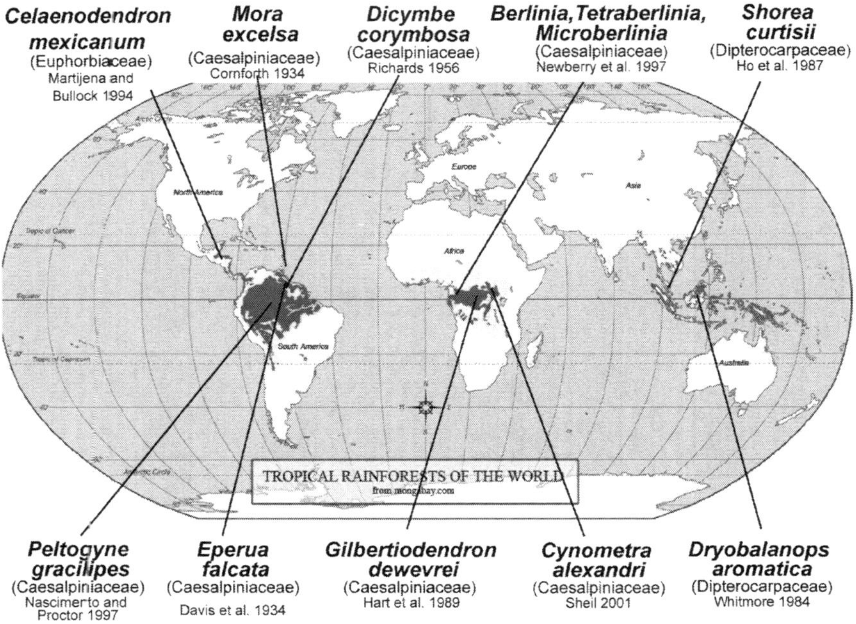

Fig. 1. Map illustrating examples of monodominant and co-dominant forests with the tree species or genus listed and the tree family in parentheses. This diagram is not an exhaustive list of monodominant species, but it illustrates how monodominant forests are found in all tropical regions.

Color image of this figure appears in the color plate section at the end of the book.

One attribute that many monodominant trees have in common is that they form ectomycorrhizal (ECM) associations. Most tropical tree species form associations with arbuscular mycorrhizal (AM) fungi, but there is a strong correlation between the dominance of tropical trees in lowland rain forests and ECM associations (Torti et al. 2001). This pattern is particularly interesting when considering how few tree species actually form ECM associations. A recent synthesis found that nearly 75% of all angiosperm plant species form AM associations, whereas only about 2% form ECM associations (Brundrett 2009). Nonetheless, the majority of tropical monodominant trees form ECM associations, suggesting that this trait somehow contributes to the success of monodominant trees. However, being ECM does not guarantee that a tree will become locally dominant, as there are numerous examples of less abundant tree species that form ECM associations, as well as monodominant trees that are AM. For this reason, Torti et al. (2001) developed a list of traits that most monodominant trees share and suggested that there must be interactions among these traits that together promote positive feedbacks leading to monodominance.

Community feedback models clearly demonstrate that negative plant-soil feedbacks lead to plant coexistence and positive feedbacks lead to dominance (Bever et al. 2010, Bever et al. 2012). These models account for individual plant responses to differing soil microbes and then scale these interactions to community-level patterns of plant diversity. The negative plant-soil feedbacks are predicted to originate from plant-pathogen interactions and the positive feedbacks are derived from plant-symbiont interactions. These theoretical models have been empirically supported by a number of studies, although the majority of the experimental work has been done as pot studies and herbaceous plant manipulations. Very few tropical studies have been performed and little is known about the functioning of plant-soil feedbacks in undisturbed systems. Tropical ECM monodominant forests are ideal model systems for exploring these feedbacks, as they offer a natural diversity gradient that has not been significantly altered by human manipulation. Since the outcome of undisturbed forest dynamics in these systems is that of single species dominance, it can be inferred that positive plant-soil feedbacks have outweighed negative feedbacks in these systems (Bever et al. 2010).

For the current review, I will be focusing on a subset of tropical monodominant tree species: those that form ectomycorrhizal (ECM) associations in lowland, aseasonal rain forests, and are not associated with extreme edaphic conditions or periodic disturbances. The reason for focusing on ECM monodominant forests is that the mechanisms operating to maintain monodominance likely include ECM-mediated feedbacks that are unique to these systems, as the ecology, evolution, and physiology of ECM fungi are different than for fungi that form arbuscular mycorrhizal (AM) associations, the dominant mycorrhizal type in most diverse tropical forests. As such, there are several attributes of ECM fungi that may confer a competitive advantage to their host trees, particularly when growing in nutrient-poor soils (Smith and Read 2008). For example, ECM fungi can grow much larger than AM fungi due to the septate nature of their cells. When ECM hyphae are broken, they can begin growing again from the last septation, enabling mycelia to grow tens and sometimes hundreds of meters in diameter. By contrast, AM fungi have coenocytic hyphae, so following cellular damage they must resume growth from the root-fungal interface. Consequently, AM fungi are not thought to grow mycelia that are more than several centimeters in diameter. This discrepancy in size places ECM fungi at an advantage in terms of nutrient foraging area and space occupancy. In fact, where ECM fungi occur, they can comprise up to 90% of the fungal biomass (Wallander et al. 2001). Also, the recent evolutionary divergence of multiple ECM fungal lineages from decomposer fungi has implications for functional differences in organic resource acquisition capabilities of AM and ECM fungi, as many ECM fungi possess and express genes that are involved in the degradation of organic polymers (Abuzinadah and Read 1986, Floudas et al. 2012). These advantages are realized for ECM trees in terms of growth, nutrient uptake, drought tolerance, pathogen resistance, and productivity.

This review explores empirical studies and hypothetical mechanisms that explicitly incorporate the unique biology of ECM fungi and focuses on a particular ECM monodominant system from Guyana, South America. Finally, these various mechanisms are synthesized into a feedback framework (Fig. 2) that integrates three non-mutually exclusive positive mechanisms derived from the ECM symbiosis that may contribute to monodominance: 1) ECM fungal mutualists preferentially benefit the ECM host tree, but donot benefit heterospecific trees, 2) positive feedbacks from ECM fungal mutualists are stronger than negative feedbacks from other soil microorganisms, and 3) ECM fungal mutualists suppress positive plant-soil feedbacks for heterospecific trees.

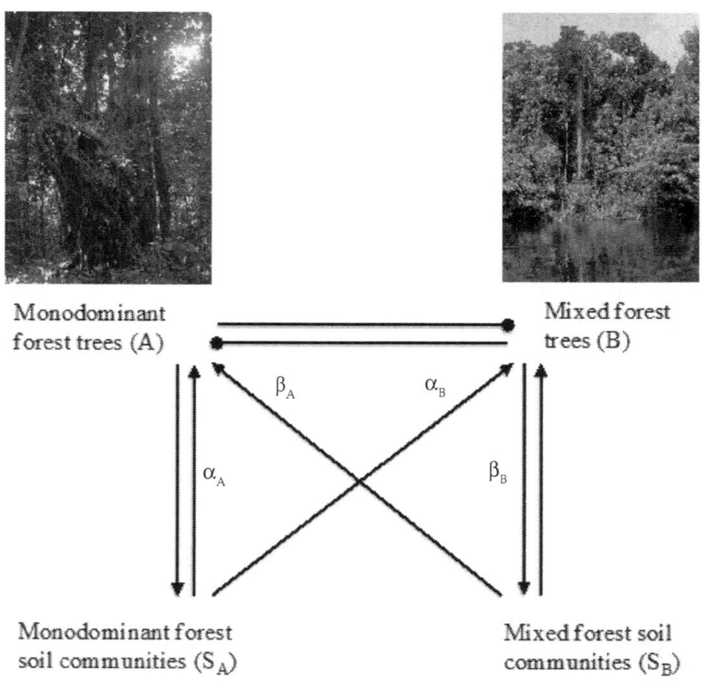

Fig. 2. Incorporation of monodominant and mixed forest trees into theoretical soil-feedback models (adapted from Bever et al. 2010, 2012). The direction and strength of the plant-soil feedbacks should correlate with the relative abundance of the tree species. The net pairwise effect can be quantified by evaluating $((\alpha_A + \beta_B) - (\alpha_B + \beta_A))$.

2. *Dicymbe corymbosa* as a Case Study

2.1 Study site

The ECM monodominant system used as a case study for exploring these feedback mechanisms is located in Guyana, South America (Fig. 3; 5°4′ N, 59°58, 700 m a.s.l.). The forest in this region is intact, aseasonal rain forest receiving approximately 4000 mm mean annual precipitation. There are extensive stands of ECM, monodominant *Dicymbe corymbosa* (Caesalpiniaceae) Spruce ex Benth. in the interior forming mosaics with higher diversity, mixed forest in which *D. corymbosa* is not found. *Dicymbe corymbosa* is an emergent tree species reaching 30–40 m and reaches dominance levels of 70–90 percent where it occurs (Henkel 2003, McGuire 2008). This monodominant system was first described by Richards (Richards 1952) and has been intensively studied by teams of ecologists and mycologists since the early 2000s. Apart from the *Gilbertiodendrondewevreii* monodominant forest in the Congo (Hart 1990, Torti et al. 2001), it has become one of the most intensively studied ECM, monodominant forests in the tropics.

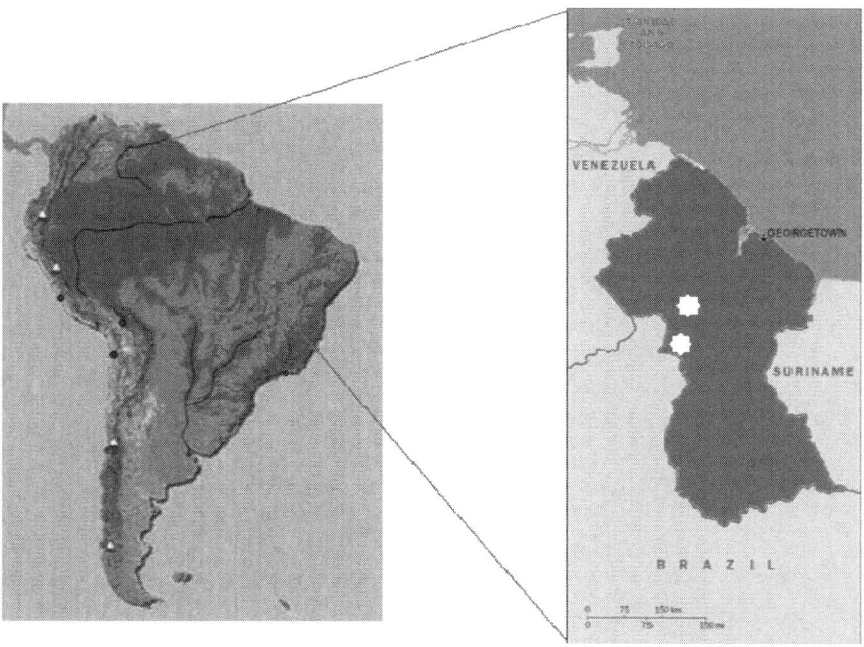

Fig. 3. Map showing the ECM monodominant *Dicymbe corymbosa* forest and mixed forest sites in Guyana.

2.2 Positive ECM-mediated feedbacks

Field experiments have revealed several positive feedbacks that are seemingly conveyed by the ECM mutualism in *D. corymbosa*, but do not benefit heterospecific trees. One of the major feedbacks is that of seedling facilitation near to conspecific adult trees in the monodominant forest (McGuire 2007a,b). That is, seedlings grow and survive better when planted near other *D. corymbosa* trees compared to when they are planted near heterospecific trees. To date, there have been two mechanisms explaining the positive distance/density-dependence of *D. corymbosa* seedling establishment. The first mechanism is that common mycorrhizal networks facilitate seedling growth and survival. This mechanism was supported by a field experiment using nylon mesh designed to exclude or allow the mycelial network to connect to the seedling roots. After one year, seedlings with access to the ECM network had significantly greater growth and survival than seedlings excluded from the network (McGuire 2007a). However, all seedlings had nearly 100% ECM root colonization, so the seedling responses were not due to limited ECM colonization, but rather access to the common mycorrhizal network. Mycorrhizal networks have been shown to influence plant community dynamics and seedling establishment in temperate ECM systems (Newman 1988, Simard and Durall 2004, Simard 2009), and detailed tracer studies using autoradiography have revealed that nutrient translocation can occur from plant to plant via these connections (Brownlee et al. 1983, Finlay and Read 1986, Wu et al. 2012). Translocation of P and C via mycorrhizal networks has been demonstrated in pine seedlings, and implies that there is the potential that monodominant ECM adult trees may support conspecific seedlings by similar pathways. The mechanism of photosynthate transfer would be particularly advantageous for seedlings in the light-limited understory of the tropical rain forest, as access to light is often the most limiting resource for seedling survival (Chazdon 1988). This positive feedback would be limited to the ECM monodominant tree in an otherwise AM rain forest community, and, therefore,fits within the plant-soil feedback framework to explain tropical ECM monodominance.

A second mechanism by which positive distance/density-dependence occurs is enhanced recruitment near to conspecific adults due to greater ECM inoculum density. This mechanism was supported in another experiment in which *D. corymbosa* seeds were planted in both the monodominant and the mixed forest, along with four other non-ECM tree species commonly found in both forests (McGuire 2007b). In the mixed forest, *D. corymbosa* seedlings had significantly lower mycorrhizal colonization after one year compared to seedlings planted in the monodominant forest. The reduced colonization was reflected in the lower survival of seedlings planted in mixed forest. Responses of the non-ECM seedlings varied, but the magnitude of survival in the monodominant forest relative to the mixed forest was much smaller than the advantage gained by *D. corymbosa* seedlings (Fig. 4). It was possible that these mixed tree species were benefitting from lower pathogen and seed predator densities, since the overall density of their conspecific adult trees was lower in the monodominant forest, although these mechanisms have not been explicitly tested. However, two seed density treatments were implemented for all tree species to test the prevalence of seed predation in both forest types, and there were no notable differences in seed predation rates in either forest type for any of the density treatments. Thus, it appeared that seedling recruitment, rather than seed survival, was more reflective of canopy diversity patterns, and that ECM inoculum density was a major factor driving these patterns for *D. corymbosa*. The reduced availability of ECM inoculum was also reflected in a molecular study that sequenced fungal DNA from forest floor samples in the monodominant and mixed forests (McGuire et al. 2010). There were significantly fewer DNA sequences aligning to ECM fungi in the mixed forest compared to the monodominant forest, which reflects the reduced colonization levels of *D. corymbosa* seedlings in the 2007 study. Whether *D. corymbosa* seedlings are subjected to pathogen attack in the mixed forest because of lower ECM inoculum has not been determined, but is a plausible hypothesis that may directly relate to the lower ECM colonization levels. In other ECM trees, reduced colonization can lead to increased vulnerability to root pathogen attack (Marx 1972, Whipps 2001, 2004), and may be a mechanism preventing the rapid spread of the ECM monodominant forest into the mixed forest.

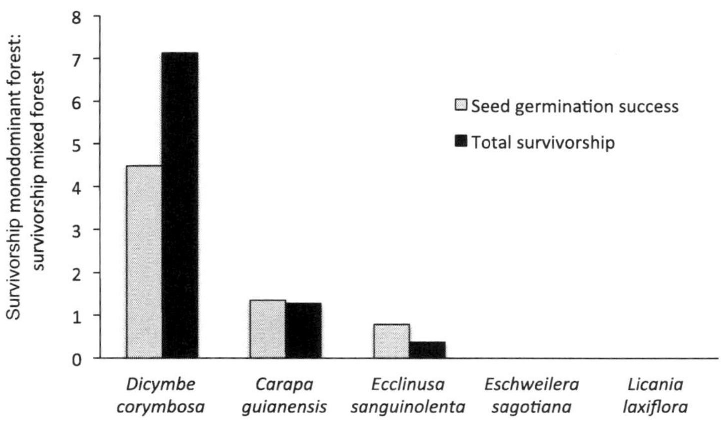

Fig. 4. Survivorship of seeds and seedlings after one year in monodominant compared to mixed forest. Seeds of *Dicymbe corymbosa* and four other species found in both forest types were reciprocally transplanted. Data are reproduced from McGuire 2007a.

An important component to plant-soil feedback theory is that greater density of microbial mutualists leads to greater positive feedback. The facilitation of seedling growth and survival near the conspecifics fits within this theoretical framework, as *D. corymbosa* seedlings are receiving greater benefit than heterospecific seedlings from the accumulated symbionts near adult *D. corymbosa* trees. There are numerous studies demonstrating this linkage in other ECM systems. For example, the highly invasive *Alliariapetiolata* in North America has been found to derive its success by decreasing ECM inoculum and root colonization of native ECM trees, which negatively affects seedling establishment of native ECM trees (Wolfe et al. 2008). Another recent study of *Populusfremontii* found that the presence of an invasive neighboring plant decreased ECM inoculum and percent colonization in the *Populus* trees, which resulted in decreased tree growth (Meinhardt and Gehring 2012). In other ECM monodominant systems, however, these mechanisms of ECM-mediated positive feedbacks have not been explicitly tested. In more diverse tropical ECM systems, such as the diverse dipterocarp forests of SE Asia, some trees seem to exhibit positive feedbacks, while others do not (Brearley 2012). However, since dipterocarp forests contain an extraordinary diversity of ECM trees, the mechanisms operating are likely to differ from an ecosystem in which only one ECM tree dominates. Also, most of the dipterocarp trees occur in low abundances, so feedback models would predict that positive feedbacks are not as dramatic as in monodominant systems.

2.3 ECM feedbacks to nutrient cycling

It has long been observed that litter decomposition in ECM forests is generally slower than in non-ECM systems, resulting in the accumulation of thick humic layers (Smith and Read 2008). There have been several reasons put forth to explain these observations. One major mechanism proposed is that that leaf litter of ECM trees is somehow more recalcitrant and resistant to decay, as has been observed in the Pinaceae (Read 1991). However, in the *D. corymbosa* system, this mechanism has not been supported. In a reciprocal leaf litter decomposition experiment, *D. corymbosa* litter decomposed at the same rate as mixed species litter in both forest types (McGuire et al. 2010). Overall, decomposition was slower in the ECM forest for all litter, suggesting that the process of decomposition was slowed by something intrinsic to the monodominant forest, rather than the decomposability of the monodominant litter. This hypothesis was further supported by two years of bi-weekly litterfall quantification, in which it was found that total biomass of litter input to the monodominant forest was less than litter input into the mixed forest. Thus, the litter accumulation in the *D. corymbosa* forest could not be attributed to greater litter production by *D. corymbosa* or to litter recalcitrance. Fungal DNA sequencing revealed that there were significantly fewer saprotrophic fungi in the monodominant forest than in the mixed forest and that the fungal communities were compositionally distinct across forest types. The conclusion was, therefore, that the slowed decomposition in the ECM forest was somehow biotically mediated.

In the ECM monodominant forest dominated by *Gilbertiodendrondewevreii* in the Congo litter decomposition was also observed to beslower than in the mixed forest (Torti et al. 2001). As in the *D. corymbosa* system, decomposition rates were not driven by differences in leaf litter recalcitrance, as mixed litter and *G. dewvreii* litter did not significantly differ in decomposition rates when they were placed in the same forest type. Further evidence refuting the hypothesis that ECM tree litter is more recalcitrant comes from a recent meta-analysis, which revealed that ECM plants do not have consistent leaf traits in terms of nutrient content and litter recalcitrance (Koele et al. 2012). Instead, it appears that the majority of studies of ECM trees have been focused on the Pinaceae and Fagales (Dickie and Moyersoen 2008), so there has been a bias in our understanding of ECM plant traits.

Slowed decomposition in ECM monodominant forests, and perhaps other ECM forests, must be driven by factors other than leaf chemistry. If litter decay rates are indeed biotically driven in these forests, then there may be some interactive feedback between ECM fungi and other decomposer microorganisms. This line of reasoning, and the fact that ECM fungi and decomposer fungi may be competing for some of the same resources and share the same spatial environment, has led to the formulation of the Gadgil hypothesis (Gadgil and Gadgil 1971, Gadgil and Gadgil 1975). This hypothesis posits that ECM fungi suppress saprotrophic microorganisms, which results in slowed decomposition, as ECM fungi are not able to decompose organic litter as quickly as free-living saprotrophs (Colpaert and van Tichelen 1996). Since ECM fungi obtain the majority of their C from their associate plant (Hobbie et al. 2002), they have lost genes associated with rapid C acquisition in the early stages of decomposition (Wolfe et al. 2012). However, ECM fungi do maintain the capacity to directly degrade other forms of organic polymers such as proteins, cellulose, and fatty acids (Abuzinadah and Read 1986, Talbot et al. 2008, Rineau et al. 2012). Therefore, the rationale for the Gadgil hypothesis is that ECM fungi may engage in some form of antagonistic relationship with saprotrophs so that they are allowed enough time to gain access to organically-bound nutrients such as N and P.

There is direct evidence supporting the Gadgil hypothesis from laboratory studies and indirect evidence from field studies. In laboratory microcosms, ECM fungi have been found to suppress both saprotrophic fungal growth and bacterial growth (Olsson et al. 1996, Lindahl et al. 2001). In one study using autoradiography, ECM fungi were found to raid the saprotrophic fungal mycelium of P, which was subsequently transferred to the ECM plant host (Lindahl et al. 1999). Allelopathy through common mycorrhizal networks has also been observed (Barto et al. 2011), which may be one mechanism by which ECM fungi directly interact with saprotrophic fungi and bacteria.

In Guyana, molecular analyses suggest that fewer decomposer fungi and potential bacteria reside in the monodominant forest compared to the mixed forest (McGuire et al. 2010), which is consistent with the predictions of the Gadgil hypothesis. Total microbial biomass was also lower in the monodominant forest and microbial community composition was distinct, suggesting that decomposer communities were indeed different in the ECM forest. Since litter decay was also slower, these various lines of evidence at least offer some indirect support for the Gadgil hypothesis. While there are numerous alternative hypotheses that may explain the patterns observed from field studies, it remains a plausible hypothesis worthy of further testing in ECM monodominant systems. A further extension of this hypothesis that is worth exploring is that ECM fungi engage in antagonistic relationships with saprotrophs foraging for more labile forms of C, N, and P, but that ECM fungi facilitate persistence of saprotrophs that directly degrade more recalcitrant compounds such as polyphenols and lignin (Wu 2011). This modification of the Gadgil hypothesis is more in line with recent findings of the loss of lignin decay genes in ECM fungal lineages (Wolfe et al. 2012), but is inevitably more difficult to test, as it requires species-specific knowledge of fungal function in the environment.

3. Containment of Monodominant Forests: Can Net Feedbacks Become Negative?

A common question put forth about monodominant forests is, "why do they not take over all lowland tropical rain forests?" When considering the ECM-mediated advantages of monodominant trees, some of the same mechanisms that facilitate local dominance may be disadvantageous for dispersal and colonization to other sites. For example, in monodominant systems where there is adult tree facilitation of conspecific seedling growth and survival (i.e., reverse Janzen-Connell mechanism), when a seed is dispersed away from a parent tree it is has a disadvantage for growth and survival, as it is denied access to the common mycorrhizal network and experiences ECM fungal inoculum limitation. It is yet unknown how long ECM inoculum can persist in the soil in the absence of an ECM tree host or what the dispersal range is for ECM fungi. In general, the dispersal capabilities of ECM fungi have not been well examined, but one recent study found that almost all of the spores of six ECM species fell within 1 m of the sporocarp cap (Galante et al. 2011). Another recent study evaluating landscape-level ECM dispersal patterns across fragments of pine "islands" showed that spore loads decreased dramatically with increasing distance from the edge of the pine fragment (Peay et al. 2012). With the exception of one ECM fungal species, most ECM fungi appeared to exhibit dispersal limitation. Dispersal limitation of ECM fungi may be an important mechanism preventing novel clumps of ECM forest from arising. The patchy nature of ECM monodominant forests is analogous to forest islands as described in the Peay et al. 2009 study (Degagne et al. 2009), and some of the same dynamics related to the theory of island biogeography may be operating with ECM fungi in these regions.

Long-term studies of ECM monodominant forest dynamics are necessary to know whether or not these forests are expanding, contracting, or in dynamic states of both. Plant-soil feedback theory is based on individual plant interactions, but in ECM monodominant forests these feedbacks may scale up to the community level. It is possible that positive plant-soil feedbacks become stronger with increasing density of conspecifics, but then begin to decline past a threshold density of monodominant trees. Pathogens or herbivores may begin to accumulate and periodic outbreaks may prevent unchecked expansion of these forests (Gross et al. 2000). A major limitation in studying *in situ* monodominant forests is that there is little opportunity to follow individual seedlings over time to observe the accumulation pathogens versus symbionts, which is the basis for the plant-soil feedback models. Theoretical modeling coupled with empirical studies will likely be necessary to formulate predictions about community threshold effects and diversity-dominance oscillations in these forests.

Future global changes due to anthropogenic activity may function to contain the expansion of monodominant forests, as there may be more frequent pathogen outbreaks or recurrent disturbance events that alter competitive dynamics and plant-soil feedbacks. Regions containing ECM monodominant forests do not experience frequent or intense disturbances (Connell and Lowman 1989, Hart et al. 1989, Torti et al. 2001), so the role of disturbance in containing the spread of ECM monodominant forests is thought to be insignificant. Nonetheless, changes in precipitation, storm surges, and other climactic factors have the potential to affect plant-fungal mutualisms (Kiers et al. 2010), and ECM fungi are known to be sensitive to disturbance (Ingleby et al. 1998, Waltert et al. 2002, Peay et al. 2009). Also, if these monodominant, ECM trees are dependent on having sufficient time for competitive exclusion to occur, more frequent disturbance events may ultimately cause a contraction of these forests.

4. Conclusions and Future Directions

It is clear that significant empirical work is needed to test many of the aforementioned hypotheses across ECM monodominant systems. Since many of the monodominant forests occur in remote locations, long-term, intensive studies are lacking in comparison to many of the more easily accessed diverse tropical forests. In the *Dicymbe corymbosa* forest in Guyana, it appears that positive plant-soil feedbacks function to maintain ECM monodominance through seedling facilitation and alterations of nutrient cycling pathways.

A major focus for future research in ECM monodominant forests should be deconstructing species-level functions of the associate ECM fungi. Feedbacks between particular species of ECM fungi and ECM trees may play some role in the competitive dominance of these trees compared to other ECM trees in the community. With the exception of some work in the Dipterocarpaceae (Brearley 2012), almost no studies have been done evaluating species-specific feedbacks, despite the strong potential for species-level differences. For example, ECM fungal diversity in the *D. corymbosa* forest is high, with more than 200 ECM fungal species estimated to coexist in this system (Henkel et al. 2012), and these individual fungal species may have different functional capacities and variable mutualistic benefits to the host. It is also possible that the ECM monodominant trees are able to support a greater diversity of ECM fungi compared to non-dominant ECM trees, which may partially explain why these trees attain such extreme dominance. With a greater diversity of symbionts, trees have access to more diverse pools of soil resources. Resource partitioning, which usually stabilizes plant species coexistence, may be functionally similar in these ECM forests, but ECM fungi may be differentially partitioning soil resources, rather than different species of plants. Different ECM species also vary in their ability to protect trees from pathogen attack (Ramachela and Theron 2010). Therefore, a major research priority for these systems should be to determine how individual ECM fungi function with respect to genetic capabilities, host pathogen protection, and resource acquisition. Since most of the ECM genera in tropical forests are the same as in temperate forests (Bâ et al. 2012), hypotheses can be formulated based on known functions of temperate fungal taxa.

Acknowledgements

We thank Krista L. McGuire and Caitlyn Gillikin for their valuable comments on this study.

References

Abuzinadah, R.A. and D.J. Read. 1986. The role of proteins in the nitrogen nutrition of ectomycorrhizal plants .1. Utilization of peptides and proteins by ectomycorrhizal fungi. New Phytol. 103: 481–493.

Bâ, A.M., R. Duponnois, B. Moyersoen and A.G. Diédhiou. 2012. Ectomycorrhizal symbiosis of tropical African trees. Mycorrhiza 22: 1–29.

Barto, E.K., M. Hilker, F. Muller, B.K. Mohney, J.D. Weidenhamer and M.C. Rillig. 2011. The Fungal Fast Lane: Common Mycorrhizal Networks Extend Bioactive Zones of Allelochemicals in Soils. Plos One 6.

Bever, J.D., I.A. Dickie, E. Facelli, J.M. Facelli, J. Klironomos, M. Moora, M.C. Rillig, W.D. Stock, M. Tibbett and M. Zobel. 2010. Rooting theories of plant community ecology in microbial interactions. Trends in Ecol. Evol. 25: 468–478.

Bever, J.D., T.G. Platt and E.R. Morton. 2012. Microbial population and community dynamics on plant roots and their feedbacks on plant communities. Ann. Rev. Microbiol. 66: 265–283.

Brearley, F.Q. 2012. Ectomycorrhizal Associations of the Dipterocarpaceae. Biotropica 44: 637–648.

Brownlee, C., J.A. Duddridge, A. Malibari and D.J. Read. 1983. The structure and function of mycelial systems of ectomycorrhizal roots with special reference to their role in forming inter-plant connections and providing pathways for assimilate and water transport. Plant and Soil 71: 433–443.

Brundrett, M.C. 2009. Mycorrhizal associations and other means of nutrition of vascular plants: understanding the global diversity of host plants by resolving conflicting information and developing reliable means of diagnosis. Plant and Soil 320: 37–77.

Chazdon, R.L. 1988. Sunflecks and their importance to forest understorey plants. Adv. Ecol. Res. 18: 1–63.

Colpaert, J.V. and K.K. van Tichelen. 1996. Decomposition, nitrogen and phosphorus mineralization from beech leaf litter colonized by ectomycorrhizal or litter-decomposing basidiomycetes. New Phytol. 134: 123–132.

Connell, J.H. and M.D. Lowman. 1989. Low-diversity tropical rain forests: some possible mechanisms for their existence. Am. Nat. 134: 88–119.

Degagne, R.S., T.W. Henkel, S.J. Steinberg and L. Fox III. 2009. Identifying Dicymbe corymbosa monodominant forests in Guyana using satellite imagery. Biotropica 41: 7–13.

Dickie, I.A. and B. Moyersoen. 2008. Towards a global view of ectomycorrhizal ecology. New Phytol. 180: 263–265.

Finlay, R.D. and D.J. Read. 1986. The structure and function of the vegetative mycelium of ectomycorrhizal plants.1. translocation of C-14 labeled carbon between plants interconnected by a common mycelium New Phytol. 103: 143–156.

Floudas, D., M. Binder, R. Riley, K. Barry, R.A. Blanchette, B. Henrissat, A.T. Martínez, R. Otillar, J.W. Spatafora, J.S. Yadav, A. Aerts, I. Benoit, A. Boyd, A. Carlson, A. Copeland, P.M. Coutinho, R.P. de Vries, P. Ferreira, K. Findley, B. Foster, J. Gaskell, D. Glotzer, P. Górecki, J. Heitman, C. Hesse, C. Hori, K. Igarashi, J.A. Jurgens, N. Kallen, P. Kersten, A. Kohler, U. Kües, T.K.A. Kumar, A. Kuo, K. LaButti, L.F. Larrondo, E. Lindquist, A. Ling, V. Lombard, S. Lucas, T. Lundell, R. Martin, D.J. McLaughlin, I. Morgenstern, E. Morin, C. Murat, L.G. Nagy, M. Nolan, R.A. Ohm, A. Patyshakuliyeva, A. Rokas, F.J. Ruiz-Dueñas, G. Sabat, A. Salamov, M. Samejima, J. Schmutz, J.C. Slot, F. St. John, J. Stenlid, H. Sun, S. Sun, K. Syed, A. Tsang, A. Wiebenga, D. Young, A. Pisabarro, D.C. Eastwood, F. Martin, D. Cullen, I.V. Grigoriev and D.S. Hibbett. 2012. The Paleozoic origin of enzymatic lignin decomposition reconstructed from 31 fungal genomes. Science 336: 1715–1719.

Gadgil, R.L. and G.D. Gadgil. 1971. Mycorrhiza and litter decomposition. Nature 233: 133.

Gadgil, R.L. and P.D. Gadgil. 1975. Suppression of litter decomposition by mycorrhizal foots of *Pinus radiata*. New Zealand J. For. Sci. 5: 35–41.

Galante, T.E., T.R. Horton and D.P. Swaney. 2011. 95% of basidiospores fall within 1 m of the cap: a field- and modeling-based study. Mycologia 103: 1175–1183.

Gross, N.D., S.D. Torti, D.H. Feener Jr. and P.D. Coley. 2000. Monodominance in an African Rain Forest: Is Reduced Herbivory Important? 1. Biotropica 32: 430–439.

Hart, T.B., J.A. Hart and P.G. Murphy. 1989. Monodominant and species-rich forests of the humid tropics: causes for their co-occurrence. American Naturalist 133: 613–633.

Hart, T.B. 1990. Monospecific dominance in tropical rain forests. Trends Ecol. Evol. 5: 6–11.

Henkel, T.W. 2003. Monodominance in the ectomycorrhizal Dicymbe corymbosa (Caesalpiniaceae) from Guyana. J. Trop. Ecol. 19: 417–437.

Henkel, T.W., M.C. Aime, M.M.L. Chin, S.L. Miller, R. Vilgalys and M.E. Smith. 2012. Ectomycorrhizal fungal sporocarp diversity and discovery of new taxa in Dicymbe monodominant forests of the Guiana Shield. Biodiv. Conserv. 21: 2195–2220.

Hobbie, E.A., N.S. Weber, J.M. Trappe and G.J. van Klinken. 2002. Using radiocarbon to determine the mycorrhizal status of fungi. New Phytol. 156: 129–136.

Ingleby, K., R.C. Munro, M. Noor, P.A. Mason and M.J. Clearwater. 1998. Ectomycorrhizal populations and growth of Shorea parvifolia (Dipterocarpaceae) seedlings regenerating under three different forest canopies following logging. For. Ecol. Manag. 111: 171–179.

Kiers, E.T., T.M. Palmer, A.R. Ives and J.F. Bruno. 2010. Mutualisms in a changing world: an evolutionary perspective. Ecology 13: 1459–1474.

Koele, N., I.A. Dickie, J. Oleksyn, S.J. Richardson and P.B. Reich. 2012. No globally consistent effect of ectomycorrhizal status on foliar traits. New Phytol. 196: 845–852.

Leigh, E.G., P. Davidar, C.W. Dick, J. Puyravaud, J. Terborgh, H. ter Steege and S.J. Wright. 2004. Why do some tropical forests have so many species of trees? Biotropica 36: 445–473.

Lindahl, B., J. Stenlid, S. Olsson and R. Finlay. 1999. Translocation of P-32 between interacting mycelia of a wood-decomposing fungus and ectomycorrhizal fungi in microcosm systems. New Phytol. 144: 183–193.

Lindahl, B., J. Stenlid and R. Finlay. 2001. Effects of resource availability on mycelial interactions and P-32 transfer between a saprotrophic and an ectomycorrhizal fungus in soil microcosms. FEMS Microbiol. Ecol. 38: 43–52.

Marx, D.H. 1972. Ectomycorrhizae as biological deterrents to pathogenic root infections. Ann. Rev. Phytopath. 10: 429–&.

McGuire, K.L. 2007a. Common ectomycorrhizal networks may maintain monodominance in a tropical rain forest. Ecology 88: 567–574.

McGuire, K.L. 2007b. Recruitment dynamics and ectomycorrhizal colonization of *Dicymbe corymbosa*, a monodominant tree in the Guiana Shield. J. Trop. Ecol. 23: 297–307.

McGuire, K.L. 2008. Ectomycorrhizal associations function to maintain tropical monodominance. *In*: Z.A. Siddiqui, M.S. Akhtar and K. Futai (eds.). Mycorrhizae: Sustainable Agriculture and Forestry. Springer, Netherlands, pp. 287–302.

McGuire, K.L., D.R. Zak, I.P. Edwards, C.B. Blackwood and R. Upchurch. 2010. Slowed decomposition is biotically mediated in an ectomycorrhizal, tropical rain forest. Oecologia 164: 785–795.

Meinhardt, K.A. and C.A. Gehring. 2012. Disrupting mycorrhizal mutualisms: a potential mechanism by which exotic tamarisk outcompetes native cottonwoods. Ecol. Appl. 22: 532–549.

Newman, E.I. 1988. Mycorrhizal Links between Plants—Their Functioning and Ecological Significance. Adv. Ecol. Res. 18: 243–270.

Olsson, P.A., M. Chalot, E. Baath, R.D. Finlay and B. Soderstrom. 1996. Ectomycorrhizal mycelia reduce bacterial activity in a sandy soil. FEMS Microbiol. Ecol. 21: 77–86.

Peay, K.G., M. Garbelotto and T.D. Bruns. 2009. Spore heat resistance plays an important role in disturbance-mediated assemblage shift of ectomycorrhizal fungi colonizing *Pinus muricata* seedlings. J. Ecol. 97: 537–547.

Peay, K.G., M.G. Schubert, N.H. Nguyen and T.D. Bruns. 2012. Measuring ectomycorrhizal fungal dispersal: macroecological patterns driven by microscopic propagules. Mol. Ecol. 21: 4122–4136.

Peh, K.S. H., B. Sonke, J. Lloyd, C.A. Quesada and S.L. Lewis. 2011. Soil Does Not Explain Monodominance in a Central African Tropical Forest. Plos One 6: 9.

Ramachela, K. and J.M. Theron. 2010. Effect of ectomycorrhizal fungi in the protection of Uapaca kirkiana seedlings against root pathogens in Zimbabwe. Southern For. 72: 37–45.

Read, D.J. 1991. Mycorrhizas in ecosystems. Experientia 47: 376–391.

Richards, P.W. 1952. The tropical rain forest. second edition edition. University Press, Cambridge, UK.

Rineau, F., D. Roth, F. Shah, M. Smits, T. Johansson, B. Canback, P.B. Olsen, P. Persson, M.N. Grell, E. Lindquist, I.V. Grigoriev, L. Lange and A. Tunlid. 2012. The ectomycorrhizal fungus Paxillus involutus converts organic matter in plant litter using a trimmed brown-rot mechanism involving Fenton chemistry. Environ. Microbiol. 14: 1477–1487.

Simard, S.W. and D.M. Durall. 2004. Mycorrhizal networks: a review of their extent, function, and importance. Can. J. Bot. 82: 1140–1165.

Simard, S.W. 2009. The foundational role of mycorrhizal networks in self-organization of interior Douglas-fir forests. For. Ecol. Manag. 258: S95–S107.

Smith, J.E. and D. Read. 2008. Mycorrhizal Symbiosis. Third Edition edition. Academic Press, San Diego (CA).

Talbot, J.M., S.D. Allison and K.K. Treseder. 2008. Decomposers in disguise: mycorrhizal fungi as regulators of soil C dynamics in ecosystems under global change. Func. Ecol. 22: 955–963.

Torti, S.D. and P.D. Coley. 1999. Tropical monodominance: A preliminary test of the ectomycorrhizal hypothesis. Biotropica 31: 220–228.

Torti, S.D., P.D. Coley and T.A. Kursar. 2001. Causes and consequences of monodominance in tropical lowland forests. Am. Nat. 157: 141–153.

Wallander, H., L.O. Nilsson, D. Hagerberg and E. Baath. 2001. Estimation of the biomass and seasonal growth of external mycelium of ectomycorrhizal fungi in the field. New Phytol. 151: 753–760.

Waltert, B., V. Wiemken, H.P. Rusterholz, T. Boller and B. Baur. 2002. Disturbance of forest by trampling: Effects on mycorrhizal roots of seedlings and mature trees of Fagus sylvatica. Plant and Soil 243: 143–154.

Whipps, J.M. 2001. Microbial interactions and biocontrol in the rhizosphere. J. Exp. Bot. 52: 487–511.

Whipps, J.M. 2004. Prospects and limitations for mycorrhizas in biocontrol of root pathogens. Can. J. Bot. 82: 1198–1227.

Wolfe, B.E., V.L. Rodgers, K.A. Stinson and A. Pringle. 2008. The invasive plant *Alliaria petiolata* (garlic mustard) inhibits ectomycorrhizal fungi in its introduced range. J. Ecol. 96: 777–783.

Wolfe, B.E., R.E. Tulloss and A. Pringle. 2012. The Irreversible Loss of a Decomposition Pathway Marks the Single Origin of an Ectomycorrhizal Symbiosis. Plos One 7.

Wu, B.Y., H. Maruyama, M. Teramoto and T. Hogetsu. 2012. Structural and functional interactions between extraradical mycelia of ectomycorrhizal Pisolithus isolates. New Phytol. 194: 1070–1078.

Wu, T.H. 2011. Can ectomycorrhizal fungi circumvent the nitrogen mineralization for plant nutrition in temperate forest ecosystems? Soil Biol. Biochem. 43: 1109–1117.

CHAPTER 11

Ectomycorrhiza in Forest Rehabilitation in Indonesia

Hesti L. Tata

1. Introduction

The total forest area[1] of Indonesia is 131.3 million ha, which is about 69.9% of the total terrestrial territory of Indonesia. However, the forest cover inside the state forest area is only 133,327.2 ha, which is only about 0.1% of designated forest area (Ministry of Forestry 2012). This implies that deforestation rates in Indonesia are very high. Based on satellite imagery, the Ministry of Forestry reported that deforestation inside the state forest area in Indonesia during 2009/2010 was 610,375.9 ha. Deforestation in Indonesia is mainly caused by fire, conversion of forest and peat-land to other land uses, mining, and forest degradation due to illegal logging and poor forest management practices (Ministry of Forestry 2012). Deforestation in the islands of Kalimantan and Sumatra was remarkably high compared with other islands in Indonesia (Holmes 2002). In the period between 1985 and 2007, forest loss in Sumatra alone was reported to be more than 70% of total forest area (Laumonier et al. 2010).

Centre for Conservation and Rehabilitation, Jalan Gunung Batu 5, Bogor 16151, Indonesia.
Email: hl.tata@gmail.com

[1]Forest area is designated by the institutional framework of forestry and includes areas normally forming part of the forest area which are temporarily cleared as a result of human intervention, but which are expected to revert to forest.

Lowland rainforests of Sumatra and Kalimantan are dominated by Dipterocarpaceae tree species, which produce high quality timber. Some species also produce non-timber products, such as illepe-nuts and dammar resin (Shiva and Jantan 1998). Most dipterocarps species naturally occur in the lowland forests of Sumatra and Kalimantan. Some species grow in peat-swamp areas (such as *Shorea balangeran* and *Shorea uliginosa*), while others can grow in higher altitude (750 m to 1200 m) montane forests (such as *Shorea platyclados* and *Shorea ovata*) (Ashton 1982, Whitmore 1984). Deforestation commonly occurs in the lowland mixed dipterocarps forests, where commercial timber trees grow well and the land is easily accessible.

Over the last three decades, many efforts have been made by the Government of Indonesia to overcome the loss of forest cover (Nawir et al. 2007). The Ministry of Forestry implemented forest rehabilitation programs in association with local communities, such as the Rehabilitation of Logged-Over Area (1996-2003), Community Based Forest Management (1994–1999), Farm Forestry (1970), Programme of Specific Allocated Funds —Reforestation Funds (2001) and the Collaborative Forest Management Project (2001). However, the programs were not successfully implemented due to the failure of the utilized approach, which was mainly based on government's instruction approach (Nawir et al. 2007).

In order to develop more integrated rehabilitation programs, the Planning Agency of The Ministry of Forestry set up a national movement to rehabilitate the state forest area (*Gerakan Rehabilitasi Nasional, Gerhan*) in 2003, totaling 3 million ha of forest in 28 provinces within 5 years.[2] However, this program failed to meet the target, owing to mismanagement of the project's implementation and policy failure (Nawir et al. 2007).

The success of forest rehabilitation is not only influenced by policy and socio-economic factors, but also technical aspects of program implementation, such as site characteristics, site-species matching, planting stocks production, and maintenance. Production of planting stocks is an important part of the reforestation and rehabilitation of degraded land. Large numbers of good quality planting stocks have to be produced in nurseries as the first step of a successful rehabilitation program.

2. Promoting Dipterocarpaceae for Rehabilitation

Enrichment planting in the area of production forests (e.g., forest areas to produce forest products for the benefit societies and industries) using native species has been regulated since 1993, under the silviculture system

[2]http://www.dephut.go.id/INFORMASI/RRL/Benih/GN_RHL.html

of Indonesian Selective Cutting and Planting (*Tebang Pilih Tanam Indonesia*, TPTI). Dipterocarpaceae is recommended for planting in the rehabilitation of degraded forest and logged areas (Soekotjo 2005), *Imperata* grassland (Adjers and Otsamo 1996) and burned forests (Anonymous 2004, Priadjati 2002, Tata et al. 2010). Despite some constraints, such as soil compactions, loss of top-soils, high temperature, and competition with weeds and fast growing tree species, Dipterocarpaceae trees are challenging to use in the rehabilitation projects (Kettle 2010, Matsune et al. 2006, Okimori et al. 2006), because they have EcM associations (Brearley 2012, Hadi 1987, Lee 1998, Smits 1994). Mutualistic symbioses between dipterocarps trees and EcM fungi is necessary for initial tree growth (Smits 1994, Turjaman et al. 2005, Turjaman et al. 2006).

3. Effect of Site Factors and Disturbance on Ectomycorrhiza

The soil type of tropical rain forests in Kalimantan island is dominated by Ultisols soils, with characteristics of high clay content, high acidity, low nutrients, and high aluminum and iron levels (Ohta and Syarif 1996), whereas soil types in Central Sumatra are dominated by Oxisols and Ultisols soils (Ketterings et al. 2000, Van Noordwijk et al. 1998). Low nutrient content but high heavy metal content in the soils may hamper the survival and growth of dipterocarp seedlings.

Mycorrhizal inoculum potential in different land use types of Jambi province, including forests rubber agroforests rubber monocultures small holder oil palm cassava field annual crops *Imperata* grasslands, has been studied. Tata (2008) reported that land use types affected EcM colonization of two dipterocarp species of *Shorea* (*Shorea selanica* and *Shorea lamellata*) in their nursery stage at 10 months after planting. All seedlings of *S. selanica* and *S. lamellata* planted in soils from different land use types were colonized by EcM fungi. The highest EcM colonization occurred on *Shorea* planted in soils from rubber agroforests (RA), while the lowest EcM colonization occurred on *Shorea* planted in soils from annual crops (CR) (Fig. 1). The soils from annual crops were categorized as fertile soils, which had higher macro nutrient contents ($P_{available}$, N, Na, Ca and Mg) compared with soils from other land uses (Tata 2008). High soil nutrient content, especially P, could inhibit colonization of EcM on the root systems (Marx et al. 1977).

Forest disturbance caused by fire affected fruiting body development of EcM fungi in East Kalimantan (Tata et al. 2003a). Two years after a wildfire in a mixed dipterocarp forest of Sungai Wain, East Kalimantan, no fruiting bodies were found on the burned forest floor. However, bioassay studies on the seedlings of *Shorea pinanga* and *Shorea stenoptera* using soil from burned dipterocarp forests have shown that the roots of both *Shorea* species were colonized with EcM fungi. EcM morphotypes encountered in the roots from different soils were varied (Table 1). This observation implies that mycorrhizal inoculum potential persists in the soils after wildfire. EcM morphotypes from unburned forests were different compared to those from burned forests (Tata et al. 2003a).

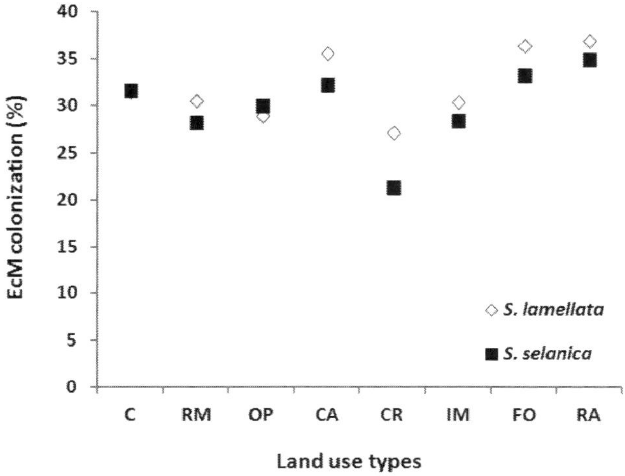

Fig. 1. Ectomycorrhizal colonization on *S. selanica* and *S. lamellata* seedlings planted on soils from different land use types in nurseries. C=sterilized soil (control), RM= rubber monoculture, OP=smallholder oil-palm, CA=cassava field, CR=annual crops, IM=*Imperata* grasslands, FO=forest, RA=rubber agroforest. Data from Tata (2008).

Table 1. Ectomycorrhizal morphotypes on *Shorea pinanga* and *S. stenoptera* seedlings grown in different soil sources from burned and unburned mixed dipterocarp forests in Sungai Wain, East Kalimantan and a reforested dipterocarp site in Haurbentes, West Java.

Tree Species	Soil sources	EcM Morphotypes	Hypha		Clamp connection
			Diameter (µm)	Color	Diameter (µm)
S. pinanga	Burned forest	Monopodial, white	4.0–4.9	hyaline	6.5–7.1
	Unburned forest	Monopodial, white	4.0–4.4	reddish brown	4.6–6.9
	Haurbentes	Monopodial, white	4.6–5.3	reddish brown	6.3–8.2
S. stenoptera	Burned forest	Monopodial, white	3.4–4.6	hyaline	5.0–8.0
	Unburned forest	Monopodial, white	4.0–4.4	hyaline	5.3–6.7
	Haurbentes	Monopodial, white	5.3–6.3	reddish brown	8.2–9.0

Data from Tata et al. (2003a)

Another study on the effect of heat treatment up to 150°C on EcM inoculum potential of forest soils from a lowland dipterocarp forest of Jambi (Sumatra) showed that heat treatment had little effect on the growth of tree cuttings from five species of dipterocarps (*Shorea selanica, Shorea leprosula, Shorea lamellata, Vatica sumatrana,* and *Hopea odorata*). Although EcM colonization in the control soils was higher than that of the soil treated with higher temperatures (150°C), the effect was smaller than expected (Fig. 2). The rapid colonization of heated soils in the experiment may imply abundant availability of EcM fungi for dipterocarp growth in field conditions. Molecular identification based on sequences from the bioassay studies of the Internal Transcribed Spacer (ITS) region of ectomycorrhizal fungi on heat-treated *S. selanica* differed from those on *S. selanica* planted in the field experiment (Tata 2008).

Logging, or timber extraction, reduces canopy cover and has been shown to affect soil EcM communities. Studies of EcM communities on the roots of *S. parvifolia* seedlings showed that the diversity of mycorrhizal types found under open canopies caused by logging was higher than that found under closed canopies. Mycorrhizal types from open canopies were thought to be pioneer fungi (Ingleby et al. 1998). Similar evidence was found in *S. lamellata* planted on different ages of rubber agroforests. EcM colonization on *S. lamellata* grown in 1 year old rubber agroforests was higher than those grown under a forest canopy. However, *S. selanica* planted under the forest canopy had the best EcM colonization (Table 2).

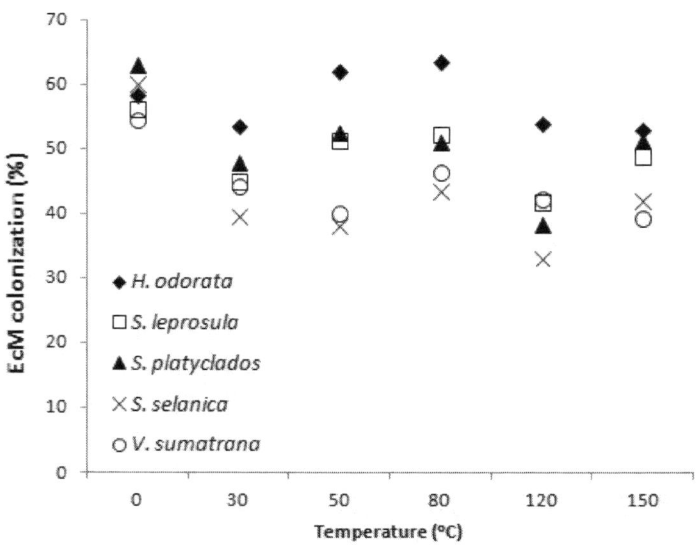

Fig. 2. Ectomycorrhizal colonization on cuttings of five dipterocarps species planted on soils that underwent a series of heat treatments. Data from Tata (2008).

Table 2. Ectomycorrhizal colonization of *S. selanica* and *S. lamellata* inoculated (+) and non-inoculated (–) with *Scleroderma columnare* in nurseries planted on different land use types.

	EcM colonization (%)			
Land uses	*S. lamellata* (–)	*S. lamellata* (+)	*S. selanica* (–)	*S. selanica* (+)
Forest	10.0 a	17.6 a	41.5 b	59.4 c
R1_forest	20.2 a	37.5 ab	27.5 a	54.1 c
R1_RAF	21.5 a	23.1 a	47.5 a	29.8 ab

Note: R1_forest: rubber garden age 1 yr derived from forest, R1_RAF: rubber garden age 1 yr derived from RAF. Values followed by same letter are not significantly different ($P< 0.05$). Data from: Tata (2008).

Complex results on the effect of *Scleroderma columnare* inoculation on growth of *S. selanica* and *S. lamellata* planted in rubber agroforests have been reported. Inoculation of *Scleroderma columnare* increased survival of *S. selanica* and *S. lamellata* seedlings one year after transplanted to the field. After two years, inoculation of EcM did not affect survival of both *Shorea* species. Seedlings growth, e.g., relative growth rate of height and stem diameter, were only increased by EcM inoculation in the second semester of two years observation (Tata et al. 2010).

4. The Importance of EcM Inoculation on Planting Stock Production

Dipterocarp trees can be generative and vegetative propagated. Planting stocks of dipterocarps are produced generatively using sown seeds in nurseries and wild seedlings collections (commonly known as wildlings) (Smits 1994, Kettle 2010), and vegetatively using shoot cutting (Smits 1992, Subiakto et al. 2005). Many studies performed during nursery stages showed that ectomycorrhizal fungal inoculation on dipterocarp seedlings increase dipterocarp seedling growth (Santoso 1988, Smits 1994, Yasman 1995, Priadjati 2002, Turjaman et al. 2005, Turjaman et al. 2006, Riniarti 2010, Tata et al. 2010) and nutrient uptake (Santoso 1988, Omon 2002, Tata et al. 2010).

In natural ecosystems, dipterocarp trees exhibit nurse effects for their seedlings by facilitating their growth and survival through hyphal networks in the soil (Read 1991, Smits 1992). Mimicking this habit, EcM inoculation can be done by planting dipterocarps seedlings in nursery using topsoils that are collected below conspecific parent trees (Smits 1992, Yasman 1995). Planting stocks from vegetative propagation, on the contrary, do not have direct contact with the hyphal networks of parent trees. Propagation techniques usually use sterilized rooting media to minimize diseases and increase survival. Hence, artificial EcM inoculation prior to transplanting is recommended (Smits 1992, Smits 1994, Yasman 1995, Omon 2002). However, Fig. 2 shows cuttings of *H. odorata, S. leprosula, S. platyclados, S. selanica* and *V. sumatrana* were colonized with EcM fungi at all degrees of heat treatment on transplanted media. Heat treatment up to 150°C might not kill mycorrhizal inocula in the soil (Tata 2008).

Efforts to rehabilitate logged forest areas in Kalimantan encountered difficulties due to the lack of EcM partners in the soil after forest disturbance. Thus, rehabilitation using dipterocarps trees in East Kalimantan requires mycorrhizal inoculation in the nursery stage (Smits 1992, Smits 1994).

5. Ectomycorrhizal Response in Field Experiment

Further studies in field conditions, however, have not been performed in detail. Some reports showed different responses of EcM fungal inoculation on the growth of dipterocarp seedlings. In a peat swamp forest in Central Kalimantan, inoculation with *Boletus* sp. and *Scleroderma* sp. increased the growth of *Shorea balangeran* (Turjaman et al. 2011). However, it is not clear whether or not EcM fungi that have been inoculated in the nursery stage, e.g., *Scleroderma* sp. and *Boletus* sp., persist in the root system of *Shorea balangeran* after transplanting to the field. Another study showed that 2 years after transplanting to a field of rubber-based agroforestry, *S. selanica* and *S. lamellata* that were inoculated in the nursery stage with *S. columnare* have different EcM communities in their root systems (Tata et al. 2010).

In logged forests in Jambi, Central Sumatra, inoculation of EcM fungi on dipterocarps seedlings (*Shorea macroptera* and *Shorea parvifolia*) planted in logged forests improved seedlings performance. Seedling growth showed a positive relation with the percentage of EcM colonization (Kikuchi 2006). This author reported that the types of EcM fungi found in planted and natural seedling were quite different.

Priadjati (2002) reported different responses of EcM inoculation in field experiments. *Shorea leprosula* cuttings planted in a secondary forest in East Kalimantan showed that EcM inoculation did not increase performance and tree growth. Non-inoculated cuttings, which were used as the control, had the best performance of height and stem diameter among other EcM inoculations (Priadjati 2002). This author found that, after transplanting to the field, control trees had a higher percentage of EcM colonization than the trees that had been inoculated with *Amanita* sp., *Russula* sp., and *Scleroderma* sp.

Another trial performed in a rubber agroecosystem showed that EcM inoculation had limited effects on the growth of *S. selanica* and *S. lamellata* (Tata et al. 2010). In the initial development of rubber agroforests, natural forest was cleared using the slash and burn technique as a low-cost establishment method (Gouyon et al. 1993, Penot 2004). No potential host trees for EcM fungi, such as trees belonging to the family of Dipterocarpaceae and Fagaceae, were still alive. Rubber seedlings were planted alongside upland paddy and other food crops. Fallow succession allowed natural regeneration until rubber trees reached a certain stem diameter that could be tapped (~8–10 years after planting; Gouyon et al. 1993, Joshi et al. 2002). Tata et al. (2010) reported that two years after planting in the rubber agroforests, the seedlings of two *Shorea* species, both inoculated and un-inoculated with *Scleroderma columnare*, were colonized by EcM fungi. Different EcM fungi colonized the seedlings that had been inoculated with *Scleroderma columnare* in nurseries. The authors implied that mycorrhizal inoculum potential persists in the soils of rubber agro-ecosystems, even rubber gardens with long histories of forest clearing. Further study was suggested to investigate whether other factors, such as squirrels and wild pigs, play roles as dispersal agents of EcM inoculums from the remaining forests (Tata et al. 2010).

In a comprehensive review, Brearley (2012) suggested that EcM fungi are mostly likely to benefit the growth of dipterocarp seedlings when they are planted in degraded areas. Therefore, inoculation of EcM seedlings at the nursery stage is recommended in the rehabilitation of degraded areas where mycorrhizal inoculum potential is limited.

6. Rehabilitation Techniques

Technically, rehabilitation of degraded dipterocarp forests can be achieved through (i) enrichment planting with desirable dipterocarp seedlings or clonal plant materials in order to increase timber productivity, and (ii) planting fast-growing light demanding species as a nurse canopy and followed by under-planting with dipterocarps (Kettle 2010). The former is established under small opening of forest canopy covers and in the presence of gaps. The latter is established under severely degraded forest conditions.

Many studies have shown techniques to rehabilitate degraded dipterocarp forests caused by logging and over-exploitation, fires, and the invasion of *Imperata* grassland (Adjers et al. 1996, Otsamo et al. 1996, Otsamo 2000, Mori 2001). Forest rehabilitation with enrichment planting can be done in gaps resulted from logging activities. To rehabilitate burned forests and *Imperata* grasslands, strip planting techniques can be adopted. Mixed planting between fast growing trees (such as *Acacia mangium*) and native tree species is the best option to rehabilitate *Imperata* grassland in South Kalimantan (Otsamo 2000). However, no EcM inoculation was mentioned in these publications.

Rehabilitation of degraded land in Haurbentes, West Java, with various Dipterocarpaceae trees, showed a successful human constructed dipterocarp forest outside their natural habitats. The establishment of a reforested dipterocarp site in Haurbentes, West Java, was initiated in order to establish a dipterocarp forest research station in 1938 (Masano et al. 1987, Subiakto et al. 2001, Mindawati et al. 2004a, b). Mindawati et al. (2004a,b) reported the growth and yield of several dipterocarp species from different origins that have been planted in Haurbentes since 1940. *Shorea leprosula, Shorea mecistopteryx, Shorea stenoptera, S. selanica, Shorea balangeran, Hopea mengerawan* and *H. odorata* demonstrated the best growth among others (Mindawati et al. 2004a,b). However, there is no report on the study of inoculation EcM fungi on dipterocarp seedlings since the early development of the Haurbentes research station in 1938. Long after the reforested dipterocarp site was established, Santoso and Turjaman (2003) reported that several species of EcM fungi were encountered on the forest floor of the Haurbentes forest research station, such as *Scleroderma* sp., *Lactarius* sp., *Boletus* sp., and *Cantharellus cibarius* (Santoso and Turjaman 2003).

Agroforestry is another option for forest rehabilitation through enrichment planting and mix-planting dipterocarp trees in agro-ecosystems. Agroforestry system in Indonesia includes taungya farming system (e.g., growing annual agricultural crops along with the forestry species during the early years of establishment of the forestry plantation), and tree-based agroforestry system, such as *Shorea javanica* and rubber (*Hevea brasiliensis*) trees. *Shorea javanica*, a dammar resin producer, has been planted in agroforestry systems in Lampung, Sumatra for many years (Michon and Bompard 1987, Torquebiau 1984). Traditional farmers who planted *Shorea javanica* (locally known as *dammar* 'mata kucing') were not aware of the technology of EcM inoculation. *S. javanica* seedlings were transplanted and domesticated in *dammar* gardens with soil from the nursery. According to farmers, transplantation is always successful, even without artificial inoculation of EcM fungi in nursery. Soils from the nursery increase chances of survival after transplantation (Michon and Bompard 1987). Nuhamara (1987) reported that several EcM fungi were found in the *S. javanica* agroforest, such as *Amanita hemibapha, Cantharellus cibarius, Lactarius* spp., *Russula* spp., and *Scleroderma* sp. This author concluded that using soils with abundant inocula, e.g., top soils with mycorrhizal networks, as sources of inocula resulted in better growth of dipterocarp trees compared with those inoculated with a single EcM fungus.

Another agroforestry system, rubber agroforests, has relatively high tree diversity (Gouyon et al. 1993, Beukema et al. 2007, Tata et al. 2008). Low management intensity allows other beneficial trees to grow, forming an agroforestry complex. EcM host trees such as *Shorea leprosula* and *Lithocarpus* spp. that grow in the rubber garden maintain EcM sporocarp diversity. *Russula* spp., *Geastrum* sp. and *Scleroderma* sp. were encountered on the rubber agroforest floor in Jambi (H.L.T. unpublished data).

The taungya system has been applied to rehabilitate burned and severely degraded forests in Bukit Suharto, East Kalimantan. *Shorea smithiana* seedlings were planted along with rubber trees and five cash crops. Three fertilizers have been used to improve soil macro-nutrients (N, P and K), but no treatment of EcM fungi inoculation was reported. One year after transplantation to the field, survival of *S. smithiana* seedlings was 85.8%, which was higher than rubber trees (80.5%) (Sutisna 2001). Although *S. smithiana* is categorized as a shade demanding species, it survived and grew in an exposed plot (Sutisna 2001). However, this author did not report the presence of mycorrhizal colonization on the root system. The soils of the site in East Kalimantan may still have considerable mycorrhizal inoculum potential, as reported by Tata et al. (2003a) in a burned dipterocarp forest in Sungai Wain, East Kalimantan.

7. Conclusions

The success of rehabilitation is determined by silvicultural and ecological factors. Ectomycorhizas play an important role in the production of planting stocks for forest rehabilitation. In the nursery stage, ectomycorrhizal fungal colonization increases early growth of dipterocarp species, which benefits planting stock production. However, after field planting, very few studies have reported success of EcM fungal inoculations on tree growth in field conditions. Inoculation of EcM fungi in the nursery stage requires skills and capital in the initial nursery development. Using topsoil collected under conspecific parent trees for planting stock productions is more applicable for EcM fungal inoculation.

8. Future Research Needs

There are several publications on the applied research of EcM in Indonesia relating to forest and land rehabilitation, published mostly in national journals and national proceedings. Most of the research was conducted with a small number of replications and inappropriate design. There has been limited basic research performed on the physiology, taxonomy, phylogeny, and ecology-community of ectomycorrhizas, has been studied that can provide evidence and support for the applied research for improvement of forest and land rehabilitation practices in Indonesia.

Another weakness of EcM studies in Indonesia is the lack of identification of EcM fungi and ectomycorrhiza. The identity of tropical EcM fungi is currently based on morphological characters of temperate EcM fungi and EcM identification book guidelines (Tata et al. 2003b), which is relatively different from sporocarps of tropical EcM fungi. Due to limited laboratory facilities, advanced technology is not currently commonly practiced.

EcM fungi promote forest rehabilitation by increasing seedling survival and tree performance. Other than that, EcMs and EcM fungi provide ecosystem services to the environment by provisioning, regulating and supporting roles. Besides silvicultural and ecological criteria, the success of forest rehabilitation also needs to consider socio-economic criteria, such as population size, market access, availability of food and fuel, and economic benefit and subsidy to local people. Factors such as increased population size, food shortages, and poverty are often the roots of forest degradation (Lamb and Tomlinson 1994). Moreover, current major drivers of deforestation in Sumatra is the development of oil palm plantations (Laumonier et al. 2010) in forested areas, which has a greater economic benefit than other crops. Therefore, forest and land rehabilitation in Indonesia should be integrated with rural development.

Rehabilitation is not only about planting trees, but also contains multiple and complex aspects and activities. Nowadays, rehabilitation and forest restoration have large targets including biodiversity conservation, biomass production, and climate change mitigation and adaptation (Ciccarese et al. 2012). Hence, to rehabilitate degraded tropical forests and lands, more efforts to integrate multiple research disciplines are required. Public participation in the rehabilitation of degraded lands in Indonesia has been increasing, such as the development of small holder timber and farm forests, using both native species and domesticated species (see, for example, Tata and Van Noordwijk 2011, Sofiyuddin et al. 2012). Forest rehabilitation may be more attractive when farmers see the alternative income, clear incentive, and policy support (Roshetko et al. 2008).

Acknowledgements

The author acknowledges Meine van Noordwijk for comments on the early draft of manuscript, anonymous reviewers for comments to improve the manuscript and Krista L. McGuire and Caitlyn Gillikin for improving the language.

References

Adjers, G. and A. Otsamo. 1996. Seedling production method of dipterocarps. *In:* A. Schulte and D. Schone (eds.). Dipterocarp Forest Ecosystems: Towards Sustainable Management. World Scientific, Singapore, pp. 391–410.

Anonymous. 2004. Development of tropical reforestation techniques. Report of joint study. Kansai Electric Power Co., Gadjah Mada University and Kanso Technos Co, Ltd p. 43.

Ashton. P.S. 1982. Dipterocarpaceae. Flora Malesiana: Vol. 9(2). MartinusNijhoffPublishers.c/o. Kluwer Academic Publishers Groups. The Netherlands.

Beukema, H., F. Danielsen, G. Vincent, S. Hardiwinoto and J. van Andel. 2007. Plant and bird diversity in rubber agroforests in the lowlands of Sumatra, Indonesia. Agrofor Syst. 70: 217–242.
Brearley, F.Q. 2012. Ectomycorrhizal associations of the Dipterocarpaceae. Biotropica 44(5): 637–648.
Ciccarese, L., A. Mattsson and D. Pettenella. 2012. Ecosystem services from forest restoration: thinking ahead. New For. 43: 543–560.
Gouyon, A., H. de Foresta and P. Levang. 1993. Does 'Jungle Rubber' deserve its name? Analysis of rubber agroforestry system in Southeast Asia. Agro for. Syst. 22: 181–20.
Hadi, S. 1987. The association of fungi in dipterocarps. *In*: A.J.G.H. Kostermans (ed.). Proceedings of the Third Round Table Conference on Dipterocarps. UNESCO. Jakarta, Indonesia, pp. 73–79.
Holmes, D.A. 2002. Where Have All the Forests Gone? Environment and Social Development, East Asia and Pacific Region. The World Bank, Washington DC. USA.
Ingleby, K., R.C. Munro, P.A. Mason and M.J. Clearwater. 1998. Ectomycorrhizal populations and growth of *Shorea parvifolia* (Dipterocarpaceae) seedlings regenerating under three different forest canopies following logging. For. Ecol. Manag. 111: 171–179.
Joshi, L., G. Wibawa, G. Vincent, D.B. Putin, R. Akiefnawati, G. Manurung, M. van Noordwijk and S. Williams. 2002. Jungle rubber: a traditional agroforestry system under pressure. International Center for Research in Agroforestry (ICRAF). Bogor, Indonesia.
Ketterings, Q.M., J.M. Bigham and V. Laperche. 2000. Changes in soil mineralogy and texture caused by slash-and-burn in Sumatra, Indonesia. Soil Science Soc. Am. J. 64: 1108–1117.
Kettle, C.J. 2010. Ecological considerations for using dipterocarps for restoration of lowland rainforest in Southeast Asia. Biodiv. Conserv. 19: 1137–1151.
Kikuchi, J. 2006. Ectomycorrhizas of dipterocarps in logged-over forests and plantations. *In*: K. Suzuki, K. Ishii, S. Sakurai and S. Sasaki (eds.). Plantation Technology in Tropical Forest Science. Springer, Germany, pp. 207–210.
Lamb, D. and M. Tomlinson. 1994. Forest rehabilitation in the Asia-Pacific Region: Past lessons and present uncertainties. J. Trop. For. Sci. 7(1): 157–170.
Laumonier, Y., Y. Uryu, M. Stüwe, A. Budiman, B. Setiabudi and O. Hadian. 2010. Eco-floristic sectors and deforestation threatsin Sumatra: identifying new conservation area network priorities for ecosystem-based land use planning. Biodiv. and Conserv. 19: 1153–1174.
Lee, S.S. 1998. Root symbiosis and nutrition. *In*: S. Appanah and J.M. Turnbull (eds.). A Review of Dipterocarps: Taxonomy, Ecology and Silviculture. Center for International Forestry Research. Bogor, Indonesia, pp. 99–114.
Marx, D.H., A.B. Hatch and J.F. Mendicino. 1977. High soil fertility decreases sucrose content and susceptibility of loblolly pine roots to ectomycorrhizal infection by *Phisolithus tinctorius*. Can. J. Bot. 55(12): 1569–1574.
Masano, H. Alrasjid and Z. Hamzah. 1987. Planting trials of dipterocarp species outside their natural distribution range in the Haurbentes experimental forest, West Java. *In:* A.J.G.H. Kostermans (ed.). Proceedings of the Third Round Table Conference on Dipterocarps. UNESCO. Jakarta, Indonesia, pp. 19–37.
Matsune, R. Soda, T. Tange, S. Sasaki and Suparno. 2006. Planting techniques and growth of dipterocarps in abandoned secondary forest in East Kalimantan. *In*: K. Suzuki, K. Ishi, S. Sakurai and S. Sasaki (eds.). Plantation Technology in Tropical Forest Science. Springer, Tokyo, pp. 221–229.
Michon, G.M. and J.M. Bompard. 1987. The dammar gardens (*Shorea javanica*) in Sumatra. *In*: A.J.G.H. Kostermans (ed.). Proceedings of the Third Round Table Conference on Dipterocarps. UNESCO. Jakarta, Indonesia, pp. 3–17.
Mindawati, N., Hendromono, M. Hiratsuka, T. Toma, Y. Morikawa and A.N. Gintings. 2004a. Tree trunk volume of *Shorea* species case study in Darmaga and Haurbentes Research Forest in West Java, Indonesia. J. Forest. Res. 1(1): 17–24.

Mindawati, N., I. Heriansyah, M. Hiratsuka, T. Toma, A.N. Gintings and Y. Morikawa. 2004b. Tree growth of dipterocarp plantation forest in Java, Indonesia. Info Hutan 1(2): 53–83.

Ministry of Forestry. 2012. Forestry Statistics of Indonesia 2011. Ministry of Forestry. Jakarta.

Mori, T. 2001. Rehabilitation of degraded forests in lowland Kutai, East Kalimantan, Indonesia . *In*: S. Kobayashi, J.W. Turnbull, T. Toma, T. Mori and N.M.N.A. Majid (eds.). Rehabilitation of Degraded Tropical Forest Ecosystems: Workshop Proceedings, 2–4 November 1999, Bogor, Indonesia. CIFOR, Bogor, Indonesia, Bogor, Indonesia, pp. 17–36.

Nawir, A.A., Murniati and L. Rumboko. 2007. Forest Rehabilitation in Indonesia: Where to After More Than Three Decades? Center for International Forestry Research (CIFOR), Bogor, Indonesia, p. 269.

Nuhamara, S.T. 1987. Mycorrhiza in agroforetsry: a case study. Biotropia 1(1): 53–57.

Ohta, S. and E. Syarif. 1996. Soils under lowland dipterocarp forests—characteristics and classification. *In*: A. Schulte and D. Schone (eds.). Dipterocarp Forest Ecosystems: Towards Sustainble Management. World Scientific, Singapore, pp. 29–51.

Omon, R.M. 2002. Dipterocarpaceae: *Shorea leprosula* Miq. cuttings, mycorrhizae and nutrients. Ph.D. Thesis. Wageningen University, the Netherlands. Tropenbos Kalimantan Series 7. Balikpapan, Indonesia.

Okimori, Y, J. Kikuchi and S. Hardiwinoto. 2006. Restoring the logged-over dipterocarps in tropical rainforests of Central Sumatra. *In*: K. Suzuki, K. Ishi, S. Sakurai and S. Sasaki (eds.). Plantation Technology in Tropical Forest Science. Springer, Tokyo, pp. 231–238.

Otsamo, A. 2000. Early development of three planted indigenous tree species and natural understoreyvegetation in artificial gaps in an *Acacia mangium* stand on an *Imperata cylindrica* grassland site in South Kalimantan, Indonesia. New For. 19: 51–68.

Otsamo, R., L. Kurniati and A. Otsamo. 1996. Dipterocarp specieson *Imperata cylindrica* dominated grasslands: a case study from South Kalimantan, Indonesia. *In*: I. Suhardi (ed.). Proceeding of Seminar on Ecology and Reforestation of Dipterocarp Forest in Gadjah Mada University, Yogyakarta, Indonesia, 24-25 January 1996. Faculty of Forestry, Gadjah Mada University and Kansai Environmental Engineering Centre, Indonesia, pp. 147–157.

Penot, E. 2004. From shifting agriculture to sustainable rubber complex agroforestry systems (jungle rubber) in Indonesia: a history of innovations production and adoption process. *In*: D. Babin (ed.). Beyond Tropical Deforestation. UNESCO/CIRAD, Montpellier, pp. 221–250.

Priadjati, A. 2002. Dipterocarpaceae: Forest fire and forest recovery. Ph.D. Thesis Wageningen University. Tropenbos-Kalimantan Series 8. Tropenbos International, Wageningen, the Netherlands.

Read, D.J. 1991. Mycorrhizas in ecosystems. Experientia 47: 376–391.

Riniarti, M. 2010. Dinamika kolonisasi tiga fungi ektomikoriza *Scleroderma* spp. dan hubungannya dengan pertumbuhan tanaman inang. Dissertation. Institut Pertanian Bogor, Bogor, Indonesia.

Roshetko, J.M., D.J. Snelder, R.D. Lasco and M. Van Noordwijk. 2008. Future challenge: A paradigm shift in the forestry sector. *In:* D.J. Snelder and R.D. Lasco (eds.). Smallholder Tree Growing for Rural Development and Environmental Services: Lessons from Asia. Advances in Agroforestry Volume 5. Springer, Berlin, pp. 451–483.

Santoso, E. 1988. Pengaruh mikoriza terhadap diameter batang dan bobot kering anakan Dipterocarpaceae. Bul Pen Hutan 504: 11–21.

Santoso, E. and M. Turjaman. 2003. Tipe-tipe struktur ektomikoriza pada *Shorea selanica, S. stenoptera, S. pinanga,* dan *S. palembanica* (Dipetocarpaceae) di Hutan Penelitian Haurbentes Jawa Barat. Bul Pen Hutan. 636: 33–37.

Shiva, M.P. and I. Jantan. 1998. Non-timber forest products from dipterocarps. *In*: S. Appanah and J.M. Turnbull (eds.). A Review of Dipterocarps: Taxonomy, Ecology and Silviculture. Center for International Forestry Research. Bogor, Indonesia, pp. 187–196.

Smits, W.T.M. 1992. Mycorrhizal studies in dipterocarp forests in Indonesia. *In*: D.J. Read, D.H. Lewis, A.H. Fitter and I.J. Alexander (eds.). Mycorrhizas in Ecosystems. CAB International. UK, pp. 283–392.

Smits, W. 1994. Dipterocarpaceae: Mycorrhizae and regeneration. Tropenbos Series 9. The Tropenbos Foundation, Wageningen.

Soekotjo. 2005. Sistem Silvikultur Intensif. Gadjah Mada University, Yogyakarta.

Sofiyuddin, M., Janudianto and A. Perdana. 2012. Potensi pengembangan dan pemasaran jelutung di Tanjung Jabung Barat. Brief 23. World Agroforestry Centre—ICRAF, SEA Regional Office. Bogor, Indonesia.

Subiakto, A., C. Sakai, S. Purnomo and Taufiqurahman. 2005. Cutting propagation as an alternative technique for mass production of dipterocarp planting stocks in Indonesia. Presentation paper at 8th Round-table Conference on Dipterocarps: Dipterocarp enhancing capacities in sustainable development and poverty alleviation. 15-17 November 2005. Ho Chi Minh City, Vietnam. [online] URL: http://www.apafri.org/8thdip/Session%202/S2_Atok.doc.

Subiakto, A., Hendromono and Sunaryo. 2001. *Ex situ* conservation of dipterocarp species in West Java and Banten. *In*: Thielges, B.A., S.D. Sastrapradja, A. Rimbawanto (eds.). *In situ* and *ex situ* Conservation of Commercial Tropical Trees. Faculty of Forestry, Gadjah Mada University and International Tropical Timber Organization. Yogyakarta, Indonesia, pp. 183–191.

Sutisna, M. 2001. Taungya experiment for rehabilitation of burnt-over forest in East Kalimantan, Indonesia. *In*: S. Kobayashi, J.W. Turnbull, T. Toma, T. Mori and N.M.N.A Majid (eds.). Rehabilitation of Degraded Tropical Forest Ecosystems. Workshop Proceedings, 2-4 November 1999, Bogor, Indonesia. Center for International Forestry Research, Bogor, Indonesia, pp. 115–122.

Tata, M.H.L. 2008. Mycorrhizae on dipterocarp in rubber agroforests (RAF) in Sumatra. Dissertation. Utrecht University, the Netherlands. Wohrmann Print Service, Zupthen, the Netherlands.

Tata, M.H.L., S. Hadi, C. Kusmana and Achmad. 2003a. Effect of forest fire on the survival of ectomycorrhizal fungi on dipterocarps. *In*: H. Aminah, S. Ani, H.C. Sim and B. Krishnapillay (eds.). Proceedings of the Seventh Round-Table Conference on Dipterocarps. 7–10 October 2002. Asia Pacific Association of Forestry Research Institutions (APAFRI). Kuala Lumpur, Malaysia, pp. 173–178.

Tata, M.H.L., S. Hadi, C. Kusmana and Achmad. 2003b. Putative ectomycorrhizal fungi at Sungai Wain Protection Forest, East Kalimantan. Proceedings the National Workshops on Conservation and Sustainable Management of Belowground Biodiversity. May 30-31, 2003. Bogor, Indonesia.

Tata, H.L., M. Van Noordwijk and M. Werger. 2008. Trees and regeneration in rubber agroforests and other forest-derived vegetation in Jambi (Sumatra, Indonesia). J. For. Res. 5(1): 1–20.

Tata, H.L., M. Van Noordwijk, M.J.A. Werger and R.C. Summerbell. 2010. Limited response to nursery-stage ectomycorrhiza inoculation of *Shorea* seedlings planted in rubber agorofrests in Jambi, Indonesia. New For. 39(1): 51–74.

Tata, H.L. and M. Van Noordwijk. 2011. Farmer participation on dipterocarp tree planting in small holder rubber plantation. *In*: E.B. Hardiyanto, S. Solberg and M. Osaki (eds.). Proceedings of International Conference on New Perspectives of Tropical Forest Rehabilitation for better Forest Functions and Management.Faculty of Forestry, Gadjah Mada University, Yogyakarta, Indonesia, pp. 38–42.

Torquebiau, E.F. 1984. Man-made dipterocarp forest in Sumatra. Agrofor. Syst. 2: 103–127.

Turjaman, M., Y. Tamai, H. Segah, S.H. Limin, J.Y. Cha, M. Osaki and K. Tawaraya. 2005. Inoculation with the ectomycorrhizal fungi *Pisolithus arhizus* and *Scleroderma* sp. improves early growth of *Shorea pinanga* nursery seedlings. New For. 30: 67–73.

Turjaman, M., Y. Tamai, H. Segah, S.H. Limin, M. Osaki and K. Tawaraya. 2006. Increase in early growth and nutrient uptake of two *Shorea seminis* seedlings inoculated with two ectomychorrizal fungi. J. Trop. For. Sci. 18: 243–249.

Turjaman, M., E. Santoso, A. Susanto, S. Gaman, S.H. Limin, Y. Tamai, M. Osaki and K. Tawaraya. 2011. Ectomycorrhizal fungi promote growth of *Shorea balangeran* in degraded peat swamp forests. Wetland Ecol. Manag. 19: 331–339.

Van Noordwijk, M., D. Murdiyarso, K. Hairiah, U.R. Wasrin, A. Rachman and T.P. Tomich. 1998. Forest soils under alternatives to slash-and-burn agriculture in Sumatra, Indonesia. *In*: A. Schulte and D. Ruhiyat (eds.). Soils of Tropical Forest Ecosystems: Characteristics, Ecology and Management. Springer-Verlag, Berlin, pp. 175–185.

Whitmore, T.C. 1984. Tropical rain forests of the Far East. 2nd edition. Clarendon Press, Oxford, pp. 352.

Yasman, I. 1995. Dipterocarpaceae: Tree-mycorrhizae-seedling connections. Ph.D. thesis, Wageningen Agriculture University, the Netherlands.

CHAPTER 12

The Controlled Ectomycorrhization Practices in Tropical Areas: Fungal Inoculum Biotechnology, Field Results and Research Perspectives

Robin Duponnois,[1,*] *Hervé Sanguin,*[1] *Amadou Mustapha Bâ,*[1,2] *Antoine Galiana,*[1] *Marc Ducousso,*[1] *Ezékiel Baudoin,*[1] *Michel Lebrun*[1] and *Yves Prin*[1]

1. Introduction

Overexploitation of forest resources resulting from excessive industrial exploitation, clearing for industrial purposes, and collection of firewood has led to a dramatic deforestation during recent decades in the Mediterranean and tropical areas (Piéri 1991). One of the detrimental effects of deforestation is soil degradation and desertification processes resulting in alterations of

[1]UMR 113 CIRAD/INRA/IRD/SUP-AGRO/UM2, Laboratoire des Symbioses Tropicales et Méditerranéennes (LSTM). TA A-82/J, Campus International de Baillarguet. Montpellier Cedex 5, France.
[2]Laboratoire Commun de Microbiologie (LCM) IRD/UCAD/ISRA, Centre de Recherche de Bel Air, BP. 1386, CP. 18524 Dakar, Sénégal.
*Corresponding author: Robin.Duponnois@ird.fr

major physicochemical and biological soil properties (Requena et al. 2001). The disturbances in plant cover largely contribute to increase soil erosion and decrease soil fertility and microbial activities (Garcia et al. 1997). Numerous studies have reported that in such conditions, indigenous inoculum levels of mycorrhizal fungi were significantly altered (Duponnois et al. 2001, Azcon-Aguilar et al. 2003). The mycorrhizal symbiotic establishment takes up and transports nutrients to plant roots, improves soil aggregation in eroded soils, and reduces water stress (Smith and Read 2008).

Mycorrhizal fungi are ubiquitous components of most ecosystems throughout the world and are considered key ecological factors in governing the cycles of major plant nutrients and in sustaining the vegetation cover (Van der Heijden et al. 1998, Requena et al. 2001, Schreiner et al. 2003). Two major forms of mycorrhizas are usually detected: the arbuscular mycorrhizas (AMs) and the ectomycorrhizas (ECMs). The AM symbiosis is the most widespread mycorrhizal association type with terrestrial plants (i.e. pteridophytes, gymnosperms and angiosperms) (Read et al. 2000). The arbuscular mycorrhizal fungi (AMF) colonize about 80–90% land plants in natural, agricultural, and forest ecosystems (Brundrett 2002). ECMs affect trees and shrubs, gymnosperms (Pinaceae), and angiosperms, and are frequently the result of the association of Homobasidiomycetes with about 20 families of mainly woody plants (Smith and Read 2008). Compared to AMs, ECMs are formed with a greater diversity of fungi, comprising 4,000 to 6,000 species of mainly Basidiomycetes and Ascomycetes (Allen et al. 1995, Valentine et al. 2004). Ectomycorrhizal fungi are not homogeneously distributed within the ecosystem, which has important implications for reforestation in areas with no native ectomycorrhizal vegetation (Marx 1991). The absence of ectomycorrhizal fungi results from natural (Terwilliger and Pastor 1999) or anthropogenic disturbance (Jones et al. 2003).

The lack of ectomycorrhizal fungal formations on root systems is a major cause of poor ectomycorrhizal plant establishment and growth in a variety of forest landscapes. Numerous studies have reported that specific ectomycorrhizal fungi can promote the survival and early growth of various tree species in the field (Castellano and Molina 1989, Kropp and Langlois 1990, Marx 1991, Castellano 1996, Roldan et al. 1996, Garbaye and Churin 1997, Duponnois et al. 2005a, 2007). Since mycorrhizal associations are estimated to occur in 95% of native undisturbed vegetation, whereas they are found in less than 1% of vegetation from disturbed sites, improving mycorrhizal soil potential may increase the benefit of the mycorrhizal effect on plant growth. This objective can be achieved by inoculating seedlings before they are transferred into disturbed sites. In order to improve the performance of afforestation, it is necessary that nurseries produce tree seedlings associated with ectomycorrhizal fungi that are ecologically compatible with the tree species and characteristics of the planting sites. According to these conditions, different methods of controlled inoculation have been identified to optimize the effect on plant growth.

Controlled mycorrhization is based on the use of mycorrhizal strains best suited to host plant species which will rapidly colonize their root systems and are well adapted to the environmental conditions of the planting site (Duponnois et al. 2007). However, this practice requires high volumes of ectomycorrhizal fungal inoculum. Hence this biotechnological process is generally associated with large investment of resources that limits its use in forest nurseries, especially in developing countries.

As the production of fungal inoculum is costly, it has been suggested that Mycorrhization Helper Bacteria in the rhizosphere could stimulate the establishment of the fungal inoculant and consequently, minimize the volume of required fungal inoculum (Duponnois et al. 1993). It has also been demonstrated that the termite mounds of *Macrotermes subhyalinus* (a litter-foraging termite frequently found in tropical areas) were colonized by a specific microflora that promotes the ectomycorrhizal fungal development (Duponnois 2006). Thus *M. subhyalinus* mound amendment could be used as a biological tool to improve the efficiency of controlled mycorrhization in forest nurseries and to minimize the technical and financial investment of this biotechnology.

The main objectives of this chapter are to describe the processes required to produce large quantities of efficient fungal inocula and to report tree growth data resulting from the use of controlled ectomycorrhization in nursery and field conditions in tropical and Mediterranean areas. In addition, innovative processes of mycorrhizal inoculation using Mycorrhization Helper Bacteria (MHB) and termite mounds of *M. subhyalinus* as catalysts for mycorrhization will be described.

2. Biotechnological Processes used in Controlled Ectomycorrhization

2.1 Criteria to adopt an inoculum formulation

Different criteria have to be taken in account to adopt an inoculum formulation. The selected ectomycorrhizal fungi must provide a positive effect on the growth and the survival of tree species used for plantation. The fungal propagules must be kept viable during storage to preserve the fungal efficiency on plant growth after inoculation. Finally ectomycorrhizal fungal inocula have to be cost-effective in order to be used in nursery practices.

2.2 Spore-based fungal inoculum

Spores from fruiting bodies can be considered as a natural source of inocula. This inoculation practice has been widely used in forest nurseries (Castellano 1994). However, this inoculation technique is only effective with fungal species that produce abundant sporocarps with large spore numbers (i.e., *Pisolithus* and *Scleroderma*).

Numerous formulations of spore inoculums have been described. Sporocarps are collected, carefully identified, and kept in paper bags. They are then brushed free of adhering soil. This inoculum type is frequently based on dry spores obtained by air drying fruiting bodies at temperatures below 35°C. The dried sporocarps are crushed in plastic bags and sieved through 200 and 500 µm sieves, respectively. The dried spores can then be used in different formulations. Spore powder can be mixed with sterilized fine sand (1:100, w/w) and used to inoculate containers by either mixing the soil substrate with this fungal mixture or by adding a small quantity of the fungal inoculum at the base of seedlings. Ectomycorrhizal inoculation can be achieved by adding spore powder into an inert carrier (i.e., clay) to form pellets that are placed at the base of seedlings (de la Cruz et al. 1990). According to Turjaman et al. 2005, in order to determine the influence of ectomycorrhizal fungi on the growth of dipterocarp species in peat soils, the sporocarps were crushed manually in plastic bags to minimize loss of spores and cross-contamination between fungi (de la Cruz et al. 1990). Crushed sporocarps and clay were mixed and formed into pellets with a ratio of 1:100 (w/w). Inoculation was made 10 days after seed germination. One hole was made in each pot and a tablet (0.4 g) of fungal inoculum was applied to a potted seedling 1 cm below the soil surface, at the proximity of the root. Another process of ectomycorrhizal spore inoculation is developed by coating seeds with a mixture of spores and a binding agent such as clay (Marx et al. 1984).

The beneficial effects of ectomycorrhizal spore inoculation on plant growth in nursery conditions have been frequently recorded (Table 1). Several disadvantages have also been reported. It is difficult to collect large quantities of fruit bodies with some ectomycorrhizal fungal species. In addition, low efficiency of the fungal inoculums due to slow germination or low spore viability is frequently recorded. Besides these critiques, this inoculation process is easy to apply and remains an efficient method for storage and transport of spores for some mycorrhizal fungal genera such as *Pisolithus* or *Scleroderma*.

Table 1. Effect of ectomycorrhizal spore inoculation on plant growth in nursery conditions.

Ref. [1]	Fungal strain	Plant species	Inoculum formulation	Effects on plant growth parameters (%) and ectomycorrhizal colonization			
				Height	Shoot biomass	Root biomass	Ecto. colonization (%)
A	*Pisolithus arhizus*	*Shorea pinanga*	Ectomycorrhizal spore pellets	+ 46.1 [2]	+ 66.7	nd[3]	87
A	*Scleroderma* sp.	*S. pinanga*	Ectomycorrhizal spore pellets	+ 41.3	+ 60.5	nd	86
B	*P. tinctorius*	*Acacia mangium*	Ectomycorrhizal spore tablets	nd	+ 29.1	+ 40.7	52
C	*Rhizopogon roseolus*	*Pinus halepensis*	Spore suspension	–30.1	nd	nd	48
C	*Suillus collinitus*	*Pinus halepensis*	Spore suspension	+ 27.8	nd	nd	75
D	*Scleroderma albidum*	*Eucalyptus globulus*	Spore suspension	+ 32.1	+ 19.7	+ 42.5	nd
D	*S. areolatum*	*E. globulus*	Spore suspension	+ 17.5	+ 9.1	+33.5	nd
D	*S. cepa*	*E. globulus*	Spore suspension	+ 30.8	+ 8.8	+ 2.0	nd
D	*S. albidum*	*E. urophylla*	Spore suspension	– 3.6	+ 3.1	–0.7	nd
D	*S. areolatum*	*E. urophylla*	Spore suspension	+ 7.2	+ 4.9	+ 4.0	nd
D	*S. cepa*	*E. urophylla*	Spore suspension	+ 6.3	+ 13.0	+ 1.4	nd
E	*P. tinctorius*	*P. halepensis*	Spore suspension	+ 37.1	+ 44.9	+ 41.7	55.4
E	*R. roseolus*	*P. halepensis*	Spore suspension	+ 38.3	+ 44.6	+ 28.6	39.5
E	*Suillus collinitus*	*P. halepensis*	Spore suspension	+ 35.1	+ 28.2	+ 30.6	28.9

A: Turjaman et al. 2005. B: Aggangan et al. 2010. C: Rincon et al. 2007. D: Chen et al. 2006. E: Torres and Honrubia 1994
[1] Ref.: Reference. [2] (mean value of ectomycorrhizal plants—mean value of the non ectomycorrhizal plants) x 100)/(mean value of the ectomycorrhizal plants). [3]nd: not determined.

2.3 Mycelium-based fungal inoculum

These inoculum formulations need the isolation of fungal symbionts in axenic conditions to obtain a pure sample of a single fungal strain. These axenic cultures are often obtained from fruit bodies of ectomycorrhizal fungi collected in the field but they can also be performed by collecting mycorrhizal roots, sclerotia, rhizomorphs and spores (Molina and Palmer 1982).

Isolating fungi from from fruiting bodies is the most successful method for obtaining pure fungal cultures. Sporocarps are of adhering soil and carefully broken in a laminar flow hood. A small amount of tissue is then removed with fine forceps and placed on a nutrient medium agar. As growth requirements vary between ectomycorrhizal fungi, different nutrient media can be used (Table 2). The fungal cultures are incubated at 25°C in the dark and subcultured until all contaminating microorganisms are eliminated. Fungal cultures are usually maintained in Petri dishes on a required nutrient agar medium at 20 and 25°C, respectively. Fresh subcultures are made every 6 to 12 weeks depending on fungal growth rates.

The formulation of mycelium-based inoculum is only feasible and economical with some fungal species that can grow in cultures. Two main cultural supports are usually used to produce fungal inocula: (i) mixture of peat/vermiculite and (ii) hydrogel (i.e., calcium alginate gel).

Glass jars (1.6 liter) containing 1.3 liter expanded vermiculite-sphagnum peat mixture (4:5 and 1:5, v:v, pH = 5.5) are heat disinfected (120°C, 20 min) (Marx and Bryan 1975). Another ratio of vermiculite-peat can be used (2:3 and 1:3) but at this ratio, the peat can release substances that are toxic for fungal growth. Hence, the first ratio mixtures are usually used. The mixture is then moistened to field capacity with 600 ml of a modified liquid nutrient medium (see Table 2). The jars are closed with lids with a 1-cm diameter hole. This hole is fitted with a 4-cm long tube filled with cotton wool. The jars are then heat disinfected a second time (120°C for 20 min). After cooling, approximately eight mycelial plugs are placed on top of the substrate. The mycelia grow down into the substrate, which is completely colonized after 6 to 10 weeks at 25°C. For faster growth, jars can be filled with a smaller quantity of substrate and shaken after mycelia have colonized a few centimeters: in this manner, mycelium is evenly distributed throughout the substrate and incubation time is shortened. This inoculum can be stored at 4°C for upto 6 months.

Dommergues et al. (1979) and Le Tacon et al. (1983, 1985) first reported the process of including fungal mycelium in polymeric gels (especially calcium alginate). This type of inoculum is more efficient than the vermiculite-peat formulation (Mortier et al. 1988) since the gel protects the mycelium from physical stresses (e.g., water stress) and from competing microorganisms.

Table 2. Composition of nutrient media commonly used for the isolation and culture of ectomycorrhizal fungi. Data from Brundrett et al. 1996.

Components	Nutrient media		
	MMN [(1)]	Pachlewski [(2)]	FDA [(3)]
Mineral nutrients (mg.l^{-1})			
$(NH_4)_2HPO_4$	250		
NH_4Cl			500
$C_4H_{12}N_2O_6$ [(4)]		500	
KH_2PO_4	500	1000	500
$MgSO_47H_2O$	150	500	500
$CaCl_22H_2O$	50	50	
NaCl	25		
Fe EDTA	20	20	
H_3BO_3		2.8	
$MnCl_22H_2O$		3.0	
$ZnSO_47H_2O$		2.3	
$CuCl_22H_2O$		0.63	
$Na_2Mo_42H_2O$		0.27	
Carbohydrate source (g.l^{-1})			
Maltose		5	
Glucose	10	20	20
Malt extract	3		
Vitamins (μg.l^{-1})			
Thiamine HCl	0.1	0.1	
Agar (g.l^{-1})	20	20	20
pH			
Adjusted pH to	5.8	5.4	5.0

[(1)] MMN: Modified Menin Norkrans medium (Marx 1969). [(2)] Pachlewski medium (Pachlewski and Pachlewski 1974). [(3)] Ferry and Das (1968). [(4)] Ammonium tartrate

Using this process, it is also possible to accurately determine the weight of mycelium or the number of living propagules contained in the inoculum. Different methods have been evaluated to measure the quantity of mycelium in the vermiculite-peat inoculum, such as the ergosterol assay (Martin et al. 1990) and chitin assay (Vignon et al. 1986), but none of them gave reliable results (Mortier et al. 1988) since the peat interferes with colorimetric measurements. With fungal strains that are able to grow from fragmented hyphae, a mycelial suspension is produced in Erlenmeyer flasks or in a fermentor filled with the liquid nutrient medium. The mycelium is then washed in tap water to remove remaining nutrients, homogenized in a

Waring blender for about 10s, and suspended in sterile distilled water. This type of fungal inoculum is quantified by measuring the fungal dry weight per ml or by counting living propagules (determination of colony forming units by spreading 1 ml of suspension on a 6-cm Petri dish with nutrient agar). The hyphal fragment suspension is then mixed (1:1, v:v) with distilled water containing 20 $g.l^{-1}$ sodium alginate and 50 $g.l^{-1}$ autoclaved dry powdered sphagnum peat. The final solution is pumped through a pipe with 2-mm holes. The drops fall into a 100 $g.l^{-1}$ $CaCl_2$ solution and form beads of reticulated calcium alginate gel (Mauperin et al. 1987). The beads are kept in $CaCl_2$ for 24h at room temperature to ensure complete reticulation. Then they are washed with tap water to eliminate NaCl and $CaCl_2$ and stored in air-tight containers at 4°C to prevent drying. This type of inoculum can be kept up to 9 months in these conditions. The beads are prepared with 1–2 g mycelium (dry weight) per liter of final solution (Mortier et al. 1988).

3. Controlled Ectomycorrhization in Glasshouse, Nursery and Field Conditions

The aims of the various technical procedures are to screen fungal isolates for their compatibility with targeted host plants and to determine plant growth promotion so the most efficient fungal isolate can be selected for use in nurseries. This topic will be mainly illustrated by some data resulting from controlled ectomycorrhization trials with Australian *Acacia* species and tropical ECM fungal strains.

Seedlings can grow in soil (disinfected or not after autoclaving) or in vermiculite-peat mixture (1:1, v:v). Several kinds of containers can be used for growing seedlings in the greenhouse depending on the aim of the experiment. The soil was then mixed with the ectomycorrhizal inoculum (10:1, v:v). The fungal inoculum control (free of any treatment) received an autoclaved mixture of vermiculite and peat moss moistened in nutrient medium at the same rate. Then the pots were arranged in a randomized complete block design. Numerous studies have reported that measuring the promoting effects of fungal inoculation on the growth and nutrient status of tree seedlings recorded in controlled conditions is the best way to select the most promising and potential mycorrhizal inoculants for large-scale inoculation programs in degraded lands in the tropics and Mediterranean areas (Table 3). Unfortunately, although several methods of inoculation can be used in practice, only some species of the genera *Laccaria*, *Hebeloma*, *Paxillus*, *Pisolithus*, *Scleroderma*, *Suillus* and *Rhizopogon* have been routinely used (Castellano 1996).

Table 3. Growth of some Australian *Acacia* species inoculated with different ectomycorrhizal fungal strains (peat-vermiculite formulation) after 4 months culturing in glasshouse conditions.

Fungal species	***Acacia* species**	**Effects on plant growth parameters (%) and ectomycorrhizal colonization**			**References**
		Shoot biomass	**Root biomass**	**Ectomycorrhizal colonization (%)**	
Pisolithus albus IR100	*A. auriculiformis*	+ 42.1 [(1)]	+ 38.6	45.2	A
P. albus IR100	*A. mangium*	+ 35.9	+ 44.1	20.1	A
P. albus IR100	*A. platycarpa*	+ 43.1	+ 9.8	31.6	A
Pisolithus sp. SL2	*A. holosericea*	+ 56.7	+ 10.6	48.3	A
P. albus COI007	*A. holosericea*	+ 55.6	+ 25.3	43.8	A
P. albus COI024	*A. holosericea*	+ 50.4	+ 17.1	10.8	A
Pisolithus sp. COI032	*A. holosericea*	+ 54.9	+ 12.6	15.0	A
P. albus IR100	*A. holosericea*	+ 57.1	+ 48.9	25.2	A
P. tinctorius GEMAS	*A. holosericea*	+ 57.8	+ 14.1	43.1	A
Scleroderma dictyosporum IR109	*A. holosericea*	+ 52.9	+ 21.0	53.4	A
S. verrucosum IR500	*A. holosericea*	+ 64.4	+ 14.1	13.8	A
P. tinctorius GEMAS	*A. crassicarpa*	+ 77.1	+ 52.3	49.4	B
P. albus COI024	*A. mangium*	+ 54.5	+ 52.1	41.7	C
Scleroderma sp. IR408	*A. holosericea*	+ 82.1	+ 89.6	13.8	D
S. dictyosporum IR412	*A. holosericea*	+ 72.9	+ 82.6	12.5	D

[(1)] (mean value of ectomycorrhizal plants—mean value of the non ectomycorrhizal plants) x 100)/(mean value of the ectomycorrhizal plants).
A: Duponnois and Plenchette 2003. B: Lesueur and Duponnois 2005. C: Duponnois et al. 2002. D: Duponnois et al. 2006

Numerous studies have reported that controlled mycorrhizal symbiosis can significantly improve the growth of tree species in degraded soils *in situ* (Smith and Read 2008). However, most of these experiments have been performed in controlled glasshouse conditions. Data on the sustainability of the positive effect of mycorrhiza on plant growth in field conditions are lacking, especially in tropical areas. Controlled ectomycorrhization experiments have been set up in Senegal using an Australian *Acacia holosericea* and the ectomycorrhizal fungus *Pisolithus albus* strain IR100 (Duponnois et al. 2007). A positive effect of mycorrhizal inoculation on *A. holosericea* development has been recorded in all of the experiments (Fig. 1). These results suggest that without suitable symbionts, this species has poor ability to scavenge for P under P-limiting conditions (Dommergues et al. 1999) and clearly shows that the controlled mycorrhization of *A. holosericea* could be a beneficial tool in improving the survival and productivity of *Acacia* species. However, in order to evaluate the possible

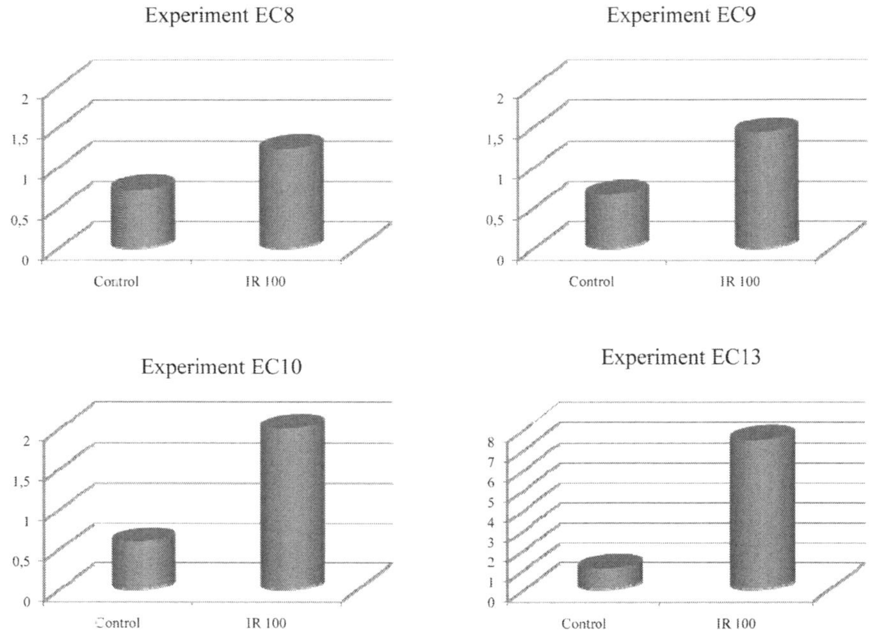

Fig. 1. Field performance of *A. holosericea* trees (expressed in tons of wood biomass per ha with 1200 planted trees per ha) inoculated or not with *Pisolithus albus* strain IR100 in field trials conducted in Senegal after 18 months of plantation. Data from Duponnois et al. 2007.

economic advantage of ectomycorrhizal inoculation, it is essential to carry out field experiments in a range of situations representative of all major types of plantations.

4. Mycorrhization Helper Bacteria (MHB)

The existence of this bacterial group was reported for the first time by Garbaye and Bowen (1989). The MHB are commonly defined as telluric bacteria enhancing the establishment of mycorrhizal symbiosis. To isolate MHB, ECMs were surface-sterilized to only isolate bacteria closely associated with the ectomycorrhizas. According to the results of Garbaye and Bowen (1989), most of the bacteria isolated from *Rhizopogonluteolus*/ *Pinusradiata* ectomycorrhizas belong to the fluorescent pseudomonads group and about 80% of the bacterial strains tested in this experiment stimulated mycorrhiza formation, whereas 20% had neutral or negative influence. Duponnois and Garbaye (1991) showed that among some bacteria isolated from sporocarps of *Laccariabicolor* and ectomycorrhizas of Douglas fir with *L. bicolor*, 14 bacterial isolates have significantly stimulated the ectomycorrhizal colonization of the Douglas fir root systems with *L. bicolor* strain S238 after 4 months of culture time in glasshouse conditions. These authors did not find any links between the bacterial effect on mycorrhiza formation and the origin of the bacteria or their taxonomic position. A bacterial strain of *Bacillus subtilis*, isolated from Douglas/*L. bicolor* ectomycorrhizas, has increased the ectomycorrhizal rate (percent of short roots mycorrhized with *L. bicolor*) to 97.3% vs. 67.3% ectomycorrhizal rate recorded in the control treatment (inoculated fungus without bacteria), while a bacterial strains isolated from a *L. bicolor* sporocarp (*Pseudomonas fluorescens* strain BBc6) exerts the same positive effect on mycorrhiza formation. This MHB effect has been recorded in different environmental conditions (glasshouse and nursery conditions) showing the high competitiveness of the bacterial inoculant against the native microflora, but also in axenic conditions that showed that the stimulation is an intrinsic property of each bacterial strain and did not result from interactions within the microbial community in the rhizosphere (Duponnois 1992).

MHB have been isolated from different plant-fungal combinations such as arbuscular mycorrhizal fungi (herbaceous plants) or ectomycorrhizal fungi (trees). To date, numerous studies have reported that many bacterial strains could stimulate either the establishment of arbuscular or ectomycorrhizal symbiosis (Table 4) (Frey-Klett et al. 2007). All these results show that the MHB concept is generic and that it is not subjected to the type of the mycorrhizal symbiosis nor to the taxonomic position of the MHB strains (Frey-Klett et al. 2007).

The MHB effect on ectomycorrhizal symbiosis has been mainly investigated with northern hemisphere tree species like *Pseudotsugamenziesii*, *Quercusrobur* (Duponnois and Garbaye 1991), *Picea abies*, *Pinus sylvestris* (Schrey et al. 2005), *Pinus radiata* (Garbaye and Bowen 1989) and some ectomycorrhizal fungi, principally *L. bicolor* (Frey-Klett et al. 2007), *Suillus luteus* (Bending et al. 2002). However, it has also been demonstrated that some bacteria (*Fluorescent pseudomonads*) could help ectomycorrhizal formation in tropical conditions with Australian *Acacia* species (Founoune et al. 2002a,b).

MHBs belonging to the fluorescent pseudomonad group have been frequently studied for their effect on ectomycorrhiza establishment in glasshouse conditions. For example, the inoculation of a MHB, *Pseudomonas monteilii* strain HR 13 has been studied on the mycorrhization of (i) an Australian *Acacia, A. holosericea,* by several ectomycorrhizal fungi and one endomycorrhizal fungus *Glomusintraradices*, and of (ii) several Australian *Acacia* species by *Pisolithusalbus* strain IR 100 under glasshouse conditions (Duponnois and Plenchette 2003). A promoting effect of HR13 on the ectomycorrhizal establishment has been recorded with all the fungal isolates (strains of *Pisolithus* and *Scleroderma*) (Fig. 2).

Table 4. MHB impacts on mycorrhiza establishment reported from the literature (From Frey-Klett et al. 2007).

Mycorrhizal fungi	MHB species	Host plant	References
Ectomycorrhizal fungi			
Amanita muscaria, *Suillus bovinus*	*Sptreptomyces*	*Picea abies*, *Pinus sylvestris*	Schrey et al. 2005
Hebelomecrustuliniforme	Unidentified bacteria	*Fagus sylvatica*	De Oliveira 1988
Laccaria bicolor	*Pseudomonas fluorescens*, *Pseudomonas* sp., *Bacillus* sp.	*Pseudotsuga menziesii*	Duponnois and Garbaye 1991
Laccaria fraterna, *Laccaria laccata*	*Bacillus* sp., *Pseudomonas* sp.	*Eucalyptus diversicolor*	Dunstan et al. 1998
Lactariusrufus	*Paenibacillus* sp., *Burkholderia* sp.	*Pinus sylvestris*	Poole et al. 2001
Pisolithus albus	*Pseudomonas monteilii* *Pseudomonas resinovorans*	*Acacia holosericea*	Founoune et al. 2002a
Pisolithus sp.	Fluorescent Pseudomonads	*Acacia holosericea*	Founoune et al. 2002b
Rhizopogon luteolus	Unidentified bacteria	*Pinus radiata*	Garbaye and Bowen 1989
Rhizopogon vinicolor, *Laccaria laccata*	*Arthrobacter* sp.	*Pinus sylvestris*	Rózycki et al. 1994
Scleroderma spp., *Pisolithus* spp.	*Pseudomonas monteilii*	Australian *Acacia* species	Duponnois and Plenchette (2003)
Suillus luteus	*Bacillus* sp.	*Pinus sylvestris*	Bending et al. 2002
Arbuscular mycorrhizal fungi			
Endogone sp.	*Pseudomonas* sp.	*Trifolium* spp., *Cucumis sativum*, *Allium cepa*	Mosse 1962
Gigaspora margarita	*Azospirillum brasilense*	*Pennidetum americanum*	Rao et al. 1985
Glomus clarum	*Azotobacter diazotrophicus*, *Klebsiella* sp.	*Ipomoea batatas*	Paula et al. 1992
Glomus deserticola	*Klebsiella pneumoniae*, *Alcaligenes denetrificans*	*Unicola paniculata*	Will and Sylvia 1990
Glomus fasciculatum	*Azotobacter chroococcum*	*Lycopersicum esculentum*	Bagyaraj and Menge 1978
Glomus fasciculatum, *Glomusmosseae*	*Rhizobium meliloti*	*Medicago sativa*	Azcón-Aguilar et al. 2003

Glomus fasciculatum, Glomusmosseae, Glomus caledonium	*Bacillus coagulans*	*Morus alba, Carica papaya*	Mamatha et al. 2002
Glomus fistulosum	*Pseudomonas putida*	*Zea mays, Solunum tuberosum*	Vosatka and Gryndler (1999)
Glomus intraradices	*Bacillus subtilis, Enterobacter* sp.	*Allium cepa*	Toro et al. 1997
Glomus intraradices	*Pseudomonas monteilii*	*Acacia holosericea*	Duponnois and Plenchette 2003
Glomus intraradices	*Rhizobium*	*Anthyllis cytisoides*	Requena et al. 1997
Glomus intraradices	*Agrobacterium rhizogenes, Pseudomonas fluorescens, Rhizobium leguminosarum*	*Hordeum vulgare, Triticum aestivum*	Fester et al. 1999
Glomus intraradices	*Streptomyces coelicolor*	Sorghum	Abdel-Fattah and Mohamedin 2000
Glomus mosseae	*Paenibacillus* sp.	*Sorghum bicolor*	Budi et al. 1999
Glomus mosseae	*Pseudomonas* sp.	*Lycopersicum esculentum*	Barea et al. 1998
Glomus mosseae	*Bradhyrhizobium japonicum*	*Glycine max*	Xie et al. 1995
Glomus mosseae	*Pseudomonas fluorescens*	*Lycopersicum esculentum*	Gamalero et al. 2004
Glomus mosseae	*Brevibacillus* sp.	*Trifolium pratense*	Vivas et al. 2003
Glomus mosseae, Glomus intraradices	*Paenibacillus brasilensis*	*Trifolium*	Artursson 2005
Complex of AMF [(1)]	*Pseudomonas putida*	*Trifolium*	Meyer and Linderman (1986)
Complex of indigenous AMF	*Pseudomonas* sp.	*Triticum aestivum*	Babana and Antoun 2005
Complex of AMF	*Bacillus mycoides*	Herbaceous plant species	von Alten et al. 1993

[(1)] AMF: Arbuscular Mycorrhizal Fungi.

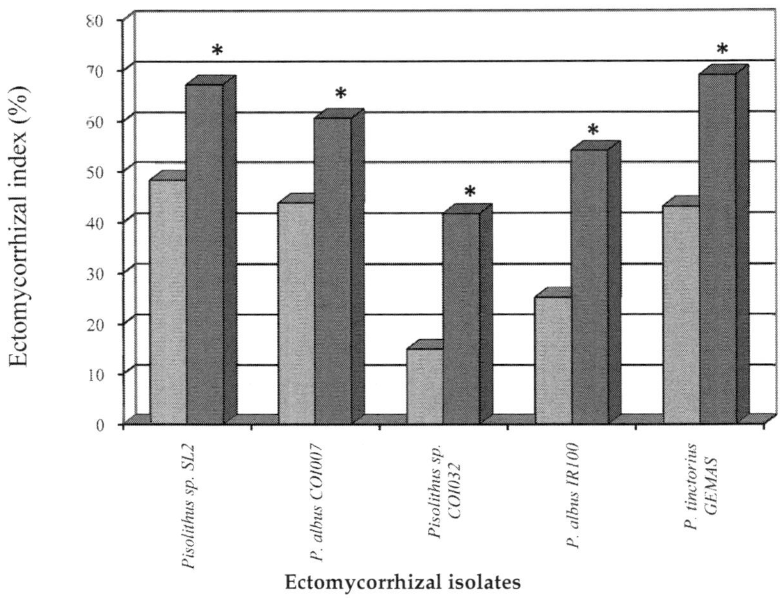

Fig. 2. Effect of *Pseudomonas monteilii* isolate HR13 on ectomycorrhiza formation of *Acaciaholosericea* with ectomycorrhizal fungi after 4 months of culturing under glasshouse conditions.

*: Significant effect of *P. monteilii* isolate on ectomycorrhizal formation for each fungal strain. ■: Fungal strain inoculated alone. ■: Fungal strain inoculated with *P. monteilii* HR13.

The MHB *P. monteilii* strain HR13 has also significantly increased the ectomycorrhizal colonization for all the Australian *Acacia* species tested in this experiment (*A. auriculiformis, A. eriopoda, A. holosericea, A. mangium* and *A. platycarpa*) (Fig. 3).

MHB have also been tested for their influence on ectomycorrhizal formations between Douglas and *L. bicolor* strain S238 in nursery conditions. One bacterial isolate (*Pseudomonas fluorescens* isolate BBc6) increased mycorrhiza formation from around 60% (per cent of short roots mycorrhized with *L. bicolor* S238N without bacterial inoculation) to 87%. This experiment was performed with a peat-vermiculite fungal inoculum. Similar positive effects were recorded when this bacterial strain was mixed together with the fungus in alginate beads (Duponnois 1992). After 4 months of plantation, BBc6 stimulated ectomycorrhizal infection from 42% to 75%. Frey-Klett et al. (1999) showed that five months after inoculation and sowing, bacterial inoculation significantly improved the percentage of ectomycorrhizal short roots on plants inoculated with two doses of fungal inoculum of *L. bicolor* (50 and 100 mg m^{-2} dry weight mycelium entrapped in alginate beads mixed into the soil at a constant dose of 1 litre m^{-2}) in a nursery (Fig. 4). The

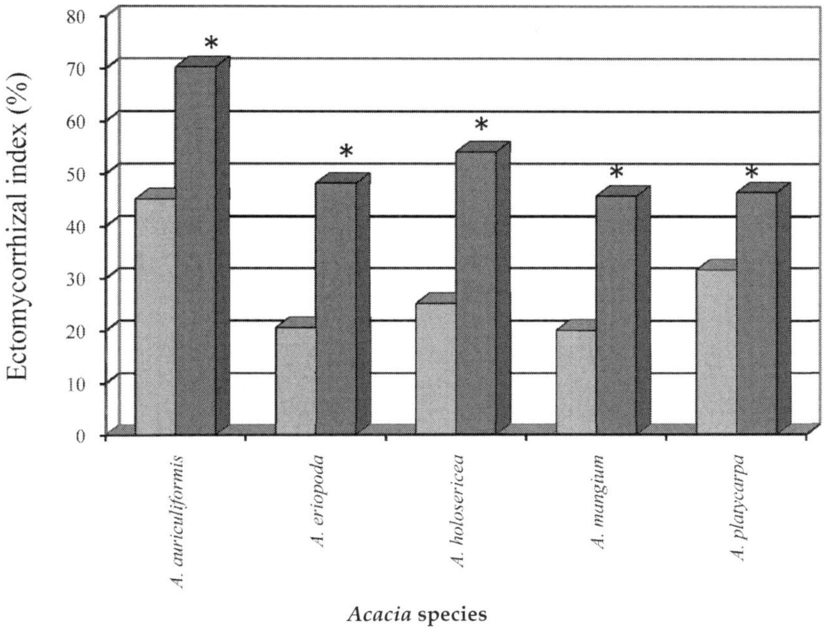

Fig. 3. Effect of *Pseudomonas monteilii* isolate HR13 on ectomycorrhiza formation with *Pisolithus albus* isolate IR100 of five Australian *Acacia* species after 4 months of culturing in glasshouse conditions. For the legend, see Fig. 2.

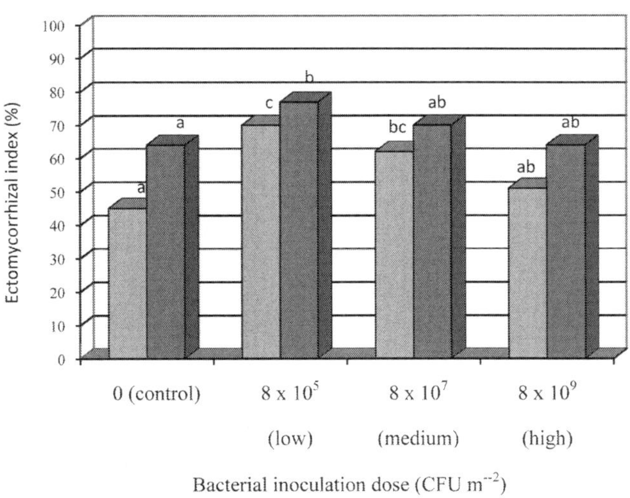

Fig. 4. Dose effect of *P. fluorescens* isolate BBcR8 on the percentage of Douglas-fir short roots mycorrhizal with *L. bicolor* S238, 23 weeks after fungal and bacterial inoculations. ■: Fungal inoculation dose: 50 mg dry weight mycelium m^{-2}. ■: Fungal inoculation dose: 100 mg dry weight mycelium m^{-2}. For each fungal inoculation dose, columns indexed by the same letters are not significantly different according to a Scheffe test. Data from Frey-Klett et al. (1999).

lowest bacterial dose increased mycorrhizal colonization from 45 to 70% in plants inoculated with the lower amount of fungal inoculum, and from 64 to 77% in plants inoculated with the higher amount of fungal inoculum. The lowest bacterial dose increased mycorrhizal colonization more than the highest bacterial dose (Fig. 4).

5. Termite Mounds of *Macrothermes subhyalinus* as a Catalyst of Ectomycorrhizal Formation: Islands of Microbial Fertility in Tropical Forest Ecosystem

Termite (*Isoptera*) mounds are frequently observed in tropical ecosystems, particularly in savanna environments. Termite activities result in the translocation of large amounts of soil from different depths (Holt and Lepage 2000). These biological disturbances strongly impact their local environment and have a major influence on soil's physical and chemical characteristics (Holt and Lepage 2000) demonstrating the termite role as ecosystem engineers.

Previous microbiological studies of termite mounds have described the microbial communities in grass-, litter- and soil-feeding termite mounds (Duponnois et al. 2005b). Fluorescent pseudomonads have only been detected in mound powder of one species of termite, *M. subhyalinus* (Table 5).

The phylogenetic analysis done on these fluorescent pseudomonads showed that these bacteria mostly belonged to *Pseudomonas monteillii* species (Duponnois et al. 2006). The dual inoculation of termite mound and ectomycorrhizal fungus (*Scleroderma* sp. isolate IR408 or *S. dictyosporum* isolate IR412) significantly enhanced root growth of *A. holosericea* seedlings after 4 months of culturing in greenhouse conditions (Table 6). In addition, termite mound amendment significantly increased ectomycorrhizal formation of *A. holosericea* with both of the fungal isolates tested in this study (Table 6). It has been also reported that an isolate of *P. monteilii* (isolate HR13) could stimulate ectomycorrhizal formation between *A. holosericea* and different fungal isolates (*Scleroderma dictyosporum*, *S. verrucosum*, *P. albus* and *P. tinctorius*) (Founoune et al. 2002a, Duponnois and Plenchette 2003) as well as AM establishment between *Acacia* species and *Glomus intraradices* (Duponnois and Plenchette 2003).

It has also been shown that another strain of *P. monteillii*, strain KR9, isolated from *M. subhyalinus* termite mounds, stimulated ectomycorrhizal formation between *S. dictyosporum* IR412 and *A. holosericea* and the growth of the host plant (Duponnois et al. 2006) (Table 7).

As the termite mound powder exerted a significant positive effect on mycorrhizal establishment and plant growth, an experiment was conducted

Table 5. Biological and chemical characteristics of *Cubitermes*, *Macrotermes* and *Trinervitermes* mound powders. Data from Duponnois et al. 2005b.

	Cubitermes sp.	*Macrotermes subhyalinus*	*Trinervitermes* sp.
NH_4^+ (µg N g^{-1} of dry mound powder)	40.9 b *	9.4 a	37.1 b
NO_3^- (µg N g^{-1} of dry mound powder)	206.9 a	3408.9 b	42.3 a
Available P (µg g^{-1} of dry mound powder)	7.7 b	3.5 a	10.2 c
Microbial biomass (µg C g^{-1} of dry mound powder)	17.0 a	22.5 ab	36.5 b
Fluorescent pseudomonads (x 10^2 CFU g^{-1} of dry mound powder)	< 1 a	79.3 b	< 1 a
Actinomycetes (x 10^2 CFU g^{-1} of dry mound powder)	37.3 b	39.5 b	22.5 a
Ergosterol (µg g^{-1} of dry mound powder)	1.381 b	0.316 a	1.717 b

*: Data in the same line followed by the same letter are not significantly different according to Student's "t" test ($p < 0.05$).

Table 6. Impacts of fungal inoculation and *Macrotermes subhyalinus* amendment on *Acacia holosericea* growth and ectomycorrhizal establishment after 4 months of culturing in greenhouse conditions.

Treatments	**Shoot biomass (mg dry weight)**	**Root biomass (mg dry weight)**	**Ectomycorrhizal colonization (%)**
Control	261 a [(1)]	33 a	0
Scleroderma sp. IR408	1458 c	318 bc	12.8 a
S. disctyosporum IR412	964 b	190 ab	14.2 a
M. subhyalinus (MS)	1288 bc	238 b	0
IR 408 + MS	1051 b	432 cd	23.6 b
IR412 + MS	1140 bc	606 d	25.8 b

[(1)] Data in the same column followed by the same letters are not significantly different according to the Student's "t" test ($p < 0.05$).

Table 7. Effect of the *Pseudomonas monteilii* strain (isolate KR9) and/or *Scleroderma disctyosporum* IR 412 on mycorrhiza formation, rhizobial development, and growth of *Acacia holosericea* after 4 months of culturing under glasshouse conditions.

	Treatments			
	Control	***P. monteilii* KR9**	***S. dictyosporum* IR412**	**KR9 + IR412**
Shoot biomass (mg dry weight)	532 a [(1)]	553 a	1236 b	1786 c
Root biomass (mg dry weight)	184 a	198 a	536 b	868 c
Number of nodules per plant	4.2 a	4.6 a	8.3 b	12.4 c
Total nodule weight per plant (mg)	6.8 a	7.1 a	15.9 b	25.3 c
Ectomycorrhizal colonization (%)	0	0	28.3 a	48.5 c

[(1)] Data in the same line followed by the same letter are not significantly different according to the Student's "t" test ($p < 0.05$).

to test a new mycorrhizal inoculation process using a "catalyser" of the mycorrhizal establishment (termite mounds of *M. subhyalinus*) to minimize the required volume of fungal inoculum added to the cultural substrate (Duponnois 2006). The effects of the termite mound powder have been explored on the mycorrhiza formation of an Australian Acacia, *Acacia holosericea*, by an ectomycorrhizal fungus (*Pisolithus albus* IR 100) and an arbuscular mycorrhizal (AM) fungus (*Glomus intraradices*). For this, seeds of *A. holosericea* were surface-sterilized with 95% sulfuric acid for 60 minutes and transferred aseptically to Petri dishes filled with 1% (w/v) agar/water medium. These plates were incubated at 25°C in the dark. The germinating seeds were used when rootlets reached 1–2 cm long. A disinfected (120°C, 40 min) sandy soil was mixed with 0, 1, 5 and 10% (v/v) of *M. subhyalinus* mound powder. Ectomycorrhizal and AM inoculation were performed by mixing the soils with either the fungal inoculum (10/1, v/v) or a control: an autoclaved mixture of moistened vermiculite/peat moss at the same rate (ectomycorrhizal control) or with non-mycorrhizal millet roots and their rhizosphere soil (AM control). Plastic containers (30 x 30 x 5 cm) filled with soil mixtures were each planted with 100 pre-germinated seeds of *A. holosericea*. Seedlings were kept in a glasshouse under natural light (daylight approximately 12 h, mean daytime temperature 30°C) and watered daily without fertilizer. After 2 months of culturing, the seedlings were uprooted from the containers and transferred into 1 liter pots (one seedling per pot) filled with the same sterilized soil but without termite mound powder amendment and fungal inoculation. Two months after the seedling transplantation, *A. holosericea* plants were uprooted and the root systems gently washed. Their growth was measured as well as the extent of ectomycorrhizal or arbuscular mycorrhizal colonization. Beside this experiment, the conventional process used for controlled mycorrhization in forest nurseries was performed by planting *A. holosericea* seedlings in 11 pots filled with the same disinfected soil previously used. For ectomycorrhizal inoculation, the soil was mixed with *P. albus* IR100 fungal inoculum (10/1, v/v) whereas the control treatment (without fungus) received an autoclaved mixture of moistened (MMN medium) vermiculite/peat moss at the same rate. For arbuscular mycorrhizal inoculation, a hole was made in each pot and filled with 1 g fresh millet root (mycorrhizal or not for the control treatment without fungus). The *A. holosericea* seedlings were kept in a glasshouse under natural light. After 4 month's culture time, *A. holosericea* plants were uprooted and their shoot and root biomass, mycorrhizal indices were measured as described earlier. The results showed that termite mound powder amendment significantly stimulated the mycorrhizal formation from both types of fungal isolates (Tables 8 and 9).

This promoting effect could result from (i) the increase of plant growth induced by nutrients added with termite mound powder (particularly root

Table 8. Influence of arbuscular mycorrhizal inoculation and termite mound powder amendment rates on *A. holosericea* growth and on mycorrhiza formation after 2 months of culturing in 1-liter pots.

Factor [(A)]	Shoot biomass (mg dry weight)	Root biomass (mg dry weight)	Mycorrhizal colonization (%)
Fungal inoculum rate (%)			
0	190.1 (22.6) [(1)] a [(2)]	70.5 (8.5) a	0
1	783.5 (41.5) b	238.1 (18.4) b	41.8 (2.5) a
5	1236.1 (48.9) c	357.1 (20.3) c	70.1 (2.6) c
10	1510.1 (73.9) d	334.5 (35.9) c	55.5 (4.3) b
Termite mound amendment rate (%)			
0	709.6 (101.1) a	175.0 (25.6) a	31.0 (5.3) a
1	1000.5 (142.9) b	274.5 (43.3) b	47.3 (7.3) b
5	1019.5 (125.2) b	294.5 (30.7) b	41.5 (6.4) b
10	988.5 (113.4) b	256.1 (29.9) ab	47.5 (6.6) b
Fungal inoculum rate (FIR)	S	S	S
Termite mound amendment rate (TAR)	S	S	S
FIR x TAR	NS	S	S

S: significant ($p < 0.05$), NS: not significant ($p < 0.05$). [(1)] Standard error of the mean. [(2)] Data in the same column and for each factor followed by the same letter are not significantly different according to the Newman-Keuls test ($p < 0.05$). [(A)] Values are means of 20 replicates for fungal inoculum and termite mound amendment rates. Fungal inoculum rate factor is for all termite mound amendment rate treatments combined; the termite amendment factor is for all fungal inoculum rate treatments combined.

growth) and (ii) the addition *via* the termite mound of a bacterial group (i.e., fluorescent pseudomonads) that could act as Mycorrhization Helper Bacteria (MHB) (Duponnois and Plenchette 2003). The data recorded from this study were similar to those obtained from the cultural practice using the conventional process of controlled mycorrhization (Table 10).

One of the main problems encountered with the conventionally controlled mycorrhization of forest planting stocks is the large quantity of fungal inoculum required for the production of high quality mycorrhizal plants in nursery conditions. In tropical and Mediterranean areas, the mycorrhizal inoculum dose added per plant to the cultural substrate is

Table 9. Effect of ectomycorrhizal inoculation and termite mound powder amendment rates on *A. holosericea* growth and on mycorrhiza formation after 2 months of culturing in pots of 1 liter.

Factor [(A)]	**Shoot biomass (mg dry weight)**	**Root biomass (mg dry weight)**	**Mycorrhizal colonization (%)**
Fungal inoculum rate (%)			
0	191.0 (27.9) [(1)] a [(2)]	74.5 (12.9) a	0
1	556.5 (61.4) b	166.1 (15.2) b	57.6 (2.1) a
5	745.5 (43.6) c	221.1 (17.4) c	61.1 (1.5) a
10	1197.5 (96.6) d	364.5 (27.1) d	61.6 (1.9) a
Termite mound amendment rate (%)			
0	501.0 (77.1) a	167.0 (27.4) a	41.1 (5.5) a
1	612.2 (78.1) ab	195.0 (30.2) ab	41.6 (5.6) a
5	858.6 (96.9) c	250.5 (27.8) b	51.2 (6.8) c
10	719.1 (135.5) bc	213.5 (33.6) b	46.3 (6.4) b
Fungal inoculum rate (FIR)	S	S	S
Termite mound amendment rate (TAR)	S	S	S
FIR x TAR	S	S	S

S: significant ($p < 0.05$), NS: not significant ($p < 0.05$). [(1)] Standard error of the mean. [(2)] Data in the same column and for each factor followed by the same letter are not significantly different according to the Newman-Keuls test ($p < 0.05$). [(A)] Values are means of 20 replicates for fungal inoculum and termite mound amendment rates. Fungal inoculum rate factor is for all termite mound amendment rate treatments combined; the termite amendment factor is for all fungal inoculum rate treatments combined.

usually of 0.1 liter per liter of soil (ectomycorrhizal inoculation) and 1 g fresh weight of mycorrhizal root per litre of soil (arbuscular mycorrhizal inoculation) (Duponnois et al. 2007). With this innovative process, these volumes are drastically lowered since, with only 450 ml of ectomycorrhizal inoculum or arbuscular mycorrhizal inoculum (mixture of spores, mycorrhizal roots, and rhizosphere soil), about 100 mycorrhized *A. holosericea* plants can be produced with the same growth and mycorrhizal colonization as mycorrhized plants produced with the conventional controlled mycorrhization procedure. Therefore the inoculation cost could be greatly decreased since termite mounds are commonly found in tropical ecosystems while requiring very low fungal amounts.

Table 10. Influence of ectomycorrhizal or AM inoculations on growth of *A. holosericea* and on mycorrhiza formation after 4 months of culturing in glasshouse conditions with the conventional process of controlled mycorrhization.

Treatments	Shoot biomass (mg dry weight)	Root biomass (mg dry weight)	Mycorrhizal colonization (%)
AM inoculation			
Uninoculated control	648.1 a [(1)]	312.2 a	0
Glomus intraradices	1834.1 b	546.3 b	59.3
Ectomycorrhizal inoculation			
Uninoculated control	550 a	290 a	0
Pisolithus albus IR100	1120.1 b	560.2 b	35.6

[(1)] For each type of fungal inoculation, data in the same column followed by the same letter are not significantly different according to the Newman Keuls test ($p < 0.05$).

6. Conclusion

Numerous studies have reported the benefits that result from using fungal inoculants to enhance the development of tree seedlings in glasshouse and nursery conditions. In the same way, numerous technical procedures have been described to produce cost-effective mycorrhizal inocula. The results presented in this chapter suggest that inoculation with ectomycorrhizal fungi can improve the early growth of the most planted tree species in tropical and Mediterranean forests and that this technique will accelerate the rehabilitation of degraded forests. Unfortunately, mycorrhizal inoculation remains underexploited in nursery practices, although the management of tree mycorrhizal status could have very important implications for tropical reforestation programs in developing countries. Controlled mycorrhization could be highly improved if the fungal inoculation is associated with Mycorrhization Helper Bacteria or termite mound powder of *M. subhyalinus*. These innovative procedures decrease the necessary amount of fungal inoculum while keeping the same mycorrhizal establishment that is usually measured with the conventional process of controlled mycorrhization. Hence the use of these cultural practices could be facilitated and encouraged in forest nurseries. Future research should prioritize the generalizability of these controlled experiments to long-term applications in tree plantations and reforestation projects.

Acknowledgements

We thank the anonymous referees for their valuable comments on this study, and Krista L. McGuire and Caitlyn Gillikin for improving the language.

References

Abdel-Fattah, G.M. and A.H. Mohamedin. 2000. Interactions between a vesicular-arbuscular mycorrhizal fungus (*Glomus intraradices*) and *Streptomyces coelicolor* and their effects on sorghum plants grown in soil amended with chitin of brawn scales. Biol. Fert. Soils 32: 401–409.

Aggangan, N.S., H.K. Moon and S.H. Han. 2010. Growth response of *Acacia mangium* Willd. seedlings to arbuscular mycorrhizal fungi and four isolates of the ectomycorrhizal fungal *Pisolithus tinctorius* (Pers.) Coker and Couch. New For. 39: 215–230.

Allen, E.B., M.F. Allen, D.J. Helm, J.M. Trappe, R. Molina and E. Rincon. 1995. Patterns and regulation of mycorrhizal plant and fungal diversity. Plant and Soil 170: 47–62.

Artursson, V. 2005. Bacterial–fungal interactions highlighted using microbiomics: potential application for plant growth enhancement. PhD Thesis, University of Uppsala, Uppsala, Sweden.

Azcón, R., R. Rubio and J.M. Barea. 1991. Selective interactions between different species of mycorrhizal fungi and *Rhizobium meliloti* strains, and their effects on growth, N_2-fixation (^{15}N) and nutrition of Medicago sativa L. New Phytol. 117: 399–404.

Azcon-Aguilar, C., J. Palenzuela, A. Roldan, S. Bautista, R. Vallejo and J.M. Barea. 2003. Analysis of the mycorrhizal potential in the rhizosphere of representative plant species from desertification-threatened Mediterranean shrublands. App. Soil Ecol. 14: 165–175.

Babana, A.H. and H. Antoun. 2005. Biological system for improving the availability of Tilemsi phosphate rock for wheat (*Triticum aestivum* L.) cultivated in Mali. Nutr. Cycling Agroeco. 72: 147–157.

Bagyaraj, D.J. and J.A. Menge. 1978. Interaction between a VA mycorrhiza and *Azotobacter* and their effects on the rhizosphere microflora and plant growth. New Phytol. 80: 567–573.

Barea, J.M., G. Andrade, V. Bianciotto, D. Dowling, S. Lohrke, P. Bonfante, F. O'Gara and C. Azcón-Aguilar. 1998. Impact on arbuscular mycorrhiza formation of *Pseudomonas* strains used as inoculants for biocontrol of soil-borne fungal plant pathogens. Appl. Environ. Microbiol. 64: 2304–2307.

Bending, G.D., E.J. Poole, J.M. Whipps and D.J. Read. 2002. Characterisation of bacteria from *Pinus sylvestris-Suillus luteus* mycorrhizas and their effects on root-fungus interactions and plant growth. FEMS Microbiol. Ecol. 39: 219–227.

Brundrett, M., N. Bougher, B. Dell, T. Grove and N. Malajczuk. 1996. Working with mycorrhizas in Forestry and Agriculture. ACIAR Monograph pp. 373.

Brundrett, M.C. 2002. Coevolution of roots and mycorrhizas of land plants. New Phytol. 154: 275–304.

Budi, S.W., D. van Tuinen, G. Martinotti and S. Gianinazzi. 1999. Isolation from the *Sorghum bicolor* mycorrhizosphere of a bacterium compatible with arbuscular mycorrhiza development and antagonistic towards soilborne fungal pathogens. Appl. Environ. Microbiol. 65: 5148–5150.

Castellano, M.A. and R. Molina. 1989. Mycorrhizae. *In*: The Biological Component: Nursery Pest and Mycorrhizae Manual, Vol. 5. T.D. Landis (ed.). Agric. Handbook 674. USDA Forest Service, Washington, DC, pp. 101–167.

Castellano, M.A. 1994. Current status of outplanting studies using ectomycorrhiza-inoculated forest trees. *In*: F. Pfleger and B. Linderman (eds.). A Reappraisal of Mycorrhizae in Plant Health. The American Phytopathological Society, St Paul, pp. 261–281.

Castellano, M.A. 1996. Outplanting performance of mycorrhizal inoculated seedlings. *In*: K.G. Mukerji (ed.). Concepts in Mycorrhizal Research. Kluwer Academic Publishers, Dordrecht, The Netherlands, pp. 223–301.

Chen, Y.L., B. Dell and N. Malajczuk. 2006. Effect of *Scleroderma* spore density and age on mycorrhiza formation and growth of containerized *Eucalyptus globulus* and *E. urophylla* seedlings. New For. 31: 453–467.

de la Cruz, R.E., E.B. Lorilla and N.S. Aggrangan. 1990. Ectomycorrhizal tablets for *Eucalyptus* species. *In*: D. Werner D and P. Muller (eds.). Fast Growing Trees and Nitrogen Fixing Trees. Gustav Fisher Verlag, Stuttgard, pp. 371.

de Oliveira, V. 1988. Interactions entre les micro-organismes du sol et l'établissement de la symbiose ectomycorhizienne chez le hêtre (*Fagus sylvatica* L.) avec *Hebeloma crustuliniforme* (Bull. ex Saint-Amans) Quél. Et *Paxillus involutus* (Batsch ex Fr.). PhD Thesis, University of Nancy. Nancy. France.

Dommergues, Y.R., H.G. Diem and C. Divies. 1979. Microbiological process for controlling the productivity of cultivated plants. US Pat No 4. 155.737, May 22.

Dommergues, Y.R., E. Duhoux and H.G. Diem. 1999. Les arbres fixateurs d'azote, Editions Espaces 34. ORSTOM, FAO, Montpellier.

Dunstan, W.A., N. Malajczuk and B. Dell. 1998. Effects of bacteria on mycorrhizal development and growth of container grown *Eucalyptus diversicolor* F. Muell. seedlings. Plant and Soil 201: 241–249.

Duponnois, R. and J. Garbaye. 1991. Techniques for controlled synthesis of the Douglas fir-*Laccaria laccata* ectomycorrhizal symbiosis. Ann. For. Sci. 48: 239–251.

Duponnois, R. 1992. Les bactéries auxiliaires de la mycorhization du Douglas (*Pseudotsugamenziesii* (Mirb.) Franco) par *Laccarialaccata* souche S238. PhD Thesis, University of Nancy, France.

Duponnois, R., J. Garbaye, D. Bouchard and J.L. Churin. 1993. The fungus-specificity of mycorrhization helper bacteria (MHBs) used as an alternative to soil fumigation for ectomycorrhizal inoculation of bare-root Douglas-fir planting stocks with *Laccaria laccata*. Plant and Soil 157: 257–262.

Duponnois, R., C. Plenchette, J. Thioulouse and P. Cadet. 2001. The mycorrhizal soil infectivity and arbuscular mycorrhizal fungal spore communities in soils of different aged fallows in Senegal. Appl. Soil Ecol. 17: 239–251.

Duponnois, R., H. Founoune and D. Lesueur. 2002. Influence of the controlled dual ectomycorrhizal and rhizobial symbiosis on the growth of *Acacia mangium* provenances, the indigenous symbiotic microflora and the structure of plant parasitic nematode communities. Geoderma 109: 85–102.

Duponnois, R. and C. Plenchette. 2003. A mycorrhiza helper bacterium (MHB) enhances ectomycorrhizal and endomycorrhizal symbiosis of Australian *Acacia* species. Mycorrhiza 13: 85–91.

Duponnois R., H. Founoune, D. Masse and R. Pontanier. 2005a. Inoculation of *Acacia holosericea* with ectomycorrhizal fungi in a semi-arid site in Senegal: growth response and influences on the mycorrhizal soil infectivity after 2 years plantation. For. Ecol. Manag. 207: 351–362.

Duponnois, R., M. Paugy, J. Thioulouse, D. Masse and M. Lepage. 2005b. Functional diversity of soil microbial community, rock phosphate dissolution and growth of *Acacia seyal* as influenced by grass-, litter- and soil-feeding termite nest structure amendments. Geoderma 124: 349–361.

Duponnois, R. 2006. Nouvelles compositions d'inocula fongiques, leur procédé de préparation et leur application à l'amélioration da la croissance des cultures. Patent N°WO/2008/012399.

Duponnois, R., K. Assigbetse, H. Ramanankierana, M. Kisa, J. Thioulouse and M. Lepage. 2006. Litter-forager termite mounds enhance the ectomycorrhizal symbiosis between *Acacia holosericea* A. Cunn. Ex G. Don and *Scleroderma dictyosporum* isolates. FEMS Microbiol. Ecol. 56: 292–303.

Duponnois, R., C. Plenchette, Y. Prin, M. Ducousso, M. Kisa, A.M. Bâ and A. Galiana. 2007. Use of mycorrhizal inoculation to improve reafforestation process with Australian *Acacia* in Sahelian ecozones. Ecol. Engineering 29: 105–112.

Ferry, B.W. and M. Das. 1968. Carbon nutrition of some mycorrhizal *Boletus* species. Trans. Brit. Mycol. Soc. 51: 795–798.

Fester, T., W. Maier and D. Strack. 1999. Accumulation of secondary compounds in barley and wheat roots in response to inoculation with an arbuscular mycorrhizal fungus and co-inoculation with rhizosphere bacteria. Mycorrhiza 8: 241–246.

Founoune, H., R. Duponnois, J.M. Meyer, J. Thioulouse, D. Masse, J.L. Chotte and M. Neyra. 2002a. Interactions between ectomycorrhizal symbiosis and fluorescent pseudomonads on *Acacia holosericea*: isolation of mycorrhiza helper bacteria (MHB) from a soudano-Sahelian soil. FEMS Microbiol. Ecol. 41: 37–46.

Founoune, H., R. Duponnois, A.M. Bâ, S. Sall, I. Branget, J. Lorquin, M. Neyra and J.L. Chotte. 2002b. Mycorrhiza helper bacteria stimulated ectomycorrhizal symbiosis of *Acacia holosericea* with *Pisolithus alba*. New Phytol. 153: 81–89.

Frey-Klett, P., J.-L. Churin, J.-C. Pierrat and J. Garbaye. 1999. Dose effect in the dual inoculation of an ectomycorrhizal fungus and a mycorrhiza helper bacterium in two forest nurseries. Soil Biol. Biochem. 31: 1555–1562.

Frey-Klett, P., J. Garbaye and M. Tarkka. 2007. The mycorrhiza helper bacteria revisited. New Phytol. 176: 22–36.

Gamalero, E., A. Trotta, N. Massa, A. Copetta, M.G. Martinotti and G. Berta. 2004. Impact of two fluorescent pseudomonads and an arbuscular mycorrhizal fungus on tomato plant growth, root architecture and P acquisition. Mycorrhiza 14: 185–192.

Garbaye, J. and J.-L. Churin. 1997. Growth stimulation of young oak plantations inoculated with the ectomycorrhizal fungus *Paxillusinvolutus* with special reference to summer drought. For. Ecol. Manag. 98: 221–228.

Garbaye, J. and G.D. Bowen. 1989. Stimulation of ectomycorrhizal infection of *Pinus radiata* by some microorganisms associated with the mantle of ectomycorrhizas. New Phytol. 112: 383–388.

Garcia, C., A. Roldan and T. Hernandez. 1997. Changes in microbial activity after abandonment of cultivation in a semi-arid Mediterranean environment. J. Environ. Qual. 26: 285–291.

Holt, J.A. and M. Lepage. 2000. Termites and soil properties. *In*: T. Abe, D.E. Bignell and M. Higashi (eds.). Termites: Evolution, Sociality, Symbioses, Ecology. Kluwer Academic Publishers, Dordrecht, pp. 389–407.

Jones, M.D., D.M. Durall and J.W.G. Cairney. 2003. Ectomycorrhizal fungal communities in young forest stands regenerating after clearcut logging. New Phytol. 157: 399–422.

Kropp, B.R. and C.G. Langlois. 1990. Ectomycorrhizae in reforestation. Can. J. For. Res. 20: 438–451.

Le Tacon, F., G. Jung, P. Michelot and J. Mugnier. 1983. Efficacité en pépinière forestière d'un inoculum de champignon ectomycorhizien produit en fermenteur et inclus dans une matrice de polymères. Ann. For. Sci. 40: 165–176.

Le Tacon, F., G. Jung, J. Mugnier, P. Michelot and C. Mauperin. 1985. Efficiency in a forest nursery of an ectomycorhizal fungus inoculums produced in a fermentor and entrapped in polymeric gels. Can. J. Bot. 63: 1664–1668.

Lesueur, D. and R. Duponnois. 2005. Relations between rhizobial nodulation and root colonization of *Acacia crassicarpa* provenances by an arbuscular mycorrhizal fungus, *Glomusintraradices* Schenk and Smith or an ectomycorrhizal fungus, *Pisolithus tinctorius* Coker & Couch. Ann. For. Sci. 62: 467–474.

Mamatha, G., D.J. Bagyaraj and S. Jaganath. 2002. Inoculation of field-established mulberry and papaya with arbuscular mycorrhizal fungi and a mycorrhiza helper bacterium. Mycorrhiza 12: 313–316.

Martin F., C. Delaruelle and J.-L. Hilbert. 1990. An improved ergosterol assay to estimate the fungal biomass in ectomycorrhizas. Mycol. Res. 94: 1069–1074.

Marx, D.H. 1969. The influence of ectotrophic mycorrhizal fungi on the resistance of pine roots to pathogenic infections. I. Antagonism of mycorrhizal fungi to root pathogenic fungi and soil bacteria. Phytopathology 59: 153–163.

Marx, D.H. and W.C. Bryan. 1975. Growth and ectomycorrhizal development of Loblolly pine seedlings in fumigated soil infested with the fungal symbiont *Pisolithus tinctorius*. For. Sci. 21: 242–254.

Marx, D.H., K. Jarl, J.L. Ruehle and W. Bell. 1984. Development of *Pisolithus tinctorius* ectomycorrhizae on pine seedlings using basidio-encapsulated seed. For. Sci. 30: 897–907.

Marx, D.H. 1991. The practical significance of ectomycorrhizae in forest establishment. Ecophysiology of Ectomycorrhizae of Forest Trees, Marcus Wallenberg Foundation Symposia Proceedings 7: 54–90.

Mauperin, C., F. Mortier, J. Garbaye, F. Le Tacon and G. Carr. 1987. Viability of an ectomycorrhizal inoculum produced in a liquid medium and entrapped in a calcium alginate gel. Can. J. Bot. 65: 2326–2329.

Meyer, J.R. and R.G. Linderman. 1986. Response of subterranean clover to dual-inoculation with vesicular-arbuscular mycorrhizal fungi and a plant growth-promoting bacterium, *Pseudomonas putida*. Soil Biol. Biochem. 18: 185–190.

Molina, R. and J.G. Palmer. 1982. Isolation, maintenance and pure culture manipulation of ectomycorrhizal fungi. *In*: N.C. Schenck (ed.). Methods and Principles of Mycorrhizal Research. The American Phytopathological Society, St. Paul, pp. 115–129.

Mortier, F., F. Le Tacon and J. Garbaye. 1988. Effect of inoculum type and inoculation dose on ectomycorrhizal development, root necrosis and growth of Douglas fir seedlings inoculated with *Laccaria laccata* in a nursery. Ann. For. Sci. 45: 301–310.

Pachlewski, R. and J. Pachlewski. 1974. Studies on Symbiotic Properties of mycorrhizal Fungi of Pine (*Pinus silvestris* L.) with the aid of the method of mycorrhizal synthesis in pure culture on agar. Forest Research Institute. Warsaw. Poland.

Paula, M.A., S. Urquiaga and J.O. Siqueira. 1992. Synergistic effects of vesicular-arbuscular mycorrhizal fungi and diazotrophicus bacteria on nutrition and growth of sweet potato (*Ipomoea batatas*). Biol. Fert. Soils 14: 61–66.

Piéri, C. 1991. Les bases agronomiques de l'amélioration et du maintien de la fertilité des terres de savanes au sud Sahara. *In*: Savanes d'Afrique, terre fertile? Actes des Rencontres Internationales, Montpellier, France, pp. 43–74.

Poole, E.J., G.D. Bending, J.M. Whipps and D.J. Read. 2001. Bacteria associated with *Pinus sylvestris–Lactarius rufus* ectomycorrhizas and their effects on mycorrhiza formation *in vitro*. New Phytol. 151: 743–751.

Rao, N.S.S., K.V.B.R. Tilak and C.S. Singh. 1985. Effect of combined inoculation of *Azospirillum brasilense* and vesicular-arbuscular mycorrhiza on pearl millet (*Pennisetum americanum*). Plant and Soil 84: 283–286.

Read, D.J., J.G. Duckett, R. Francis, R. Ligrone and A. Russell. 2000. Symbiotic fungal associations in "lower" land plants. Philos. Trans. R. Soc. Lond. Ser. B-Biol. Sci. 355: 815–830.

Requena, N., I. Jimenez, M. Toro and J.M. Barea. 1997. Interactions between plant-growth-promoting rhizobacteria (PGPR), arbuscular mycorrhizal fungi and *Rhizobium* spp. in the rhizosphere of *Anthyllis cytisoides*, a model legume for revegetation in mediterranean semi-arid ecosystems. New Phytol. 136: 667–677.

Requena, N., E. Perez-Solis, C. Azcon-Aguilar, P. Jeffries and J.M. Barea. 2001. Management of indigenous plant–microbe symbioses aids restoration of desertified ecosystems. Appl. Environ. Microbiol. 67: 495–498.

Rincon, A., M.R. de Felipe and M. Fernandez-Pascual. 2007. Inoculation of *Pinus halepensis* Mill. with selected ectomycorrhizal fungi improves seedling establishment 2 years after planting in a degraded gypsum soil. Mycorrhiza 18: 23–32.

Roldan, A., I. Querejeta, J. Albadalejo and V. Castillo. 1996. Growth response of *Pinus halepensis* to inoculation with *Pisolithus arhizus* in a terraced rangeland with urban refuse. Plant and Soil 179: 35–43.

Rózycki, H., M. Kampert, E. Strzelczyk, C.Y. Li and D.A. Perry. 1994. Effect of different soil bacteria on mycorrhizae formation in pine (*Pinus sylvestris* L.). Folia Forest. Polon. series A (Forestry) 36: 91–102.

Schreiner, R.P., K.L. Mihara, K.L. McDaniel and G.J. Bethlenfalvay. 2003. Mycorrhizal fungi influence plant and soil functions and interactions. Plant and Soil 188: 199–209.

Schrey, S.D., M. Schellhammer, M. Ecke, R. Hampp and M.T. Tarkka. 2005. Mycorrhiza helper bacterium Streptomyces AcH 505 induces differential gene expression in the ectomycorrhizal fungus Amanita muscaria. New Phytol. 168: 205–216.

Smith, S. and D. Read. 2008. Mycorrhizal Symbiosis, 2nd edn. Academic Press, London.

Terwilliger, J. and J. Pastor. 1999. Small mammals, ectomycorrhizae, and conifer succession in beaver meadows. Oikos 85: 83–94.

Toro, M., R. Azcón and J.M. Barea. 1997. Improvement of arbuscular mycorrhiza development by inoculation of soil with phosphate-solubilizing rhizobacteria to improve rock phosphate bioavailability (^{32}P) and nutrient cycling. Appl. Environ. Microbiol. 63: 4408–4412.

Torres, P. and M. Honrubia. 1994. Inoculation of containerized *Pinus halepensis* (Miller) seedlings with basidiospores of *Pisolithus arhizus* (Pers) Rauschert, *Rhizopogon roseolus* (Corda) and *Suillus collinitus* (Fr) O Kuntze. Ann. For. Sci. 51: 521–528.

Turjaman, M., Y. Tamai, H. Segah, S.H. Limin, J.Y. Cha, M. Osaki and K. Tawaraya. 2005. Inoculation with the ectomycorrhizal fungi *Pisolithus arhizus* and *Scleroderma* sp. improves early growth of *Shorea pinanga* nursery seedlings. New For. 30: 67–73.

Valentine, L.L., T.L. Fieldler, A.A. Hart, C.A. Petersen, H.K. Berninghausen and D. Southworth. 2004. Diversity of ectomycorrhizas associated with *Quercus garryana* in southern Oregon. Can. J. Bot. 82: 123–135.

Van der Heijden, M.G.A., J.N. Klironomos, M. Ursic, P. Moutoglis, R. Streitwolf-Engel, T. Boller, A. Wiemken and I.R. Sanders. 1998. Mycorrhizal fungal diversity determines plant biodiversity ecosystem variability and productivity. Nature 396: 69–72.

Vignon, C., C. Plassard, D. Moussain and L. Salsac. 1986. Assay of fungal chitin and estimation of mycorrhizal infection. Physiol. Végét. 24: 201–207.

Vivas, A., A. Marulanda, J.M. Ruiz-Lozano, J.M. Barea and R. Azcón. 2003. Influence of a *Bacillus* sp. on physiological activities of two arbuscular mycorrhizal fungi and on plant responses to PEG-induced drought stress. Mycorrhiza 13: 249–256.

von Alten, H., A. Lindemann and F. Schönbeck. 1993. Stimulation of vesicular-arbuscular mycorrhiza by fungicides or rhizosphere bacteria. Mycorrhiza 2: 167–173.

Vósatka, M. and M. Gryndler. 1999. Treatment with culture fractions from *Pseudomonas putida* modifies the development of *Glomus fistulosum* mycorrhiza and the response of potato and maize plants to inoculation. Appl. Soil Ecol. 11: 245–251.

Will, M.E. and D.M. Sylvia. 1990. Interaction of rhizosphere bacteria, fertilizer, and vesicular-arbuscular mycorrrhizal fungi sea oats. Appl. Environ. Microbiol. 56: 2073–2079.

Xie, Z.P., C. Staehelin, H. Vierheilig, A. Wiemken, S. Jabbouri, W.J. Broughton, R. Vogeli-Lange and T. Boller. 1995. Rhizobial nodulation factors stimulate mycorrhizal colonization of nodulating and nonnodulating soybeans. Plant Physiol. 108: 1519–1525.

CHAPTER 13

Biodiversity and Sustainable Use of Wild Edible Fungi in the Sudanian Centre of Endemism: A Plea for Valorisation

Nourou Soulemane Yorou,[1,2]* *N'Golo Abdoulaye Koné,*[3] *Marie-Laure Guissou,*[4] *Atsu Kudzo Guelly,*[5] *Dao Lamèga Maba,*[2,5] *Marius R.M. Ekué*[6] and *André De Kesel*[7]

1. Introduction

Wild Edible Fungi (WEF) play an important role in the livelihood of local inhabitants (Boa 2004). In tropical Africa, picking wild edible mushrooms is a lucrative activity and involves hundreds of rural women (Boa 2004, Degreef et al. 1997, Buyck 1994a). Over 300 edible mushrooms are recorded in sub-Saharan Africa (Rammeloo and Walleyn 1993, Walleyn and Rammeloo 1994). Still, mycophagy greatly varies from one country to another, even within ethnic groups of the same area (Yorou and De Kesel 2002, Guissou et al. 2008). Annual consumptions of over 30 kg per habitant have been recorded in Central and East Africa (Degreef et al. 1997). In rural tropical Africa, WEF are important in terms of species richness and consumed quantities, but also as a potential source of proteins, vitamins, and minerals (Degreef et al. 1997, Adewusi et al. 1993, Ogundana and Fagade 1981, 1982, De Kesel and Malaisse 2010).

Authors' affiliations given at the end of the chapter.

In tropical Africa, ethnomycological knowledge appears to be most elaborate and widespread in the Zambesian miombo forests where Degreef et al. (1997) and De Kesel and Malaisse (2010) enumerated 42 edible species in Katanga (Democratic Republic of Congo). Other important ethnomycological contributions from that region are available for Tanzania (Härkönen 1992, Härkönen et al. 1994, 2003), Zambia (Pegler and Piearce 1980, Piearce 1987), Malawi (Morris 1984, 1994), and Burundi (Buyck 1994a, Buyck and Nzigidahera 1995). The contributions from Eyi-Ndong (2009), Eyi-Ndong and Degreef (2009, 2010), Eyi-Ndong et al. (2011), combined with those from Mossebo and Amougou (2002), Mossebo et al. (2002), and Malaisse et al. (2008), provide an extensive documentation of WEF in the Guineo-Congolese area. As far as West Africa is concerned, investigations on WEF along with subsequent ethnomycological knowledge have been initiated only recently in a few countries such as Benin (De Kesel et al. 2002, Yorou 2000, 2010, Yorou and De Kesel 2002, Yorou et al. 2002a,b), Burkina Faso (Guissou et al. 2005, 2008), Senegal (Bâ et al. 2011), Niger (Hama et al. 2010), and Ivory Coast (Koné et al. 2010a,b). Though fungi are widespread and commonly used in rural areas of West Africa, numerous investigations demonstrated that women and in some cases children are the traditional harvesters, sellers of WEF, and keepers of ethnomycological information (Buyck 1994a, De Kesel et al. 2002, Guissou et al. 2008). In WA, local mycological knowledge is diverse and full of instructions, including patterns of collecting, drying, processing, and cooking techniques (Yorou and De Kesel 2002, Guissou et al. 2005, 2008). Yorou and De Kesel (2002) and De Kesel et al. (2002) demonstrated that folk classification of WEF by Nagot people in Central Benin challenges modern taxonomy, whilst local nomenclature carries important information related to the ecology (either saprotrophic or symbiotic), growth habit (solitary or gregarious), morphology/size, colour, flesh context, and taste of the mushroom. Unfortunately, local usage and subsequent ethnomycological know-how are actively eroding and/or disappearing from one generation to another (Guissou et al. 2008), mostly due to urbanization and modern schooling.

In tropical Africa, WEF occur predominantly in the Zambesian and Sudanian Centres of Endemism (Ducousso et al. 2002, De Kesel et al. 2002, Härkönen et al. 2003, De Kesel and Malaisse 2010). Still, the dense evergreen forests of the Guineo-Congolese domain host an important diversity of WEF (Eyi-Ndong et al. 2011). Though a few species may occur throughout the year, the occurrence of edible ectomycorrhizal (EcM) species is seasonal. Whilst numerous species appear just after the first rains, many others need persistent rains over a longer period. In seasonal forests of tropical Africa, natural production of WEF is related to the presence, health, and photosynthetic performance of partner tree species. The latter are mainly *Isoberlinia* spp., *Berlinia* spp., *Uapaca* spp., and *Anthonotha* spp., among others (Bâ et al. 2012, Ducousso et al. 2002). Unfortunately, many such EcM trees are disappearing at an alarming rate (Neuenschwander et al. 2011), either through large scale deforestation or selective logging of timber species. The process of deforestation is due mostly to shifting cultivation, timber exploitation, land conversion to pastures, and increasing urban demands of wood fuel (charcoal). Shifting cultivation usually results in the slashing, burning and conversion of large forest areas into croplands, whilst timber exploitation consists primarily of selective logging the economically important aforementioned EcM trees. In any case, the removal and destruction of partner tree species not only erodes the diversity of the area, but also severely hampers the natural production of locally exploited WEF. In Benin, about 28 WEF are threatened with extinction (Yorou and De Kesel 2011), chiefly because of forest fragmentation, habitat disturbance, and timber exploitation for charcoal. In a similar way, numerous fungi-rich WA ecosystems are undergoing an unrestrained process of fragmentation (Hahn-Hadjali et al. 2010, Wegmann et al. 2010), along with the disappearance of economically important timber species and their associated WEF. This process deeply affects the lives of the people living in surrounding villages and cities. Due to such uncontrolled deforestation, the forest ecosystems of WA are not only losing their potential as providers of Non Timber Forest Products, but also their ability to improve the livelihood of rural communities. As WEF are diverse and of local economic importance, the installation of an economic sector based on WEF can serve as a model to promote awareness among rural people, forest officers/administration, and decision makers. The main goal of establishing an economic sector of WEF is to face poverty and to conserve the hosts and the ecosystems of fungal species all alike. In the present contribution, we are attempting (1) to explore the diversity of WEF and fungal hotspots in West Africa and (2) to highlight the natural productions, phenology, nutritional, and socio-economic exploitation of WEF in West Africa. Data are assembled from intensive mycological investigations in some West African countries (Benin, Togo, Burkina Faso…) over the last 10 years, but also from extensive mycological exploration in various ecosystems from other countries like Ghana, Mali, Ivory Coast, and Guinea (started in 2009). Using a case study from Benin, we will analyse different steps towards the establishment of an economic sector based on WEF in West Africa.

2. Fungal Hotspots and WEF-rich Ecosystems in West Africa

WEF are generally composed of two major biological groups, notably the symbiotic species and the saprotrophic ones. In West Africa, there is a north-south gradient in diversity, ecological groups (symbiotic, parasitic, or saprotrophic), and natural productions of WEF. The dense southern forests of the coastal provinces of countries like Benin, Togo, Ghana, and Ivory Coast exhibit a rather low diversity of edible, mostly saprotrophic fungi. Most of the native edible taxa are members of *Pleurotus*, *Marasmiellus*, *Lentinus*, *Cookeina*, and *Armillaria*. Plantations of exotic tree species, man-made ecosystems, fallows, and to some extent disturbed habitats generally harbour few edible fungi of the genera *Chlorophyllum*, *Macrolepiota*, *Volvariealla*, and *Agaricus*. Towards the North, i.e., entering the ecological zone where woodlands and gallery forests of the Guineo-Sudanian type start to dominate (White 1986), a high diversity of edible EcM fungi is observed (Bâ et al. 2011, De Kesel et al. 2002, Ducousso et al. 2002, Yorou et al. 2002a,b). For landlocked countries such as Burkina Faso, Niger, and Mali, WEF are most diverse in the southern Sudanian woodlands (Guissou 2005, Sanon et al. 1997). The Sudanian Centre of Endemism (SCE) is home to hundreds of WEF including notably representatives of *Russula*, *Lactarius*, *Lactifluus*, *Amanita*, and Boletus. Numerous termites' fungi (*Termitomyces* spp.) are present in West Africa. These taxa are considered an important resource and they play a seasonal but substantial role in the local diet, medicine, and cash incomes for native villagers. The SCE is a narrow vegetation band stretching from East (Soudan) to West (Senegal) between 9° and 11° North latitudes (White 1986). Vegetation is composed of a mosaic of woodlands, savannahs and gallery forests, mostly dominated by tree species belonging to the Caesalpiniaceae, Phyllantaceae and Dipterocarps. In this area, some timber species such as *Afzelia* spp., *Isoberlinia* spp., *Uapaca* spp., *Berlinia grandiflora* and *Anthonotha* spp. form EcM symbiosis with numerous WEF. Ectomycorrhizal fungi are predominant not only in terms of species richness, but also in terms of fresh biomass, a large proportion of which is consumed by local inhabitants (Table 1). Though woodlands form vast areas in the SCE, they are actually fragmented and regressing annually (FAO 2010). More conserved and semi-natural woodland ecosystems are located in numerous forest reserves. Due to their conservation status as natural or semi-natural ecosystems, Caesalpinioid and Phyllantioid-dominated woodlands and gallery forests of the SCE emerge as fungi-rich ecosystems in West Africa. The presence in their flora of important EcM tree species such as *Berlinia* spp. and *Uapaca* spp. makes them remarkable WEF-rich ecosystems, generally acting as refuge ecosystems for numerous unique and rare species (Yorou and De Kesel 2011). Their generally fragmented size (Natta et al. 2002, Sparovek et al. 2002) coupled with the unique and rare mycoflora they host make the gallery forests important fungal hotspots with top conservation priority. Because of the presence in its flora of numerous EcM trees such as *Anthonotha* spp., *Gilbertiodendron* spp. *Paramacrolobium coeruleum* (to quote only a few), Guinea harbours the highest diversity of potential WEF that have been scantily documented ethnomycologically.

Table 1. Contrasting relative dominance of fungal flora between two dense forests (FD = dense forest and FR = riparian forest) lacking EcM trees and 4 woodland ecosystems (SA = shrub savannah, SB = wooded savannah, FC = Woodland of Uapaca spp. and FI = woodland of *Isoberlinia* spp.) dominated by EcM trees in Sudanian area.

	FD	FR	SA	SB	FC	FI
Total Biomass (kg/ha/year)	8.6 ± 5.2	26.7 ± 4.7	241.7 ± 13.88	223.7 ± 15.4	293.9 ± 16,1	285.7 ± 16
% Ectomycorrhizal fungi from the total biomass	6	6.8	88.3	98.9	95.0	90.0
% Saprotrophic species from the total biomass	92.0	93.2	11.7	1.1	4.0	10.0
% Wild edible Fungi from the total biomass	2.3	4.1	85.4	87.7	73.1	78.9

Source: Yorou NS (2000), unpublished data.

3. Diversity of WEF in West Africa with Special Emphasis on the Most Abundant Genera

The specific diversity of WEF in a given country depends on a range of factors, including its surface area and the diversity in ecosystems and floristic composition. West Africa counts conservative estimates of about 55 to 70 edible fungal species (Bâ et al. 2011). Whilst monographs are progressing in a few countries such as Benin (De Kesel et al. 2002) and Burkina Faso (Guissou 2005), numerous other countries have never been investigated ethnomycologically. De Kesel et al. (2002) recorded about 50 edible species in Benin. In Burkina Faso, Guissou et al. (2008) reported on about 40 useful wild fungi of which 21 are consumed by various ethnic groups. Over 70 EcM fungi were recorded in Ivory Coast by Bâ et al. 2011, a large proportion of which are edible. Similarly, about 110 EcM fungi, composed mostly of edible ones, are present in various ecosystems of Guinea (Rivière et al. 2007). According to Bâ et al. (2011), a non exhaustive list of edible species from Senegal includes 45 species. The total list of 72 species whose edibility is confirmed in Benin, Burkina Faso, Ivory Coast and Togo is presented in Table 2.

In general, genera *Lactarius s. l.* (including *Lactifluus*), *Amanita, Termitomyces* and *Russula* are the most common WEF in WA (Fig. 1).

All four genera *Lactifluus, Lactarius, Russula* and *Amanita* form EcM symbioses with the above-mentioned tree species. *Termitomyces* are not ectomycorrhizal, but they form mutualistic symbiosis with many termite species belonging to the Macrotermitinae subfamily, the so-called fungus growing termites.

Table 2. List of WEF confirmed as edible in some West African countries

Fungal Species	**Biological group**	**Countries**			
		Benin	**Burkina Faso**	**Ivory Coast**	**Togo**
Afroboletus luteolus (Heinemann) Pegler & Young	Symbiotic	+			+
Afrocantharellus platyphyllus (Heinem.) Tibuhwa	Symbiotic	+			+
Agaricus sp1	Saprotrophic		+		
Agaricus sp2	Saprotrophic		+		
Agaricus goossensiae Heinem.	Saprotrophic	+	+		
Agaricus subsaharianus L.A. Parra, Hama & De Kesel	Saprotrophic	+	+	+	+
Agaricus volvatulus Heinem. & Goos	Saprotrophic	+			
Amanita aff. *craseoderma Bas*	Symbiotic	+			
Amanita crassiconus Bas	Symbiotic	+			
Amanita massasiensis Härk. & Saarim.	Symbiotic	+	+		+
Amanita strobilaceovolvata Beeli	Symbiotic	+			
Amanita subviscosa Beeli	Symbiotic	+	+		+
Amanita cf. *rubescens* (Pers.: Fr) S. F. Gray	Symbiotic	+			+
Amanita cf. *xanthogala* Bas	Symbiotic	+			
Auricularia cornea Ehrenb.	Saprotrophic	+			+
Boletus loosii Heinem.	Symbiotic	+			
Cantharellus congolensis Beeli	Symbiotic	+			+
Cantharellus addaiensis Heinem.	Symbiotic	+	+		+
Cantharellus aff. *rufopunctatus* var *rufopunctatus* (Beeli) Hein	Symbiotic	+		+	
Cantharellus pseudofriesii Heinem.	Symbiotic			+	
Macrolepiota cf. *rachodes*	Saprotrophic	+	+		
Cookeina sulcipes (Berk.) Kuntze	Saprotrophic	+			+
Marasmiellus inoderma (Berk.) Singer	Saprotrophic	+			+
Gymnopus sp.	Saprotrophic		+		
Hygrophoropsis aurantiaca (Wulfen) Maire	Saprotrophe			+	
Hygrophoropsis sp.	Saprotrophic			+	
Lactarius baliophaeus Pegler	Symbiotic	+			
Lactifluus denigricans (Verbeken & Karhula) Verbeken	Symbiotic	+			
Lactifluus densifolius (Verbeken & Karhula) Verbeken	Symbiotic	+			

Table 2. contd....

Table 2. contd.

Fungal Species	Biological group	Countries			
		Benin	Burkina Faso	Ivory Coast	Togo
Lactifluus edulis (Verbeken & Buyck) Buyck	Symbiotic	+			+
Lactifluus flammans (Verbeken) Verbeken	Symbiotic	+			
Lactifluus gymnocarpus (Heim ex Singer) Verbeken	Symbiotic	+		+	+
Lactiflluus gymnocarpoides (Verbeken) Verbeken	Symbiotic	+	+	+	+
Lactifluus luteopus (Verbeken) Verbeken	Symbiotic	+	+		+
Lactarius pseudogymnocarpus (Verbeken) Verbeken	Symbiotic	+			+
Lactifluus meduase Verbeken	Symbiotic				+
Lactarius saponaceus Verbeken	Symbiotic	+			
Lactarius tenellus Verbeken & Walleyn	Symbiotic	+			+
Lactifluus sudanicus nom. prov. (MD105)	symbiotic				+
Lentinus squarrosulus Mont.	Saprotrophic	+		+	+
Lentinus aff. *atrobrunneus* Pegler	Saprotrophic			+	
Lentinus aff. *brunneofloccosus* Pegler.	Saprotrophic			+	
Pleurotus tuber-regium (Fr.) Fr.	Saprotrophic	+	+	+	+
Leucoagaricus cf. *bresadolae* (Schulz.) Bon & Boiffard	Saprotrophic	+			
Leucocoprinus cretatus Locq. ex Lanzoni	Saprotrophic	+	+		
Macrocybe lobayensis (R. Heim) Pegler & Lodge	Saprotrophic	+	+		+
Marasmius heinemaniannus V. Antonin	Saprotrophic	+			
Phlebopus sudanicus (Har. & Pat) Heinem.	Symbiotic	+	+	+	+
Pleurotus sp1	Saprotrophic			+	
Pleurotus sp2	Saprotrophic			+	
Pleurotus cystidiosus Muller	Saprotrophic	+			
Psathyrella tuberculata (Pat) A. H. Sm	Saprotrophic	+	+	+	+
Russula sp1	Symbiotic			+	
Russula cellulata Buyck	Symbiotic	+			
Russula congoana Pat.	Symbiotic	+	+		+

Table 2. contd....

Table 2. contd.

Fungal Species	Biological group	Countries			
		Benin	Burkina Faso	Ivory Coast	Togo
Russula oleifera Buyck	Symbiotic	+			+
Russula aff. *rubroalba* (Singer) Romagnesi	Symbiotic	+			+
Russula sesenagula Beeli	Symbiotic		+		
Schizophyllum commune (L.) Fr	Saprotrophic	+			
Termitomyces clypeatus R. Heim	Symbiotic	+			+
Termitomyces eurrhizus (Berk.) R. Heim	Symbiotic	+		+	+
Termitomyces fuliginosus R. Heim	Symbiotic	+	+	+	+
Termitomyces letestui (Pat.) R. Heim	Symbiotic	+		+	+
Termitomyces medius R. Heim et Grassé	Symbiotic	+		+	+
Termitomyces microcarpus (Berk. & Broom) Heim	Symbiotic	+			+
Termitomyces robustus (Beeli) Heim	Symbiotic	+			+
Termitomyces schimperi (Pat.) R. Heim	Symbiotic	+		+	+
Termitomyces striatus (Beeli) Heim	Symbiotic	+		+	+
Tubosaeta brunneosetosa (Singer) E. Horak	Symbiotic			+	
Volvariella volvacea (Bull.) Singer	Saprotrophic	+		+	+
Volvariella earlei (Murr.) Shaffer	Saprotrophic	+		+	+
Volvariella acystidiata Pathak	Saprotrophic			+	

Source (De Kesel et al. 2002, Guissou et al. 2008, Yorou et al. 2002b). Symbiotic species include ectomycorrhizal and termite fungi.

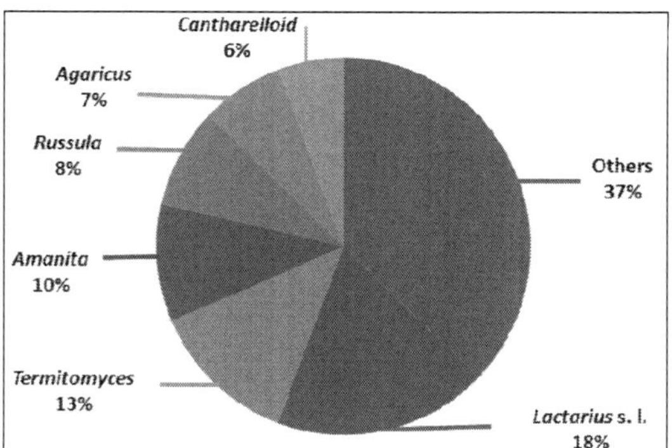

Fig. 1. Frequency of some WEF genera in West Africa (data from Table 2).

3.1 Common ectomycorrhizal edible genera

3.1.1 The genera Lactifluus and Lactarius

In tropical Africa and in West Africa in particular, the milkcaps (*Lactifluus* and *Lactarius*, see Verbeken et al. 2011, 2012) play an important role in the growth and regeneration of numerous timber species and ecosystems. Both genera contain approximately 90 described species in tropical Africa for an estimated total richness of 150, of which 60 species might be common to the Sudano-Zambezian domains (Verbeken 1996, 1998a,b, 2000, Verbeken and Walleyn 2000, 2010). In both ecozones, many (if not all) *Lactifluus* and *Lactarius* species are used by local inhabitants as food (Härkönen et al. 2003, Verbeken and Walleyn 2010). In WA, *Lactifluus* and *Lactarius* species are the most common, making up 16% of the total number of WEF species. A total of 22 *Lactifluus* and *Lactarius* species, composed mostly of edible ones, is recorded in Benin alone (van Rooij et al. 2003). In this country, a higher diversity may be expected since the sampling effort during repetitive investigations was biased toward central and western Benin. In Benin (Yorou and De Kesel 2011, van Rooij et al. 2003) and in Burkina Faso (pers. observation), many rare and edible species are confined and may be ecologically specific to gallery forests dominated by *Berlinia grandiflora*, including notably the one in Bassila (Central Benin), the gallery forests of Kou, and the Forest Reserve of Mouhoun in South-western Burkina Faso. Many *Lactarius* and *Lactifluus*, notably *Lactifluus edulis*, *Lactifluus gymnocarpoides*, *Lactifluus medusae* and *Lactifluus flammans*, number among the most appreciated WEF in Benin, Togo and in West Africa in general. Some representatives of edible *Lactifluus* and *Lactarius* from West Africa are presented in Fig. 2.

3.1.2 The genus Amanita

Though the genus *Amanita* is worldwide reputed to encompass many poisonous species, representatives of this genus are commonly used as food in West Africa and in tropical Africa (Buyck 1994a, Härkönen et al. 2003, Boa 2004). In the Democratic Republic of Congo, for example, Parent and Thoen (1977) mentioned that the genus *Amanita* ranks second in terms of annually consumed biomass whilst Morris (1984) reported on *Amanita loosii* as the most sold edible species in local markets. According to current records, a total of 7 *Amanita* species are consumed in West Africa, i.e. 10% of all WEF. Three species (*A. subviscosa*, *A. masasiensis*, *A.* cf. *xanthogala*) are well known and common in the culinary history of the Nagot people in central Benin. In the neighbouring country of Burkina Faso, Guissou et al. (2008) reported three edible species. Härkönen et al. (1994) mentioned a few cases

Fig. 2. Some edible *Lactarius* species from West Africa. a. *Lactarius flammans* (Kou Forest Reserve, Burkina Faso), b. *L. gymnocarpus* (Baro Forest Reserve, Upper Guinea), c. *L. foetens* (Mouhoun Forest Reserve, Burkina Faso), d. *L. medusae* (Fazao National Park, Togo), e. *L.* cf. *pelliculatus* (Forest Guinea), f. *L. tenellus* (from Benin).

Color image of this figure appears in the color plate section at the end of the book.

where temperate *Amanita* species (introduced in tropical Africa together with partner *Pinus* trees) have been misidentified by local people and caused fatal deaths. In many West African countries, exotic *Pinus* plantations are absent and it is almost impossible to record temperate species. In Benin, Togo, and even in central Africa (De Kesel et al. 2002, Eyi-Ndong et al. 2011), one species related to the temperate and boreal species *A. rubescens* occurs. Molecular data are needed to clarify the genetic affinity between tropical African specimens and temperate (and boreal) specimens of *A. rubescens*. Our ethnomycological investigations reveal that this sibling species of *A. rubescens* is eaten by local people without trouble. *Pinus* plantations are established in the mountains of the Fouta Djallon area in Guinea. So far there is no known report about poisonings due to misidentifications. It is actually very difficult to accurately discuss the diversity and edibility of *Amanita* species in West Africa. Not only is mycophagy rarely investigated in many West African countries, but the genus itself suffers from numerous taxonomical ambiguities (Bas 1969, Härkönen et al. 1994, 1995). *Amanita* is probably more diverse than actually reported. A total of 55 species are described and accepted for conservative estimates of 70 Amanita species in tropical Africa.

3.1.3 The genus Russula

Russula is probably one of the most ecologically important fungal groups in tropical Africa, where it emerges as the most diverse EcM genus with conspicuous fruit bodies. All species are known to form EcM symbiosis with native forest trees in Sudano-Zambezian woodlands and dense evergreen forests as well (Buyck 1994b,c, 1997, Diédhiou et al. 2010). Reportedly, *Russula* includes about 160 identified/described species from a conservative estimate of 290 species (Buyck 1994b,c, 1997, Härkönen et al. 1994). In general, the majority (if not all) of *Russula* species are edible though many species are not consumed by local people due to a repulsive smell or taste. A total of about 40 species are consumed in tropical Africa according to Buyck (1994). Six (6) edible species have been recorded so far in West Africa (see Table 2 above), including notably *R. congoana* and *R. oleifera* (Fig. 3), making it 9% of the total WEF so far inventory in this

Fig. 3. Some edible *Russula* species from West Africa. a. *Russula congoana* (Moussaya Forest Reserve, Upper Guinea), b. *Russula* sp1. (Ziama National Park, Guinea), c. *Russula oleifera* (Bui national Park, Ghana).

region. In Benin, De Kesel et al. (2002) mentioned 4 edible species. Three edible species are known from Togo whilst Guissou et al. (2008) reported only two from Burkina Faso.

3.1.4 The genus Cantharellus

A total of 46 *Cantharellus* (including the new genus *Afrocantharellus* according to Tibuhwa et al. 2012) species are known from tropical Africa (Tibuhwa et al. 2008, 2012, Buyck et al. 2013, De Kesel et al. 2011, Eyssartier and Buyck 1998). Conservative estimates of 60 species are suggested by Verbeken and Buyck (2001) for the whole of tropical Africa. *Cantharellus* encompasses several well-known and highly esteemed edible species (Eyssartier 2003). In the Zambesian ecozone, numerous species are widely collected and huge quantities are frequently sold at local markets and along the roadsides (Buyck 1994a, Malaisse 1997, Buyck et al. 2000, De Kesel and Malaisse 2010). According to Parent and Thoen (1977), *Cantharellus* ranks first in terms of diversity and annually harvested biomass. A total of 11 species are well known edible species in Western Burundi (Buyck and Nzigidahera 1995) whilst Malaisse (1997) reported over 16 species. Eyi-Ndong et al. (2011) reports a total of 8 species in central African rain forests. Unlike in the Zambesian ecozone, the genus *Cantharellus* is much less diverse in Sudanian woodlands. This is probably due to the absence in West Africa of 14 dominant tree species in *Brachystegia* and two more in the genus *Julbernardia*. Both these genera are important and typical partners of *Cantharellus* in the Zambesian area. Because of the low species diversity, their scarcity in natural areas, and the confinement to specific habitats (Yorou and De Kesel 2011), the use of *Cantharellus* species as food is rather unpopular in West Africa. Only one *Cantharellus* species (*Cantharellus addaiensis*) is widespread in *Isoberlinia*-dominated Sudanian woodlands (De Kesel et al. 2002, Yorou et al. 2002a,b). In contrast, *Berlinia*- and *Uapaca*-dominated forest galleries host a larger diversity of species (Yorou and De Kesel 2011); including notably *Afrocantharellus platyphyllus*, *Cantharellus congolensis*, *C. conspicuus* and *C. solidus* (De Kesel et al. 2002, 2011, Eyssartier et al. 2002) in Benin and Togo. There is no doubt that gallery forests of the neighboring countries—Ghana, Ivory Coast, Burkina Faso, Senegal and more importantly Guinea, host more and possibly different *Cantharellus* species. A few representatives of edible *Cantharellus* species from tropical Africa are given in Fig. 4.

Fig. 4. Some edible *Cantharellus* species from West Africa. a. *Cantharellus addaiensis* (Wari-Maro Forest Reserve, Benin), b. *Cantharellus congolensis* (Aledjo Forest Reserve, Togo), *Afrocantharellus platyphyllus* (Kota gallery Forest, Benin), d. *Cantharellus* cf. *conspicuus* (Bassila gallery forest, Benin).

Color image of this figure appears in the color plate section at the end of the book.

3.2 Special focus: Termitomyces as a delicacy in West Africa and elsewhere

The associated fungi of termites (*Termitomyces*) exhibit the highest food value among WEF in tropical Africa (Parent and Thoen 1977, Degreef et al. 1997, Malaisse 1997). *Termitomyces* have a paleo-tropical (tropical Africa and Asia) distribution. Molecular studies including both partners gave strong evidence of its unique African rain forests origin (Aanen and Eggleton 2005, Nobre et al. 2011). They display a large range of morphological features, including great variability in their size (small, medium and large), pseudorhizes (absent or not), and perforatorium (shape, size, see Fig. 5). Whilst species such as *T. microcarpus* exhibit small-sized-fruit bodies (1–1.5 cm cap diameter, stipe 1–3 cm long x 1–2 mm diameter), numerous others such as *T. schimperi* and *T. letestui* present larger fruit bodies (cap up to 20 cm diameter, stipe 15 cm long x 2–4 cm diameter). A total of about 20 *Termitomyces* species have been recorded throughout tropical Africa (Heim 1977, Mossebo et al. 2002). The largest Termitomyces in tropical Africa, *Termitomyces titanicus*, is not yet recorded in West Africa. Buyck (1994) mentioned 4 species in western Burundi, Pegler and Piearce (1980)

Fig. 5. A few edible *Termitomyces* from West Africa. a. *Termitomyces* sp. (Fazao National Park), b. *Termitomyces microcarpus*, c. *Termitomyces schimperi* (Adjengré region, Togo), d. *Termitomyces letestui* (Pèrèré region, Benin).

enumerated 6 species in Zambia, and Morris (1986) reported 8 species for Malawi. Mossebo et al. (2002) inventoried 14 species in Cameroon. West African ecosystems host numerous *Termitomyces* species, many of which are still waiting for descriptions, as attested by numerous specimens from Togo and Guinea with deviating anatomo-morphological features (pers. observ.) than those already described (De Kesel et al. 2002, Mossebo et al. 2002). With a total richness of 9 species so far recorded, the genus *Termitomyces* makes about 20% of WEF in Benin (De Kesel et al. 2002, Yorou et al. 2002b). It represents about 13% of all WEF recorded for West Africa. Seven *Termitomyces* species are recorded in the southern part of Ivory Coast (Koné et al. 2010a,b). Repetitive mycological investigations generated more than 8 *Termitomyces* species in Togo (A. Guelly, unpublished data). According to

Koné (2013) and taking into account all available data on fungus growing termites and their symbiotic fungi, a conservative diversity of at least 20 species can be estimated in West Africa. In addition, the published sequence data (COI and ITS rDNA sequences) for both fungus-growing termites and their symbiontes allocate the centre of termite agriculture to western Africa (Nobré et al. 2011, Koné et al. 2011).

Termitomyces range among the top 3 most appreciated WEF in tropical Africa (Rammeloo and Walleyn 1993). They remain a delicacy in West Africa where many species are highly appreciated because of their taste (De Kesel et al. 2002, Heim 1977). In West Africa, *Termitomyces* sporophores are commonly used as food, but highly appreciated species are commonly traded at local and regional markets (Fig. 6), contributing to secure cash incomes for local people. In Ivory Coast for example, fruit bodies of *Termitomyces letestui* are intensively collected and sold, with approximate annual harvests amounting 7000 kg per year. Seasonal income (i.e., February to April each year) of participants in the marketing chain varies from US$58.38 to US$341.62, thus playing a crucial role for the livelihood of rural communities (Koné et al. 2010b).

Fig. 6. Trade of *Termitomyces letestui* in Ivory Coast. a. from harvesters to dwellers, b and c. Trade by women along roadside.

4. Natural Productions and Phenology of WEF in West Africa

Sudanian woodlands and gallery forests display not only a high diversity of WEF, but also harbour large quantities of fungal biomass. Quantitative investigations on the natural production of WEF in West Africa are rare, and very little (if not at all) is known about effective annual harvests and consumption by rural communities. However, taking into consideration the high diversity of WEF and that West Africa woodlands generally cover hundreds of thousands of hectares each, there is no doubt that annual natural productions are far higher than seasonal harvests. Parent and Thoen (1977) mentioned annual harvests of 1000 kg in Upper-Shaba region (Democratic Republic of Congo). In a recent study, Yorou et al. (2002a) measured an annual natural fresh biomass of up to 220 kg/ha for a total of 30 WEF in woodlands of Central Benin. As the study was executed during three months (covering 15th June to 15th September), and thus it covers only a part of the fruiting period, effective natural productions of over 300 kg per hectare may be expected. However, not all vegetation types produce such large fungal biomass. Significant difference ($F_{(5,107)} = 179,13$, $p < 0,01$) was recorded within investigated ecosystems (Yorou et al. 2002a), with annual productions ranging from 8.6 ± 5.2 kg/ha (for dense forests poor in EcM trees) to 293.9 ± 16.1 kg/ha (for Caesalpinioid-dominated woodlands). The study suggests that Caesalpinioid- and Phyllantioid-dominated woodlands harbour larger biomass of EcM WEF (up to 99%) in comparison to Caesalpinioid-poor dense forests (Fig. 7). Natural productions are mostly dominated by species members of *Lactarius* (+ *Lactifluus*), *Russula* and *Amanita*. Genera *Lactarius*

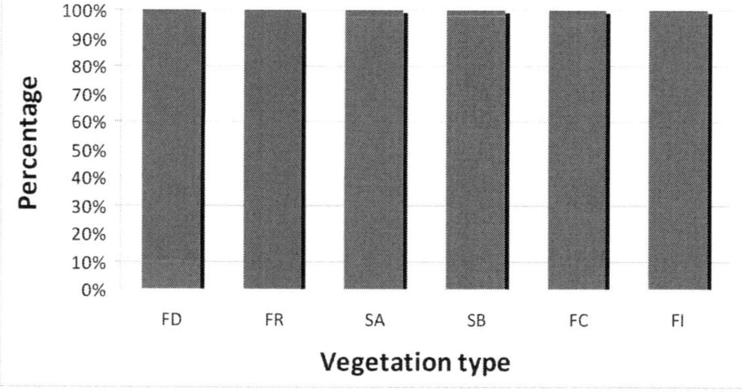

Fig. 7. Proportion of EcM versus saprotrophic WEF biomass within six different vegetation types. Caesalpinioid-dominated woodlands (SA, SB, FC, FI) are dominated (up to 99%) by EcM edible fungi in comparison to Caesalpinioid-poor dense forests (FD and FR), where saprotrophic fungi are predominant. Source Yorou: 2000 (unpublished).

and *Lactifluus* are not only the most common in term of species richness, but the most dominant in woodlands and gallery forests as well (Yorou et al. 2002a). In Sudanian ecozones, a single hectare of woodlands dominated by *Isoberlinia* spp. and/or *Uapaca* spp. may produce over 6000 sporophores, and up to 121 kg fresh biomass of *Lactifluus gymnocarpoides* annually, making it the most dominant WEF in Sudanian woodlands.

The Sudanian ecozones are characterised by a rainy season of up to 5–6 months (May to October) which contrasts strongly with a long severe dry season. While the dry season is generally the harvest period of many crops, the rainy season is particularly a difficult time for numerous farmers because of the lack of harvested crops. The beginning of the rainy/wet season (May to early July) represents a difficult period for local inhabitants. It is known as a shortage period because stocks are empty or sold out and new crops still unavailable. It is during this period that WEF play an important role as suppliers of food (De Kesel and Malaisse 2010, Yorou and De Kesel 2002) and essential proteins (Adewusi et al. 1993, Guissou et al. 2005). Yorou et al. (2002a) demonstrated that more than 70% of total annual productions of WEF are obtained at the beginning of the raining season (May to early July), which coincides with the aforementioned shortage period, thus contributing substantially to improve the food security of vast numbers of WA people. Natural productions of WEF significantly vary throughout the fruiting period ($F_{(10,107)}$ =157,57; p<0,01) and fit with an J-inverse curve (Fig. 8). The first production peak is obtained with the first intense rains (usually in May and June). Ecologically, the first rains generally break

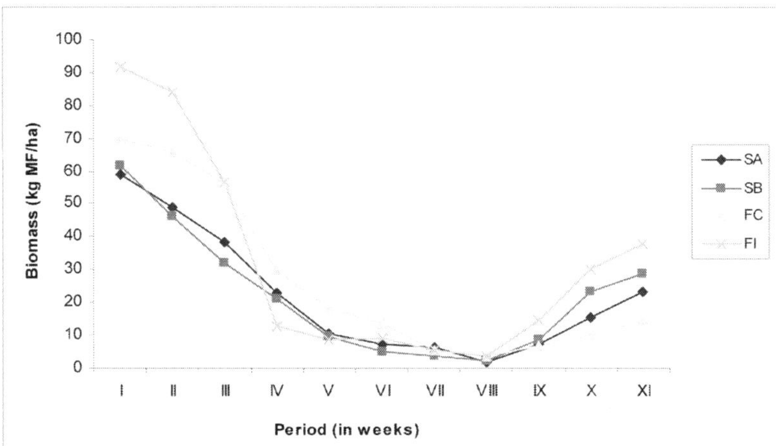

Fig. 8. Temporal variability ($F_{(10,107)}$ =157,57; p<0,01) over 6 months (11 sampled weeks I to XI, 2-week interval from May till October) of total natural productions of ectomycorrhizal WEF in Sudanian woodlands (SA, SB, FC and FI). Natural production fits a J-inverse curve with two peaks of productions. Source: Yorou 2000 (unpublished).

water stress of the myco-rhizosphere enabling the re-growth of below-ground fungal propagules (ectomycorrhizal root tips, mycelium, sclerotia, spores.....). Water availability within the myco-rhizosphere, coupled with the re-growth of EcM mycelium and the intensification of photosynthesis of partner trees provoke intensive fructifications of numerous species at the same time. As the soil becomes more humid in July, August, and September, natural productions of WEF decline. Another smaller production peak is obtained again towards the end of the rainy season, generally ending September to early October when soils gradually become drier. Temporal variability of natural production may also be related to specific preferences and phenology of both fungal and host species. Though species such as *Cantharellus addaiensis* and *Amanita* cf. *craseoderma* occur throughout the rainy season with no significant variability ($F_{(10, 121)} = 1.74$, $p = 0.07$ and $F_{(10, 121)} = 1.5$, $p = 0.15$ respectively), many other species members of *Lactifluus*, *Lactarius*, *Russula* and *Amanita* occur predominantly at the beginning of the rainy season and become less frequent towards the end of it. More information on the phenology of EcM species is needed to accurately explain temporal variability of natural productions and the phenology of individual species.

5. Steps Towards Establishing an Economic Sector based on WEF in West Africa: Challenges from a Case Study in Central Benin

Since more than 70 % of natural productions are available in a relatively short span of time and that local uses are so far lower than natural availability, it is essential to implement policies that maximize the use of WEF throughout the year. This is particularly justified by the fact that (1) the use of WEF as food is a common practice (Fig. 9), (2) it is also common to observe WEF being sold at local markets and/or at roadsides, contributing hereby to secure cash incomes for villagers, and importantly (3) to promote awareness among villagers, foresters and decision makers about the potential of using WEF as an income generating activity. An economic sector based on WEF may act as an incentive to make villagers gain interest in forest conservation and sustainable use. In West Africa, WEF are harvested, exploited and sold in a rather traditional way. Assistance is needed in order to help villagers get a maximum profit from these forests. Many aspects need to be addressed, i.e. ranging from the ones related to the basic and sustainable harvesting towards transformations and the end-users of mushrooms located either in the forest villages or in and around cities.

Fig. 9. Use of wild mushrooms as food by local people in Ivory Coast. a. Processing dried mushrooms before cooking, b and c. Cleaning *Termitomyces letestui* for cooking.

To examine the effectiveness of an economic sector based on WEF in Benin (West Africa) and its relevance to improve the livelihood of villagers, a case study was implemented in Bassila region in central Benin. Because of both the variability of vegetation types and the presence of several ethnic groups, the Bassila region was regarded as suitable for such a study. Through this case study, we would like to illustrate (1) the different steps we followed in this study and (2) address some of the constraints we faced. A simplified diagram will be suggested for scientists willing to establish a value chain of WEF in rural communities.

5.1 Ethnomycological surveys to select priority WEF in Bassila region

Ethnomycological surveys were conducted according to Yorou and De Kesel (2002). A total of 6 ethnic groups were considered in the study. The aim is to identify species that are appreciated by local people and are sold at local markets and that can be promoted in a value chain according to various considerations/criteria. During the first step of this study, forest walks were undertaken with villagers to monograph and identify all WEF they used as food. Repeated interviews were conducted in order to check the reliability of the information and to eliminate incoherent stories. The purpose was to get a reliable dataset enabling extraction of specific ethnic profile lists of WEF. We recorded a total of 21 edible species (for all ethnic groups). The diversity of WEF varied from one group to another, i.e., 7 species for Peuhl people to 16 species for Nagot people. In the second step, each ethnic group was requested to make a simple and matrix ranking (Rastogi 1999, Martin 1995) of the species they commonly use. Simple ranking is performed by asking each group to rank the top 10 species according to their nutritional importance (gustative appreciation). A matrix ranking (on a 1 to 8 points scale) of all top 10 species identified from simple ranking is performed against 4 criteria (local price per kg, preference for selling or own use, spatial and temporal availability in nature, and drying requirement of the species). Matrix ranking allowed the identification of the top 8 priority species, notably *Termitomyces letestui, T. schimperi, T. clypeatus, Lactifluus flammans, Lactifluus gymnocarpoides, Lactifluus medusae, Cantharellus* aff. *rufopunctatus* and *Psathyrella tuberculata.* These 8 species have been unanimously pointed out as interesting target species to act as a backbone for a regional economic sector based on wild edible fungi.

5.2 Identification of WEF-rich habitats and assessment of natural productions and phenology

The purpose of this activity is to identify the habitats of the identified priority species, phenology, and natural availability. This is of great importance since knowing target species' preferred habitat, distribution, and phenology will greatly reduce harvest effort of villagers. Except for *Psathyrella tuberculata,* all above mentioned species include termite fungi (*T. letestui, T. schimperi* and *T. clypeatus*) and EcM ones (*Lactifluus flammans, L. gymnocarpoides Lactifluus medusae* and *Cantharellus* aff. *rufopunctatus*). Ecologically, Bassila region harbours Caesalpinioid- and Phyllantioid-dominated woodlands for which quantitative data on natural production and species distribution are available (Yorou et al. 2002a). Annual productions of *Termitomyces* are difficult to assess since their fructification is sporadic and relies upon many environmental factors. Our numerous attempts to assess natural productions of *Termitomyces* within permanent plots were unsuccessful. However, Koné et al. (2010a,b) reported an annual estimation of over 7000 kg for *T. letestui* in southern Ivory Coast. *Cantharellus* aff. *rufopunctatus* is a species that is confined to gallery forests dominated by *Berlinia grandiflora.* Annual productions of this species amounts 178 kg/ha, on the basis of direct weekly biomass measure of 8.9 kg during 5 fruiting months. In summary, woodlands and gallery forests dominated by *Isoberlinia* spp. and *Uapaca* spp. were identified during the study as important target habitats to deploy sustainable harvesting in the context of a WEF-based local economical activity (in Central Benin).

5.3 Market surveys, identifying potential end users and assessing economic balance: The challenges

Market surveys are aiming at identifying species that are jointly appreciated by the harvesters and the end users as well. Market surveys should allow making an economic examination of the sector, taking into account harvesting effort, but also costs analysis related to mushroom drying, labelling, and transportation from the forest villages to the location of potential users. Such factors, coupled with the variability of environmental factors, remain the most challenging aspects in the establishment and continuity of a regional economic sector based on WEF. To be efficient and visible, this economic activity not only needs to identify good target species, but also needs to be available throughout the year. It is clear that both fresh and dried specimens should be available and that they should meet the users' expectation. This goes from personal criteria and requirements about transformed fungi (dried, in oil) to the way the edible fungi are packed, labelled and presented. In addition, costs for collecting in the forest, drying and transportation from rural to urban markets should be as low as possible to allow a rapid flow at reasonable price.

In this particular case (Bassila region), though we were able to identify priority species along with their production and distribution, a successful mushroom-based sector could not be initiated because of (1) the lack of potential buyers within and outside the villages, (2) lack of assistantship to promote awareness and to coach villagers from harvesting to the selling of WEF and finally (3) consumers' preferences of fresh mushrooms over dried ones. Because mushrooms are available in their immediate environment, no villager from the study area can imagine buying mushrooms except the very rare and highly prized species such as *Termitomyces*. Many villagers harvest the little quantities of WEF needed for their own personal uses, with very limited lucrative ambition. As WEF are collected only for personal use, and that they are mostly appreciated in fresh conditions, there is no need for collecting and drying more than daily required. We hardly observed villagers drying surplus specimens for a long term use (neither for personal consumption nor for sales). However, drying mushrooms seems not to be a major worry in tropical rural areas. Wild edible mushrooms can be either dried under the sun or smoked together with bush meat (Fig. 10). In the neighbouring country Burkina Faso, the highly appreciated *Phlebopus sudanicus* is harvested and commonly dried under the sun by Bobo people for long term exploitation (Guissou et al. 2005).

Fig. 10. Drying wild edible mushrooms. a and b. Drying of *Psathyrella tuberculata* and *Volvariella volvacea* under sun, c. Smoking of *Termitomyces letestui* together with bush meat.

Another problem we faced during our investigations was the paucity of highly esteemed species in natural areas. *Termitomyces* species usually present a high economical output, as can be confirmed in the study areas where several women were observed selling *T. fuliginosus* specimens. Whenever available in large quantities, *Termitomyces* species could successfully support a large scale sector of WEF in West Africa, as the appreciation and demand is high and most end users (either in villages or in cities) are prepared to buy them at a relatively high price (Koné et al. 2010b). Exactly the same phenomenon is observed with *Termitomyces letestui* in the termite rich miombo areas of Lubumbashi (DR Congo, De Kesel pers. obs.). Unfortunately, because of their dependence on termites species their fructification being moderated by various environmental factors (Koné 2013), *Termitomyces* species are not constantly available in nature to justify a long lasting value chain. Unless grown under controlled conditions, it will remain a moderately predictable but still highly seasonal edible product from the forest.

One important challenge in the study area, and probably in many West African cities, is to revive culinary habits of citizens so as to introduce WEF in their daily dishes. Though WEF are valuable by villagers throughout the study areas, they hardly gained attention of citizens of the neighbouring cities. Generally, there is a decline in the use of WEF from one generation to another and when moving from forest villages to the cities (Yorou and De Kesel 2002, Degreef et al. 1997). This is probably a general fact in the whole of West Africa (Guissou et al. 2008), and it may negatively influence the promotion of value chain of WEF. Reversing, changing, or widening culinary habits of citizens is probably the most tedious task in the actual West Africa context, as firstly, many imported foodstuffs are actually regarded as the best, and secondly, wild fungi are usually considered to be poisonous. If *Termitomyces* can attract the interest of many villagers and citizens, much effort is needed to increase the attention of the people in cities towards EcM edible mushrooms. This can be successfully performed through large scale awareness-raising programme about the culinary and nutritional benefits of WEF, and importance of WEF to improve food security among villagers and citizens.

5.4 Recommended protocol for the establishment of an economic sector based on WEF from tropical Africa.

For an enterprise to be successful, basic studies and some considerations should be made. We propose a simplified diagram (Fig. 11) including important steps and studies to be considered for the establishment of a value chain of WEF in tropical Africa. Most required studies include:

1) Studies of the diversity, distributions and natural production of WEF: this will allow the generation of annual availability of WEF as well as identification of WEF-rich habitats,
2) Ethnomycological investigations: to record patterns of local uses but also to identify priority species,
3) Studies on different ways to locally dry and store WEF. It is evident that these techniques should be adapted to the local needs and possibilities. They should be economic, efficient, low-tech, affordable, safe and reliable.
4) Market studies to demonstrate the need for and the efficiency of the sector. What is the demand and from where comes the demand. The market potential is qualitative, quantitative and surely not evenly

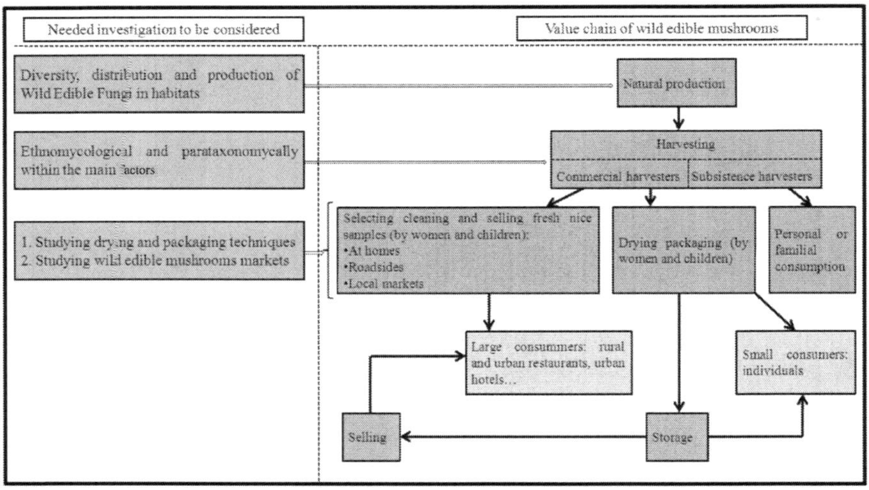

Fig. 11. Simplified diagram showing different steps and most needed investigations to be considered in erecting an economic sector based on WEF.

distributed. Is the sector solid enough to overcome drawbacks in production due to seasonal fluctuations? A thorough analysis is needed to know who is winning and who is losing. There is no doubt that a small scale pilot project will be needed in order to extrapolate and simulate a large scale market.

6. Concluding Remarks

Wild edible fungi are mostly EcM fungi and they are a seasonal but abundant and important source of food. In parts of West Africa, they represent a modest but stable income for rural people. The socio-economic potential of these wild edible fungi is high because our studies have indicated that the naturally produced quantities largely surpass the demand and local consumption. Wild edible fungi are under-utilized in West Africa. This potential is however currently being hampered or even annihilated by wide scale and devastating human activities. Let there be no doubt that the massive production of charcoal has become the biggest threat to all EcM forests in the whole of tropical Africa. Although forest products, including edible fungi, are often highly valued by local inhabitants, much effort is still needed to promote awareness among people that harvesting and marketing of edible fungi can only become an income generating activity if EcM forests are maintained and preserved. Multi-disciplinary investigations including taxonomists, ecologists, food scientists, food technologists, and economists should be implemented in order to secure a better exploitation of EcM forests and their generally associated edible fungi. There is very little doubt that sustainable use and preservation of West African forests, rather than cutting them down for charcoal,will yield a long-term improvement in the livelihood of millions of rural people.

Key words: Wild Edible Fungi, diversity, natural productions, phenology, local uses, economic sector, West Africa

7. Aknowledgement

We are grateful to local communities who provided valuable ethnomycological information. Yorou NS is grateful to the German Research Foundation for financial supports (DFG project Ag7/19-1 and Yo 174/2-1). Koné NA is undebted to the International Foundation for Science (Grant D/4982-1). The authors acknowledge Krista L. McGuire and Caitlyn Gillikin for improving the language.

8. References

Aanen, D.K. and P. Eggleton. 2005. Fungus-growing termites originated in African rain forests. Curr. Biol. 15: 851–855.

Adewusi, S.R.A., F.V. Alofe, O. Ademeyi, O.A. Alofabi and O.L. Oke. 1993. Studies on some edible wild mushrooms from Nigeria. I. Nutritional, teratogenic and toxic considerations. Plant Foods Human Nutri. 43: 115–121.

Bâ, A.M., R. Duponnois, M. Diabaté and B. Dreyfus. 2011. Les champignons ectomycorrhiziens des arbres forestiers en Afrique de l'Ouest. Méthodes d'étude, diversité, écologie, utilisation en foresterie et comestibilité. Editions IRDpp. 264.

Bâ, A.M., R. Duponnois, B. Moyersoen and A.G. Diédhiou. 2012. Ectomycorrhizal symbiosis of tropical African trees. Mycorrhiza 22: 1–29.

Bas, C. 1969. Morphology and subdivision of *Amanita* and a monograph of its section *Lepidella*. Persoonia 5: 285–579.

Boa, E. 2004. Wild Edible Fungi. A Global Overview of Their Use and Importance to people. Non-Wood Forest Products. FAO, Rome 17: 1–147.

Buyck, B. 1994a. Ubwoba, champignons comestibles de l'Ouest de Burundi. Administration Générale Coopération pour le Développent, Agriculture volume 34: 123.

Buyck, B. 1994b. *Russula* I (*Russulaceae*). Flore Illustrée des Champignons de l'Afrique Centrale 15: 335–408.

Buyck, B. 1994c. *Russula* II (*Russulaceae*). Flore Illustrée des Champignons d'Afrique Centrale 16: 411–542.

Buyck, B. 1997. *Russula* III (*Russulaceae*). Flore Illustrée des Champignons de l'Afrique Centrale volume 17: 545–597.

Buyck, B. and B. Nzigidahera. 1995. Ethnomycological notes from Western Burundi. Belgian J. Bot. 128: 131–138.

Buyck, B., G. Eyssartier and A. Kivaisi. 2000. Addition to the inventory of the genus *Cantharellus* (*Basidiomycota, Cantharellaceae*) in Tanzania. Nova Hedwigia 73: 3–4.

Buyck, B., F. Kauff, C. Cruaud and V. Hofstetter. 2013. Molecular evidence for new *Cantharellus* (Cantharellalles, Basidiomycota) from tropical African miombo woodlands and a key to all tropical African Cantherelles. Fungal Diversity 58: 281–298.

De Kesel, A. and F. Malaisse. 2010. Edible wild food: Fungi. *In*: F. Malaisse (ed.). How to Live and Survive in Zambesian Open Forest (Miombo Ecoregion). Gembloux, Presses agronomiques: 422pp. + CD rom, pp. 41–56.

De Kesel, A., N.S. Yorou and B. Buyck. 2011. *Cantharellus solidus*, a new species from Benin (West-Africa) with a smooth hymenium. Cryptog. Mycol. 32: 277–283.

De Kesel, A., J.T.C. Codjia and N.S. Yorou. 2002. Guide des champignons comestibles du Bénin. Cotonou, République du Bénin, Jardin Botanique National de Belgique et Centre International d'Ecodéveloppement Intégré (CECODI). Impr. Coco-Multimedia, pp. 275.

Degreef, J., F. Malaisse, J. Rammeloo and E. Baudart. 1997. Edible mushrooms of Zambezian woodland area: A nutritional and ecological approach. Biotechnology Agronomy Society and Environment 1: 221–231.

Diédhiou, A.G., M.A. Selossé, A. Galiana, M. Diabaté, B. Dreyfus, A. M. Bâ, S. Miana and G. Béna. 2010. Multi-host ectomycorrhizal fungi are predominant in a Guinean tropical rainforest and shared between canopy trees and seedlings. Environ. Microbiol. 12: 2219–2232.

Ducousso, M., A.M. Bâ and D. Thoen. 2002. Les champignons ectomycorrhiziens des forêts naturelles et des plantations d'Afrique de l'Ouest: une source de champignons comestibles. Bois For. Trop. 275: 14.

Eyi-Ndong, H. 2009. Etude des champignons de la forêt dense humide consommés par les populations du nord du Gabon. PhD thesis, Université Libre de Bruxelles, pp. 271.

Eyi-Ndong, H. and J. Degreef. 2009. Inventaire des champignons consommés par les Pygmés du nord du Gabon. XVII Congrès AETFAT, 28 Fév-2 Mars 2007, Yaoundé Cameroun.

Eyi-Ndong, H. and J. Degreef. 2010. Diversité des espèces de *Cantharellus*, *Lentinus* et *Termitomyces* consommés par les Pygmés du Nord du Gabon. *In*: J. van der Burgt, J. van der Maesen and J.M. Onana (eds.). Systématique et Conservation des plantes Africaines. Kew Royal Botanic Garden, pp. 133–141.

Eyi-Ndong, H., J. Degreef and A. De Kesel. 2011. Champignons comestibles des forêts denses d´Afrique Centrale. Taxonomie et identification. ABC taxa, Bruxelles 10: 255.

Eyssartier, G. 2003. The genus *Cantharellus*. PhD thesis, Muséum National d´Histoire Naturelle, Paris.

Eyssartier, G., B. Buyck and A. Verbeken. 2002. *Cantharellus conspicuus* sp. nov. Crypto. Mycol. 23: 95–102.

Eyssartier, G. and B. Buyck. 1998. Contribution à la systématique du genre *Cantharellus* en Afrique tropicale: Etude de quelques espèces rouges. Belgian J. Bot. 131: 139–149.

FAO. 2010. FAOSTAT, FAO Statistical Databases. http,//faostat.fao.org/, dernier accès le 21 janvier 2010.

Guissou, K.L.M. 2005. Les *Macromycètes* du Burkina Faso. Thèse de Doctorat, Université de Ouagadougou, Burkina Faso.

Guissou, K.M.L., P. Sankara and S. Guinko. 2005. *Phlebopus sudanicus* ou la viande des Bobos, un champignon comestible dans le Département de Satiri au Burkina Faso. Crypto. Mycol. 3: 195–204.

Guissou, K.M.L., A.M. Lykke, P. Sankara and S. Guinko. 2008. Declining wild mushrooms recognition and usage in Burkina Faso. Economic Bot. 62: 530–539.

Hahn-Hadjali, K., R. Wittig, M. Schmidt, G. Zizka, A. Thiombiano and B. Sinsin. 2010. La végétation de l´Afrique de l'Ouest. *In*: B. Sinsin and D. Kampmann (eds.). Biodiversity Atlas of West Africa, volume I: Benin. Cotonou/Frankfurth/Main pp. 78–85.

Hama, O., E. Maes, K.M.L. Guissou, D. Ibrahim, M. Baragé, L.A. Parra Sánchez, O. Raspé and A. De Kesel. 2010. *Agaricus subsaharianus*, une nouvelle espèce comestible et consommée au Niger, au Burkina Faso et en Tanzanie. Crypto. Mycol. 31: 221–234.

Härkönen, M. 1992. Wild mushrooms, a delicacy in Tanzania. Univ. Helsingensis 2: 29–31.

Härkönen, M., T. Saarimäki and L. Mwasumbi. 1994. Edible and poisonous mushroom of Tanzania. The African J. Mycol. Biotech. 2: 99–130.

Härkönen, M., T. Saarimäki and L. Mwasumbi. 1995. Edible mushrooms of Tanzania. Karstenia 35 (supplement): 1–92.

Härkönen, M., T. Niemelä and L. Mawasumbi. 2003. Edible, harmful and other fungi. The Finnish-Tanzanian Friendship Society. Norrlinia 10: 200.

Heim, R. 1977. Termites et champignons. Les champignons termitophiles d'Afrique Noire et d'Asie méridionale. Paris.

Koné, N.A. 2013. Symbiose termite-champignon: Origine, co-diversification et fructification saisonnière du symbiote fongique (*Termitomyces* spp.). Thèse Unique de Doctorat de l'Université Nangui Abogoua, Côte d'Ivoire : 47p.

Koné, N.A., K. Dosso, S. Konaté, Y.J. Kouadio and K.E. Linsenmair. 2011. Environmental and biological determinants of *Termitomyces* species seasonal fructification in central and southern Côte d'Ivoire. Insectes Soc. 58: 371–382.

Koné, N.A., D. Koné and P. Nicot. 2010a. State of knowledge of fungal diversity in Côte d'Ivoire. *In*: S. Konaté and D. Kampmann (eds.). Biodiversity Atlas of West Africa Tome III, pp. 172–177.

Koné, N.A., S. Konaté and E.K. Linsenmair. 2010b. Socio-economic importance of *Termitomyces* in Côte d'Ivoire. *In*: S. Konaté and D. Kampmann (eds.). Biodiversity Atlas of West Africa Tome III, pp. 177–178.

Malaisse, F. 1997. Se nourrir en forêts claires africaines. Approches écologiques et nutritionnelles. CTA. Wagenigen, pp. 384.

Malaisse, F., A. De Kesel, G. N'Gasse and G. Lognay. 2008. Diversité des champignons consommés par les Bofi de la Lobaye (République centrafricaine). Geo-Eco-Trop 32: 1–8.

Martin, G.J. 1995. Ethnobotany: A methods manual. London. Champman & Hall. Chapter anthropology.

Morris, B. 1986. Notes on the genus *Termitomyces* Heim in Malawi. Soc. Malawi J 39: 40–49.

Morris, B. 1984. Macrofungi of Malawi. Some ethnobotanical notes. Bull Br. Mycol Soc 18: 48–57

Morris, B. 1994. Bowa: Ethnobotanical notes on the macrofungi of Malawi. *In*: J.H. Senya and A.C. Chikuni. Proceedings of the XIIIth Plenary meeting. AETFAT, Malawi, pp. 635–793.

Mossebo, D.C. and A. Amougou. 2002. Contribution à l'étude du genre *Termitomyces* (*Basidiomycètes*) au Cameroun: écologie et systématique. Bull. Soc. Mycol. Fr. 118: 195–249.

Mossebo, D.C., A. Amougou and R.E. Atangana. 2002. Contribution à l'étude du genre *Termitomyces* (*Basidiomycètes*) au Cameroun : écologie et systématique. Bull. Soc. Mycol. Fr. 118: 195–249.

Natta, A.K., B. Sinsin and L.J.G. van der Maesen. 2002. Riparian forests, a unique but endangered ecosystem in Benin. Notulae Florae Beninensis 4. Botanische Jahrbücher 124: 55–69.

Neuenschwander, P., B. Sinsin and G. Georgen. 2011. Protection de la nature en Afrique de l'Ouest: Une Liste rouge pour le Benin. Nature Conservation in West Africa: Red List for Benin. International Institute of Tropical Agriculture, Ibadan, Nigeria, pp. 365.

Nobré, T., N.A. Koné, S. Konaté, K.E. Linsenmair and D.K. Aanen. 2011. On the origin and co-diversification of fungus-growing termites and their fungal symbionts. Mol. Ecol. 20: 2619–2627.

Ogundana, S.K. and O.E. Fagade. 1981. Nutritional value of some Nigerian edible mushrooms. Mushroom Sc. 11: 123–131.

Ogundana, S.K. and O.E. Fagade. 1982. Nutritional value of some Nigerian edible mushrooms. Food Chem. 8: 263–268.

Parent, G. and D. Thoen. 1977. Food value of edible mushrooms from Upper-Shaba Region. Economic Bot. 31: 436–445.

Pegler, D.N. and G.D. Piearce. 1980. The edible mushrooms of Zambia. Kew Bull. 35: 475–491.

Piearce, G.D. 1987. The genus *Termitomyces* in Zambia. Mycologist 1: 111–116.
Rammeloo, J. and R. Walleyn. 1993. The edible fungi of Africa south of the Sahara. Scripta Bot. Belg. 5: 1–62.
Rastogi, A. 1999. Methods in Applied Ethnobotany. Lessons from the field. Discussion papers n° MNRR 99/1. Kathmandu. International center for Integrated Mountain Development.
Rivière, T., A.G. Diedhiou, M. Diabate, G. Senthilarasu, K. Hatarajan, A. Verbeken, B. Buyck, B. Dreyfus, G. Bena and A.M. Bâ. 2007. Genetic diversity of ectomycorrhizal *Basidiomycetes* from African and Indian tropical rain forests. Mycorrhiza 17: 415–428.
Sanon, K.B., A.M. Bâ and J. Dexheimer. 1997. Mycorrhizal status of some fungi fruiting beneath indigenous trees in Burkina Faso. For. Ecol. Manag. 98: 61–69.
Sparovek, G., S.B.L. Ranieri, A. Gassner, I.C. de Maria, E.E. Schnug, R.F. dos Santos and A. Joubert. 2002. A conceptual framework for the definition of the optimal width of riparian forests. Agr. Ecosyst. Environ. 90: 169–175.
Tibuhwa, D., B. Buyck, A. Kivaisi and L. Tibell. 2008. *Cantharellus fistulosus* sp. nov. from Tanzania. Cypto. Mycol. 29: 129–135.
Tibuhwa, D., S. Sanja, L. Tibell and A.K. Kivaisi. 2012. *Afrocantharellus* gen. stat. nov. is part of African *Cantharellaceae*. IMA Fungus 3: 25–38.
van Rooij, P., A. De Kesel and A. Verbeken. 2003. Studies in tropical African *Lactarius* species (*Russulales, Basidiomycota*) 11. Records from Benin. Nova Hedwigia 77: 221–251.
Verbeken. A. 1996. New taxa of *Lactarius* (*Russulaceae*) in tropical Africa. Bull. J. Bot. Nat. Belg. 65: 197–213.
Verbeken, A. 1996. Biodiversity of the genus *Lactarius* Pers. 5. A synopsis of the subgenus *Lactifluus* (Burl.) Hesler & A.H. sm. Emend. Mocotaxon 66: 363–386.
Verbeken, A. 1998a. Studies in tropical African *Lactarius* species. 5. A synopsis of the subgenus *Lactifluus* (Burl.) Hesler & A.H. Sm. Emend. Mycotaxon 66: 363–386.
Verbeken, A. 1998b. Studies in tropical African *Lactarius* species. 6. A synopsis of the subgenus *Lactariopsis* (Henn.) R. Heim Emend. Mycotaxon 66: 387–418.
Verbeken, A. 2000. Studies in tropical African *Lactarius* species. Persoonia 17(3): 377–406.
Verbeken, A. and R. Walleyn. 2000. Studies in tropical African *Lactarius* species. 7. A synopsis of section *Edules* and a review on the edible species. 132: 175–184.
Verbeken, A. and B. Buyck. 2001. Biodiversity and ecology of tropical African ectomycorrhizal fungi in Africa. *In*: R. Watling, J.C. Franckland, A.M. Ainsworth, S. Isaac and R. Robinson (eds.). Tropical Mycology: Macromycetes, volume 1,. CAB International, Wallingfort, UK, pp. 11–24.
Verbeken, A. and R. Walleyn. 2010. Fungus Flora of Tropical Africa, National Botanic Garden of Belgium Volume 2. Monograph of *Lactarius* in Tropical Africa, 54 planches, pp. 160.
Verbeken, A., J. Nuytinck and B. Buyck. 2011. New combinations in *Lactifluus*. 1. L. subgenera *Edules, Lactariopsis*, and *Russulopsis*. Mycotaxon 118: 447–453.
Verbeken, A., K. van De Putte and E. De Crop. 2012. New combinations in *Lactifluus*. 3. *Lactifluus* subgenera *Lactifluus* and subgenera *Piperati*. Mycotaxon. 120: 443–450.
Walleyn, R. and J. Rammeloo. 1994. The poisonous and useful fungi of Africa south of the Sahara: a literature survey. Script. Bot. Belg. 10: 1–56.
Wegmann, M., M. Machwitz, M. Schmidt and S. Dech. 2010. Fragmentation de la forêt tropicale humide-biodiversité en danger. *In*: B. Sinsin B and D. Kampmann (eds.). Biodiversity Atlas of West Africa, volume I: Benin. Cotonou/Frankfurth/Main pp. 86–91.
White, F. 1986. La végétation de l'Afrique. Paris, ORSTAM-UNESCO: pp. 384.
Yorou, N.S. 2000. Biodiversité, écologie et productivité des champignons supérieurs dans diverses phytocénoses de la forêt classée de Wari-Maro au Bénin. Thèse d'Ingénieur Agronome, FSA/UNB: 122 pages + annexes.

Yorou, N.S. 2010. Larger fungi. *In*: B. Sinsin and D. Kampmann (eds.). Biodiversity Atlas of West Africa, volume I: Benin. Cotonou/Frankfurth/Main pp. 324–331.

Yorou, N.S. and A. De Kesel. 2002. Connaissances ethnomycologiques des peuples Nagot du centre du Bénin (Afrique de l'Ouest). Proceeding of XVI the AETFAT congress, Brussels 2000. Syst. Geogr. Plants 71: 627–637.

Yorou, N.S. and A. De Kesel. 2011. Champignons supérieurs. Larger fungi. *In*: P. Neuenschwander, B. Sinsin and G. Goergen (eds.). Protection de la Nature en Afrique de l'Ouest: Une Liste Rouge pour le Bénin. Nature Conservation in West Africa: Red list for *Benin*. International Institute of Tropical Agriculture (IITA), Ibadan, Nigeria, pp. 47–61.

Yorou, N.S., A. De Kesel, B. Sinsin and J.T.C. Codjia. 2002a. Diversité et productivité des champignons comestibles de la forêt classée de Wari-Maro (Bénin, Afrique de l'Ouest). Proceedings of XVIth AETFAT Congress, Brussels 2000. Syst. Geogr. Plants 71: 613–625.

Yorou, S.N., A. De Kesel, J.T.C. Codjia and B. Sinsin. 2002b. Biodiversité des champignons comestibles du Bénin. Proceedings of the Symposium-Workshop on Biodiversity in Benin. Abomey-Calavi (Benin) October 30th to November 18th 2002, pp. 231–240.

[1]Department of Natural Resources Management, Faculty of Agronomy, University of Parakou, BP 123, Parakou, Benin.

[2]Department Biology I, Ludwig-Maximilians-Universität München, Menzinger Str. 67, 80638, Munich, Germany.

[3]URF des Sciences de la Nature et de l'Environnement, Station d'Écologie de Lamto, Université Nangui Abrogoua (Côte d'Ivoire), BP 28 N'Douci.

[4]École Normale Supérieure, Université de Koudougou, BP 376, Koudougou, Burkina Faso.

[5]Département de Botanique et Écologie Végétale, Faculté des Sciences, Université de Lomé, 081 BP 1515 Lomé, Togo.

[6]Laboratoire d'Écologie Appliquée, Faculté des Sciences Agronomiques, Université d'Abomey-Calavi, 01 BP 526, Cotonou, Bénin.

[7]Department of Cryptogamy (Bryophyta & Thallophyta), National Botanic Garden of Belgium, Domein van Bouchout, B-1860, Belgium.

*Corresponding author: n.s.yorou@bio.lmu.de

Index

Color Plate Section

Chapter 1

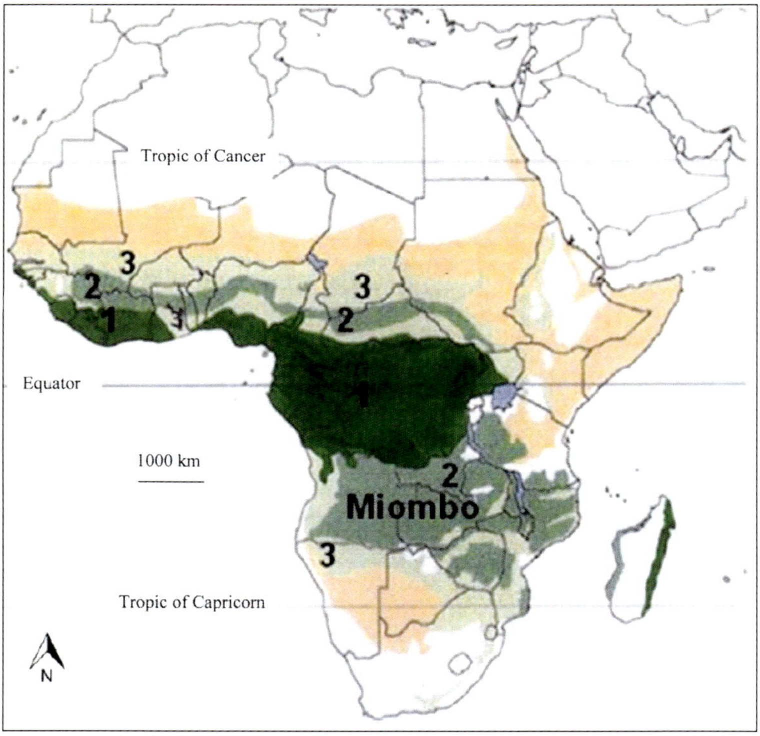

Fig. 1. Distribution of EcM trees in tropical Africa: (1) rainforests in the Guinea-Congo region; (2) open forests in the Sudanian and Zambezian regions; (3) savanna woodlands in the Sudanian and Zambezian regions (Bâ et al. 2012).

Chapter 2

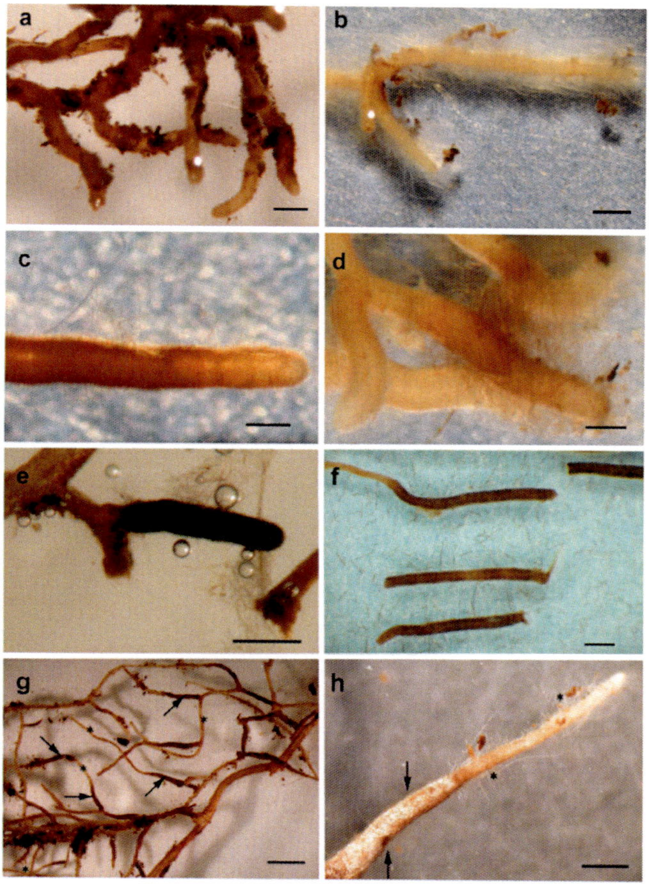

Fig. 1. Micrographs of mycorrhizas of *Neea* species 1 (a-e), *Neea* aff. *floribunda* (g-h), *Pisonia* sp. (g-h), a. *Russula puiggarii* – *Neea* species 1, b. *Russula-Lactarius* – *Neea* species 1, c. *Tomentella-Thelephora* species 1 – *Neea* species 1, d. *Tomentella-Thelephora* species 2 – *Neea* species 1, e. Ascomycete – *Neea* species 1, f. *Tomentella-Thelephora* species 3 – *Neea* aff. *floribunda*, g.,h. . *Tomentella-Thelephora* species 3 – *Pisonia* sp., g. overview: long roots partially with hyphal mantle (arrows), partially with root hairs and no hyphal mantle (*), h. one root shown at higher magnification: distal portion with root hairs (*), proximal portion with hyphal mantle (arrows). Scale bars: a.,b.,f.,h. 1 mm; c.-e. 0,5 mm; g. 5 mm. Data from Haug et al. 2005.

Chapter 6

Fig. 3. (A) *Podoserpula miranda* and (B) *Cantharellus garnieri*. Photos by courtesy from Ducousso Marc, CIRAD.

Fig. 4. *Pisolithus albus* from New Caledonia. A: *Pisolithus albus* MD07-117 from the Koniambo massif; B: *Pisolithus albus* MD07-228 from the Ouen-Toro, Noumea; C: cross section of *Pisolithus albus* MD07-166 from Pindjen water-fall and D: globose spores (8.77 to 9.62 µm) of *Pisolithus albus* MD06-379 from Poum, erected spines (1.2 µm) are clearly visible. From Jourand et al. (2010a).

Fig. 9. *Eucalyptus globulus* seedlings after 12-weeks growth. A and A′ mycorrhizal; B and B′: non-mycorrhizal (controls). A and B: no nickel added; A′ and B′ seedlings treated with Ni. From Jourand et al. (2010b).

Chapter 7

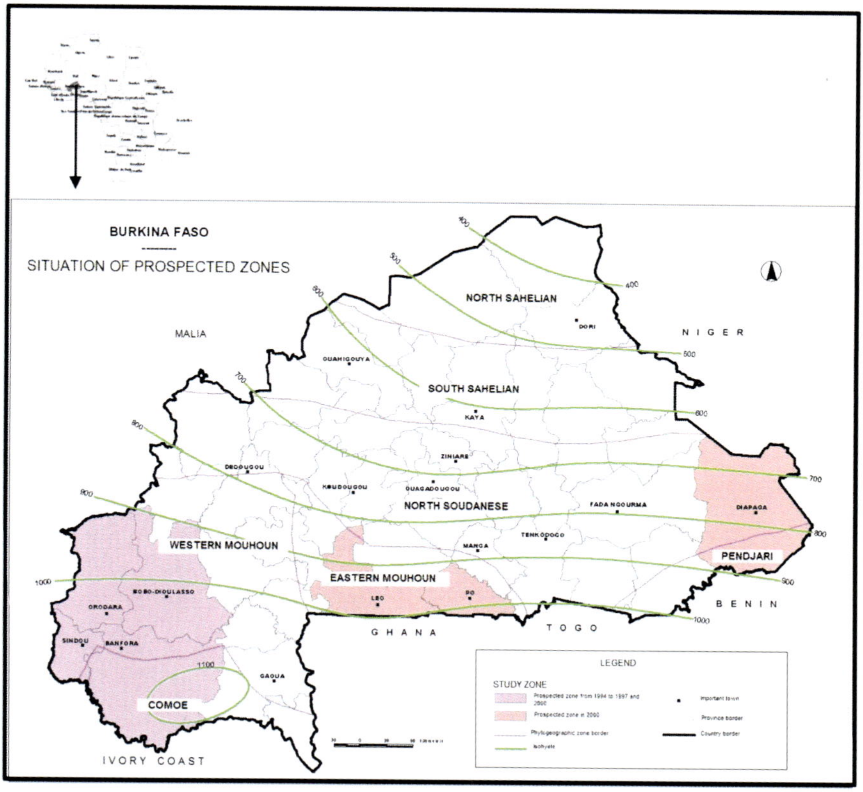

Fig. 1. Climatic regions of Burkina Faso and localization of prospected areas. Data from Sanon et al. (2009a).

Chapter 8

Fig. 1 Sporocarps of *S. sinnamariense*. Bar is 2.5 mm.

Fig. 3. Yellow *Gnetum* EM root tip formed with *S. sinnamariense*, showing extensive extramatrical mycelia. Bar is 0.3 mm.

Chapter 9

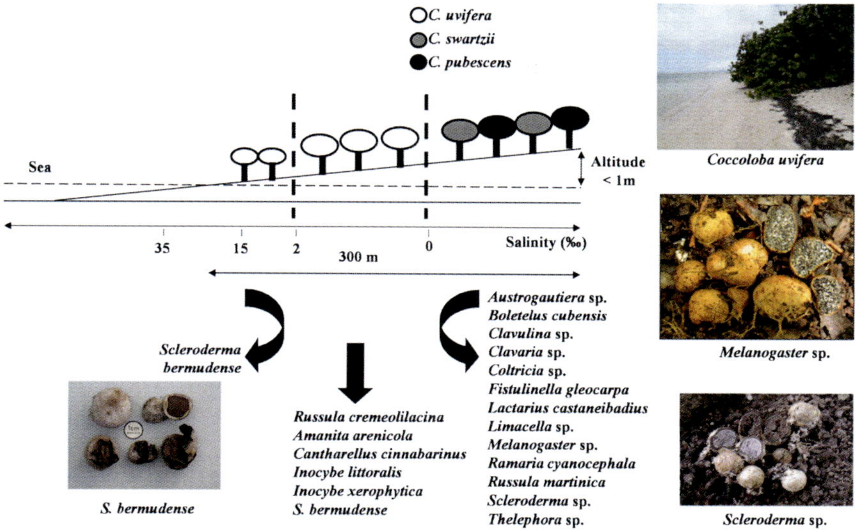

Fig. 1. Distribution of three *Coccoloba* species (*Coccoloba uvifera, C. swartzii* and *C. pubescens*) and their putative ectomycorrhizal fungi along a salinity gradient in Martinique and Guadeloupe (Lesser Antilles).

Chapter 10

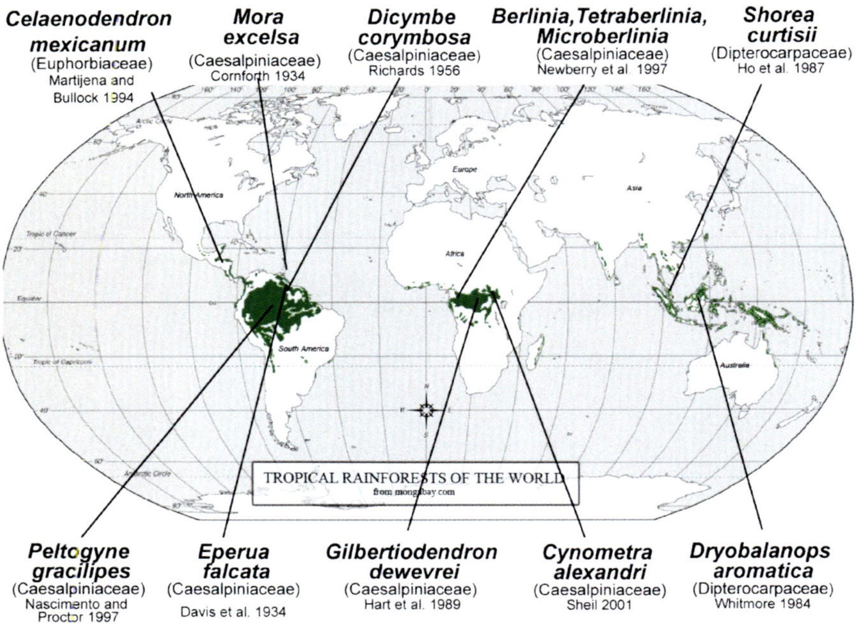

Fig. 1. Map illustrating examples of monodominant and co-dominant forests with the tree species or genus listed and the tree family in parentheses. This diagram is not an exhaustive list of monodominant species, but it illustrates how monodominant forests are found in all tropical regions.

Chapter 13

Fig. 2. Some edible *Lactarius* species from West Africa. a. *Lactarius flammans* (Kou Forest Reserve, Burkina Faso), b. *L. gymnocarpus* (Baro Forest Reserve, Upper Guinea), c. *L. foetens* (Mouhoun Forest Reserve, Burkina Faso), d. *L. medusae* (Fazao National Park, Togo), e. *L.* cf. *pelliculatus* (Forest Guinea), f. *L. tenellus* (from Benin).

Fig. 4. Some edible *Cantharellus* species from West Africa. a. *Cantharellus floridulus* (Wari-Maro Forest Reserve, Benin), b. *Cantharellus congolensis* (Aledjo Forest Reserve, Togo), *Afrocantharellus platyphyllus* (Kota gallery Forest, Benin), d. *Cantharellus* cf. *conspicuus* (Bassila gallery forest, Benin).